MANAPEPIG

MANAPEPIG

BARTROLOMEW MC INNCEL

Contents

Prologue

Dearest explorers

If you are reading this parable
you are here now stepping
into a realm of most terrible
and wonderful beckoning

Between these page furls
your party will need
to uncover old worlds
so few accept to seek

You are offered these visions
that only those elite
with the noblest of intentions
dared whisper or speak

Now, in you, secrets rumble
like the mysteries of the sea
They are linked through mists of jungle
and this knowledge you will need

1

I -- GUY WITH A LITTLE GRAPEFRUIT HEAD

I knew this guy that had a little grapefruit head.

His name was Josh and his face was kind of normal size but because of his undersized head it made his mouth look extra big like those goombas in the Mario movie.

He always had a short buzz haircut cut too, as if he needed to make that area of his body even smaller. He was a bully though, in primary elementary school and he loved to wear these big thick sweater hoodies and baggy clothes.

All the kids lived in fear of Josh because you never knew when he was going to strike next. So, I waited until he was in the cafeteria one day eating lunch and I glided up behind him in full stealth mode like I was on a Marty McFly hover board, and as gently as docking a space shuttle, I placed an entire sloppy chili-dog right down into his hoody and one of those little carton milks with the mouth part opened up. His clothes were so thick he didn't even notice.

I wish I could have been there when the chili slop or chocolate milk started to soak through to his back, or when some other student noticed the dumpster trash smell coming from the sopping wet bulge in his hood.

What if he actually found out by pulling his hood onto his head? Onto his little head. "Aaaaaaaah... WHAT the'fffff..."

It looked like he was always wearing the same clothes though. What if he never even took them off and just slept in them like a weirdo and as soon as he laid down at night the milk would start gushing out?

Silly Josh. That's what happens when you pick on people smaller than you. Life finds a way.

Oh! Hi, there. We haven't been properly introduced yet. My name is blehhehhehh, like you care, and I was a real-life genuine kid detective while I was growing up. I've been solving mysteries since I was blehhehhehh, like you give a shit.

I've been a kid detective my whole life. I still am. Well. Technically I'm an adult kid detective at this point. In so many years, however, I have uncovered a menagerie of truly extraordinary spectacles, somewhat like an archeologist who travels the world in a trusty Aussie hat hunting for legendary artifacts and searching for lost relics while getting in fist fights constantly and hitting people with whips.

One fine day, as it happened, the realization suddenly dawned upon me of what tremendous importance it was to the world, this vast collection of rare gems of insight I had unearthed. In this moment I was at once set upon by the unavoidable knowing that I must share no less than an essential morsel of these secrets here now with those of you having the good taste to listen. I would be thoroughly remiss to take any such course otherwise.

Truly remiss.

What's that you said? Oh! Not sure what the word "remiss" means? Haha. That's perfectly reasonable. I can assure you. No one knows. And we never will.

Alas, I have compiled this text of incredible discoveries and awe inspiring wonders on the order that it may be set adrift on the ocean of time. Often, as it happens, we are not afforded the luxury of being present at the right perfect moment that those effects of our good deeds are felt. But whomever you are in the world and in whatever time you find yourself, I hope one day the contents of this book are able to spill over you like a fine mixture of chunky bean sauce and coagulating cream.

And finally, to Josh. I know this was all a long time ago now, but didn't I tell you, golf ball head? What did I say? If you mess with the full-size head kids, you're gonna find out.

2

II -- WE DON'T TAKE KINDLY 'ROUND HEEYUH

First off, we don't need no simpleton nonsensical types comin' 'round heyuh tryin'uh cause up a whole mess uh trouble why we tryin'uh have a good honest sensical talk.

We're good science fearin' folks 'roun' these parts an' we don't take kindly to no not-science fearin' folks 'roun' these parts.

Matter fact, if you're in some kind'a closed minded, makin' shit up cult then go on and GIT! Yuh hear?! Go on back to your kind! You belong with them! I can't take care you no moe!!

Go on! GIT!

DAMNIT! I SAID GIT! **D,X**

I ain't cryin'! YOU'RE --> cryin'!

DAMNIT, I CAIN'T!!!

I cain't do it! I'm sorry, goddamnit! I cain't.

GET over here, religgy. Come 'eer, boy. I swear I promise I'll never leave you again, boy. Who's a good boy?! Who's a good boy?! Who wants to read a fun book about science?! Who wants to read a fun book about science?!

Ok! But it's gonna cost you at least two handshakes. Nothing's free. This book costs at least two handshakes. Okay. Shake hands! Goooooood! Now shake the other hand! No. Other hand. No. No... No. Other hand. No. You already gave me that hand. Now I want this one. This one. C'mon, shake. GOOOOOOOD JOB! GOOOOOOD BOOOOOOOY!

Alright, alright. Let's talk business. No, I'm not having a stroke.

I got some questions for religgies then. I "got" some questions — I went out and retrieved them, not simply have them in my possession.

Anyway, why do you think God and evolution are mutually exclusive? Why do you think those two concepts are like Robbin Hood and that other guy trying to cross a log bridge at the same time and there's not enough room for the both of them and inexplicably they can't just take turns crossing, so then they must enter into mortal combat with their bow staffs for no goddamn reason?

Isn't it one of your beliefs that God made conscious man because he was lonely or something and he wanted other sovereign minds to share the universe with? Well, why wouldn't he also enjoy watching the mechanisms of nature like sunsets and fireflies and Tasmanian devils that are always screaming, and fat little baby seals and hummingbirds and that one fish that lives in the anus of the other fish?

Why wouldn't he also want them to entertain and dazzle him the same way you and your consciousness do? Aren't you supposed to be growing and making mistakes and learning from them? So, why wouldn't nature herself get to do that too? Or is it just all about you?

You don't think watching the adaptive arms races of thousands of species is as breathtaking and wondrous as watching you just living from chair to chair with your cankles, simply existing to eat and score points in petty office drama?

Man! How will nature ever compete with that?!

You don't think he likes a few surprises every now and again? You think birthday parties are just only for KIDS?! You think just cuz he's a grown-up he doesn't have feelings too?!

Everything is about you, you, you, ISN'T IT?! You know, he had a life and dreams before you came along.

He works his ass off all day to make sure you don't get hit by meteors, and make sure that none of those things you just saw in that horror movie end up crawling into this dimension and into your house for real, like that old grieving widow that crawls around on the ceiling and then down your wall and then puts her face right over your face while you're sleeping but the rule is she can only get you if you're sleeping with your mouth open and then she gets to dip down into your mouth and lick your whole tongue and then you wake up in the morning like "What the fuck is in my mouth?!"

Who do you think is protecting you from her most of the time?

And then he has to work with the public all day taking customer orders across the universe.

"Yeah, can I get a perfect life with no problems ever no matter how it effects other people and then the two enchilada plate with no sides... Right. Yeah. No sides on that. That's fine. Right. I understand. Yeah, that's why I said it. Yeah, and then the three fish tacos and a side of sour cream just in case I change my mind later and... What? You what? Wait. No sour cream? You guys don't have ANY sour cream? At all?! Not like for other purposes or...? There's not like an industrial thing or...? No?! Okay. Uuuh. Wow. Uuuh. I didn't see that comin'. I guess, theeeeeeen. Man. I guess I'll just... I can't believe you guys don't... Uuuh, yeah, I don't even know, uuuuuh, just give me the fish tacos anyway and a large fiesta horchata drink with no ice, and that's it. Yep. Thanks. Oh! And a little side of mayo. So, do you guys know of any places close to here that would have sour cream?... ...What?... ...That place?! Isn't that a jewelry store? Why would a jewelry store have sour cream?"

Then he comes home after a long day of work so excited to see you and all your smiling faces, and all he wants is for you kids to just be some good, him-fearing people and drop to your knees and act like a bunch of groveling henchmen for a mean villain who gets bored and orders you to sing and dance for his amusement, and you're like "But myaffter, we don't know how" and he's like "DO IT!" So, you start doing your best and he starts laughing and you start crying.

You know, as all loving parents do.

"Hey guys, I'm home! Oh, is this glued pasta picture for me? Aww, I love it. It says "number one dad." You guys are the best kids ever, NOW ON YOUR KNEES!!! -->"

That's got love written all over it.

By the way, why does it gotta be this weird master-slave relationship too? Why can't it just be like a mutual respect and acknowledgement of an alignment with goodness and compassion or something? Why would he have to control-freak micro-manage every little molecule in nature when he could just lead it in a dance and work with its natural rhythms?

Oh, and I'd also like to extend an olive branch about taking names in vain or whatever. It's not *my* fault that the culture imprinted theologically incorrect exclamatory statements on me as an innocent wee child.

And besides, who do you think has the biggest sense of humor of all? Why do you think sometimes people who are stone-cold dead and already shipped off to the walk-in frig suddenly wake up from an insane afterlife experience and scare the crap out of hospital staff? Why do you think problem children and royal pains in the britches still exist? Do you really think those rascals survived in the gene pool this long solely on their own account?

Think again.

Either way, if the process of evolution was ever starting to go off the rails and beginning to get too out of hand I'm sure he could just go back and tweak something, like Dr. Strange. Maybe in the original iteration of Earth the first species to make it to intelligent tool maker was the anus fish and God was like "Sooooo, like, I don't *hate* this outcome, per se or whatever, buuuuut liiiiiiike why don't we just keep having *fun*, yuh know? Isn't this so much fun? Why don't we do it again?! Oh my God, we should totally do it again! :D Let's DO it!

Okay! Go ahead! Yep. Reshuffle. Yeah. It's fine. They'll never know."

Or maybe you think evolution takes too long and God gotta be places and he had tuh gyitt'er done. So, you're gonna be with him in heaven for an endless succession of zillions and jillions of years, as you attest, but somehow he just really had to get you all slapped together the same way that you had to get that semester project finished the night before it was due?

"Shoot, shoot, shoot! I only have infinity of time left! I need to get these human things done so they can worship me and fear me constantly! Quick, uh, how should I make'm? I don't have time to start from scratch. Let's just uuuuuse a monkey aaaaaand, what else? Let's add some pig aaaaaand... fuck it, that's good enough. Mrs. McClusky will never know."

I know these arguments won't faze most of you religgies. You say God can't work through evolution because there wouldn't be any proof of his existence or something. Well, maybe the proof isn't supposed to be in the physical world thus making it into some kind of logically inconsistent and schizophrenic acid trip where completely random things can just happen all the sudden out of nowhere. That would be torture. It's truly a blessing that our world is scientifically consistent and we get to use our wonderful minds to explore and study it.

And anyway, what kind of test would this even be if God's existence was totally obvious and there wasn't any plausible deniability? You wouldn't need church or whatever. You could just look in the encyclopedia and it would be like "Yeah, we proved God that one time, so you better not." It would just be a scientific fact that you better be good and not be a cheater. And half of you would run off and start a snake church and obsess over it all day and never get anything done and totally defeat the whole purpose of being here. How would that be a test of our character? It wouldn't. Thank God there's no obvious sign of God. You gotta dig a little deeper if you want that. Either way, I wouldn't be so sure that the mechanisms of evolution and nature are not another part of his grand design.

Besides, if all of the magical holy stuff was more empirical, then scientists would always be trying to measure it, and then at some point a chariot would crash and the government would recover the angel bodies but deny it, even after the black and white grainy footage surfaced showing the haunting images of autopsies on dead angel bodies face up and naked on cold steal dissection tables with their long blonde hair strewn everywhere plastered to the sides of their faces, just staring up blankly at the ceiling with their mouths open and wings all mangled from the wreck, and for some reason there's no genitals apparently. So, it's like, well, why don't they even have genitals?

Is that what you want? Is that the kind of thing you want to see happen?! Is that what YOU WANT for the world?! **D,X**

Maybe we're just here in this dimension on vacation for a little bit and we're not supposed to be on the phone all day calling home constantly like a ninny. And maybe God isn't the antithesis of science; maybe he is the ultimate science. And you think you're being cool by making fun of the chess club, but God is actually the president of the chess club and as soon as he takes off his glasses and lets down his ponytail you'll realize how hot he was all along.

You know, in western civilization there has been one religion that has dominated the culture for millennia. You know which one I'm talkin' about -- the one that loves to put the symbol of the fish on the back of their cars. The head guy lives in Rome and he has that huge hat. Well, his hat is actually symbolic of a fish too. You know the guy I'm talkin' about. The space pimp.

If that's your thing but you can still retain the ability to acknowledge the critical importance of science and reason, then okay I guess. I can still work with that. But keep all this fishy business in mind for later. The plot thickens. Dun, dun, duuuun.

I've been pickin' on religgies, but what about all the atheez out there that think I'm a gullible unstable mess for all of this mumbling God talk and erratic gibberish. Well, atheez, why don't you try this? Just stop being so smug and cocky for a second and go ask the universe directly whether it's actually conscious or not. Then wait. I don't mean wait five minutes. I mean just wait. And ask polite. The universe reserves the right to refuse service. Then after you've done all that, your subconscious will serve you up some wishful thinking placebo to help fuel your delusions and comfort you.

Not really. Just wait.

Now you think I'm REALLY crazy. It's okay. Maybe I am crazy.

So crazy it just might work.

I haven't been the same since that time that the entire world flooded with water and we few lucky ones were forced to live aboard floating barges to survive, and I met that one traveler on the open ocean and I offered to trade him some real genuine paper I had from the Before Time, and I proved to him it was real paper by taking it out of its canister and rubbing it around under my nose all over my lips and softly whispering "pyepper... ...pyeeeepper... ...have you EVer seen real PYEpper?"~

He only offered me thirty minutes alone with one of the women on his boat for my two pages of REAL paper, and I was like "PFFFFF! I could get half a DOZEN women for this PYEPPER!"

Oh, and by the way, don't nobody try to give me any of that "God's actually a woman" nonsense. The job of being God is a REAL job running the entire universe, not making me a sandwich. C'mon. Obviously a dude. A gay dude, maybe. But female? PFFFFF!

Don't worry. There aren't any of them in here reading at this level. They're off doing dishes or doing one of these play jobs we made for them so they can feel like they're a real part of society doing real people stuff. Although, it might be backfiring at this point. Now they're out here asking for more pay, and that's on top of the original expense we already have to pay to keep them surrounded all day with all of these male actors pretending to be customers and co-workers.

Remember the good ol' days when we could just stick'm all in one room banging away on a bunch of typewriters "for the war effort?" Those were the days. And we gave them those little uniforms and even awarded them sometimes, as if you could get an award for typewriting.

It used to be cute but now they're getting all cocky about it. We're just gonna have to tell'm at some point.

Not to mention all of the things we have to allow now to occur in order to make all of this seem legit, like how we have to let some of them be "doctors" and let them take on real-life female patients so they don't suspect anything, but now those female patients haven't actually been getting any real medical attention for years.

Dear God. What have we done?

Operation Make Believe was supposed to help the world.

)X

3

III -- COME AT ME BRO

So, what exactly the hell is a ManApePig?

Well, it's you. You're the ManApePig.

WOE WOE WOE!! Relax!! I mean it scientifically, dude! Chill, dude!

Besides, I'm just a book. It's not like you can intimidate me into compliance. The words are already written. Put that broken beer bottle down, yuh nut!

Jeez, dude.

Just pull your stool back up to the bar and keep reading, never ending story style... at the bar... where people normally read books.

What was I even saying before your violent outburst?

Oh, yeah. Yeah. So, if you just give me a chance to explain for a second

I'll explain exactly why it is that you are a filthy ManApePig.

Eeeeasy...

Scientifically filthy.

So, what do I mean by that exactly? I mean that you are literally a member of the ape family of animals that has been modified like a hairless pig (Eeeeasy...) and endowed with a big ol' man brain by a very specific type of evolution that you likely haven't even heard of.

That's "man" as in all mankind, to be clear — or peoplekind, if you prefer.

You might be thinking "Who gives a hoot how I evolved? I'm here now, ain't I?" Well, first of all I find it strange that you talk like that, but let me ask you this. What made you so incredibly special that you were able to out-compete every other hominin group like the Denisovans with their enhanced altitude stamina and the Neanderthal with their incredible brute strength?

We didn't out-class Neanderthal by trying to beat them at their own game. We out-proliferated them by our intelligence — an intelligence we gained and maintained by doing something fundamentally different from them. It was something so seemingly simple that it is *nearly always* overlooked.

The key critical deviations for us that made all the difference in the world between you and I and the Neanderthal were dissimilarities in our diets and in the foods we ate.

Hopefully you know by now that diet is important. You can't just eat whatever's clever. There aren't any fancy little British kids with magical wands flowing through your veins using sorcery to convert all the garbage junk you eat into key essential nutrients.

So, what was the secret sauce in our diet? What gave it that touch of magic to transform us into the most transcendent hominin on the planet?

Before we go there, don't get me wrong about one thing, our physical strength still matters. We always need to be healthy, stealthy and spry. But even our most hulked out diesel Chad ancestors were still weak little nerds in the animal kingdom and we got picked on all the time. And by picked on I mean they ate our monkey ass. Ate like... uhh... not like... you know what I mean.

Yes, I know not all monkeys are apes, STEVE! You freakin' nerd. Get out of here before I EAT your monkey ass! STEVE!

Not like... uh... YOU know what I mean! Freakin' Steve.

We all know that chimps are more than 98% the same as us genetically. Well, at least I think we all know that. Chimpanzees are the closest animal to us and share a common ancestor, but they're actually FIVE TIMES stronger than us pound for pound. Yeah, I know, they have weird little bodies and stubbed out legs like you could just feed them skittles and hide them from your mom in the stuffed animals in your closet until NASA suspects something and then they run away and you finally find them laying next to a cold river in the woods all gross and dying. But if that little guy tried to phone home and accidentally squeezed your device too hard he would crush it into a sputtering mess of gizmos and battery acid. They are stronk — especially together!

It's perfectly healthy and natural for us however to intellectually explore our own bodies and mentally touch ourselves to the thought of us running through the forest in a leopard skin toga, leaping off a cliff onto the back of a wooly mammoth and stabbing it in the carotid artery with a bone knife, but that was actually a Neanderthal hunting strategy that constantly got their asses killed. It was incredibly dangerous. We, on the other hand, figured out things like using dart spears and atlatls to take them down at a safe distance.

Really, the sexiest part about us is that twinkle in our eye when we are working things out and testing the fences. No challenge is safe from us when we set our sights and lock on, like an Australian outback dude slowly turning to whisper "Cleverrr girllll..."

Now, just let yourself relax for a moment. Forget about squatty aliens. Forget about Steve. Forget about velociraptor.

Go back deep into the past to a time indeed quite long ago. Go back far beyond ancient Rome, in fact folding it several hundred times over.

Along a seaside bluff calmly sits a young ape mother. She stares in quiet wonder at the endless water before her. Their whole troop looks on in awe, gazing over the rolling ocean. Never in their lives have they seen this and it's unlike anything they have ever fathomed.

They have reached the end of a journey where the river meets the sea. From here there is nowhere left to go. There is no further place downstream. Their eyes study each horizon and every nose turns into the wind carefully sifting through these new aromas and searching for familiar scents.

Across the expanse of the planet and over the immensity of Earth, there will be places of unique opportunity where the fresh water meets the surf.

This is where they first saw seabirds out searching amongst the rocks and collecting nutritious beach food from the sand and pools and out-crops.

In this unassuming first vital moment they had discovered a new form of sustenance. In their own time they would come to adopt it and un-leash its potential like a flood in torrent.

This marked the humble beginnings which became the first flint strike ignition to catapult them hurdling forward through geological epochs and ages.

The Earth's surface would change and the continents would slowly shift as their lifetimes continued passing, forever adapting to their new environment. For thousands of generations, they would look out at that same ocean, while time and selective forces worked patiently to mold and forge them.

On and on like this it went until their descendants finally emerged with a mind so creatively potent it would explore every corner of the Earth.

Each voyaging migration into unforgiving climates and every weapon against hunger they used to form their diets were like one great human story hand painted on the rock face over a deeply ancient history in stone layers that still know us.

4

IV -- NOSE UNDERWATER AND LIARS IN THE SCIS

You wanna see some freak shit?

Start with this. Go look up proboscis monkeys and check out the images and come back when you're done.

Did you see? Whatch'yuh think?

"Eww, gross. They look gross. Eww. Someone kill them. Ewww."

Actually, they're pretty awesome. It's only the oldest adult males that have those extra big floppy gonzo noses. The females and younger males have noses much more similar to a human's... or a house elf's.

It's generally assumed the extra large size nose in the older males is a prowess display like a lion's mane or something. And, yes, their noses look exactly like what you were thinking they look like, you CHILD.

Switching gears, do you remember those sweet James Bond films where the villain had a secret underground lair within a volcano on an island and the only way to get inside it as you approached the island was you had to go underwater through a submarine tunnel until you surfaced inside?

This idea is also similar to a beaver dam or to the experiment where you flip a cup upside down and lower it into the water and it keeps the pocket of air trapped in it and stays dry inside.

Now, imagine a normal cave with a typical front opening like a doorway that just faces straight outward towards the world, and imagine this cave is on the side of a steep mountain near the ocean. When the tide rises up to the cave the water will just rush into it. By the time the tide reaches the top of the cave opening the entire cave will be submerged underwater.

Now imagine another cave nearby where the outside front of it actually looks like a large human nose like the famous carvings on the side of Mt. Rushmore. For this cave the nostril hole of the nose is the entrance that you have to go up vertically through to get in, and the nostril hole opening is a little bit lower down than the floor of the cave.

So, because of this covering over the front of the cave, it makes the cave itself like the upside down cup in the water. The nostril hole opening of nose is just like the cup's mouth opening facing downward. For the cup upside down in the water, you can go up into it through the bottom and into the air inside. Likewise, you can go up into the nose cave through the nostril hole the same way.

It should be obvious at this point, but what do you think happens to this nose cave when the tide rises? Well, as the ocean level raises up past the nostril opening and conceals the *outside* of the nose cave under the water, the water level *inside* the cave just stops right where the rim of the nostril is and the cave stays dry inside just like the cup upside down in the water. So, from inside the cave, when you look at where the nostril hole entrance is, it now just looks like a pool of water. Now it looks like the cool amphibious docking area inside of the secret volcano island lair.

I'm sure you have already guessed where this is going. This isn't just how a nose cave by the sea would be - this is how *your* nose is built. Your nose is a protective hood that extends over the sinus holes in your face, and it exists precisely for the purpose of enabling you to go underwater without having your sinus cavity immediately flooded with water gushing into your nasal passages and down the back of your throat.

Your nose protects you.

The other apes do not have this sinus hood feature we call a nose. Their nostrils are just like the first cave opening that is facing straight forward. If they get their face submerged, then the water will just pour straight into their nasal cavity. It is fairly well known that apes famously hate bodies of water and try to avoid them. Surely now you can better understand much of why that is.

Our nose is sometimes referred to as a "dive nose." This is to say it is for the act of "diving" as in swimming underwater -- not "diving" as in jumping into water head first.

There are some fun-loving zoo people who have trained various ape species to swim in pools. When they do, they prefer to keep their heads above water. But you can see how the ones trained to dive swim with their heads below water will always keep their head facing straight downward like they are snorkeling. This is because they need to make their *whole head* like that upside down cup of trapped air in the water. Other apes that have been trained to dive deeper down in the water will instead keep one of their hands on their nose, pinning it shut, for reasons which are now completely obvious to you.

Now, we can return to the proboscis monkeys. These monkeys are really quite fascinating. They share several extremely unique traits with us humans. Proboscis monkeys are called the most prolific swimmers amongst monkeys. They will swim and dive for the better part of an hour in fresh water and sea water, just as humans do.

In looking at the female and younger male proboscis monkeys one can see the obvious resemblance of their noses to ours. And now you already know precisely why they have those noses exactly what they are for -- diving and swimming.

There are also lots of little types of monkeys that are not quite the expert swimmers that big proboscis monkeys are but are still good underwater swimmers in their own right. Diving macaques are a good example. You can see with them as well the protruding nose feature and the length of their nose bridge.

These diving noses we have are streamlined and aerodynamic to help our face not be such a blunt instrument, as it is with the other apes, and this of course helps us slide through the water a little better.

If you go and look through some of the pictures of diving macaques you can see a number of them with eerily human looking faces.

Spoo'oo'ooky!

There is actually one theory out there that has been gaining some traction and states that macaques are really an Illuminati clone ex-periment where they crossbred and mixed the DNA of hipsters from Seattle with the DNA from humans.

We really do have to pause, however, and just appreciate the fact that the very species of primate that swim and dive are also the exact same ones with these noses, which of course includes us. The utility of the nose is painfully obvious at this point. But when you hear blustery scoffing Savannah rebuttals from the fancy official scientists, just stop and ask yourself this question. Honestly, what are the odds of this perfect correlation happening? What are the chances of this perfect correspondence between the primates that swim underwater and the development of these unique noses? Seriously. The only miracle is the fact they still try to deny this.

It is also worth noting that there are a few other types of land animals that swim a lot with their heads below water - animals like a type of camel and the iconic moose. These *long snout* diving land animals have flappy looking nostrils that actually collapse and seal shut underwa-ter. This tends to be the logical choice with long snouts where their nostrils are on *top*, and so, it doesn't make any sense to have an extra nose up on top of their nose. In the seaside cave analogy, their cave entrance is up on the top of the cave, and so the only thing that works is to have a sealing doorway.

The only other option for swimming mammals is to just have a blow hole, which consists of specialized muscles and valves to seal the nostril completely shut. But that is a much more involved and so-phisticated option reserved exclusively for the far more aquatically advanced species like dolphins and whales that live and sleep in the water full time — not for monkeys and moose that have one foot in both worlds.

Alas, when I was a school boy, not all that long ago, they tried to feed us some cockamamie gibberish that humans had gotten their covered pointy noses from the need to "heat air" while living in cold environments. Uuuuuuuuuh what?! First off, if we were THAT adapted to cold environments we would be covered in body hair and we would not need clothes to survive there.

The only real cold adaptations that some of us have been getting have been lighter skin to absorb scarce sunlight and also adapting *smaller* average nose sizes. Noses are a liability in cold weather. They are just frost bite waiting to happen and they don't provide you with any advantage in the cold at all.

The Golden Snub Nose monkeys of the Himalayas are the most cold-adapted monkeys on Earth, with big beautiful fur coats. They also have basically no nose to speak of and look like some trippy little skeleton faces, but they are still quite cute though. That much everyone agrees on.

The Golden Snub Nose monkeys also have more rounded features. This is because in cold environments you basically want to have less surface area and be smaller and fatter and rounder so you radiate away less heat. What you don't want are a bunch of long skinny pointy body parts which increase your surface area and radiate away more of your precious body heat — you know, body parts like pointy protruding noses.

Our noses are NOT a cold weather feature. Nature has made this very clear.

Our style of nose is quite unique in nature and is unmistakably a feature specifically for diving animals with flat faces. That's it. It's a diving nose.

If we were not an intermediately adapted swimming and diving animal then we would have absolutely no need whatsoever for this big superfluous beak jutting out form the front of our face. There is not a single solitary other purpose for having this pecker or nature would have told us about it loud and clear.

And now you know. This is why you don't look like a skeleton face guy peeling off sweet guitar licks on a death metal poster or even look like that Michael Jackson skin disease guy in the children's wizard movie -- Lord Voldemmmmmm'actually... the one we're not supposed to talk about.

Well, I would love to be a mature adult and just leave the argument here, but I know there will be some anthropology and paleontology cunts out there that will say in whiny little cunt voices "Oh yeah, well, what about nose fetishes?"

Here we go.

So, what do they mean by that exactly?

You know what? Who cares what they mean by that? They suck. They're just mad
because they NEVER win any of the medieval reenactment battles even though they keep trying to sneak in real weapons.

And they keep getting away with it because the COUNCIL of Belthea keeps GRANTING them immunity grace because Spell Binder PORTIUS of the council has a FANCY for Phillandrew of Albhor's older sister.

That's right, Portius! Everyone doth hath knowing already! At the next vote of the realm a fortnight hence, you and your secret yearnings in your loins will be laid bare for all to see and we will RID the council of you and your TWISTED appetites.

Anyway, what the fetish argument means is that often times species and animals will develop traits that serve no practical purpose other than the opposite sex really likes them and keeps selecting for it. In other words, they have a fetish for it.

There are abundant examples such as birds of paradise with fancy tails so massive that they can barely fly. Peacocks as well have enormously large and extravagant tails that don't benefit the males in any utilitarian way and actually are somewhat of a liability and a burden.

You often see these fetish selections in males in big flashy ways. The main problem is that the ladies don't know when to quit. They will just keep selecting for something until it gets so big that it becomes maladaptive and overburdening. You can imagine cases such as the bird of paradise who got the mutation that made his tail just a little bit bigger and it finally broke the camel's back and made him just too heavy to fly. Well, then no female would be selecting him because he would just starve to death.

These fetish selections will max out practical proportions. But they won't exceed sizes that are hard limited by, well, by reality. You don't see male lions with manes so large that they're tripping over them like Rapunzel the lion.

So, what's the short answer to how I know our human noses didn't become hooded and pointy because of fetishes? Well, because they are nowhere near a practical limit in size. They could be way longer. We could have some serious Pinocchio stuff going on at least a full hand's length out from the front of our face. We could have the schnoz version of Mexican pointy boots.

Now that's how big our nose could easily have gotten if we were strictly a land animal and big noses were being fetishistically selected for. Air is far less dense than water, and so the length limit for a nose is much larger. In an aquatic environment, however, a full hand's length of nose would make life much more difficult and it would make turning your head and course correction *far* more cumbersome.

Even if the long nose was a fetish, then it is exactly the water environment that would keep it limited to its current size. It's only swimming and diving that would limit it to these proportions.

Besides, why would women select for men with this fragility right in the middle of their face? Why would they want this thing that is so easy to break and bloody? We have been an extremely violent species. Our pointy human nose bones are a huge liability in a fight. People get their noses broken all the time. That's not sexy.

"Oo'oo'oo. Check out that dandy. Meeeyow. I heard his bones break easy. Mmmmm. Mommy like."

You wanna know the other way I know human women don't fetishize huge unicorn noses? It's this really advanced technique called we can just ask women what they like. Our human women possess the power of speech and they are able to just communicate that information directly to us, once we get past the stage of them getting really mad at us for not already being able to just read the information straight out of their mind and already know exactly what they were thinking.

Besides, studies have already been done to find out what people are most attracted to. It is obviously of great interest to us. Things like A/B studies have been used to reveal that men and women actually have their instinctive preference for a nose and face which is the un-weighted *average* of their population group, not the *biggest*. In other words, we are not into bigger and bigger and bigger and bigger noses. If someone showed up with an elephant nose down below their chin we would recoil in horror, not get all hot and bothered by it.

Nature made the nose for a very clear and obvious reason, and we just rolled with it in terms of attraction. The hot boy was the stud that could surface out of the waves with the water streaming off of his rip-pling hairless muscles and flowing romance novel mane while using his hands to hold a spear and a fish; it wasn't the nerd coming up coughing and gagging using one hand to plug his front sinus hole. The nose was useful just like big muscles are useful, and as such it became attractive to the ladies and they kept choosing it over and over and over and over...

The human nose is only the tip of the iceberg that we will be deep div-ing down to in order to explore its titanic mass and its many breath-taking outcroppings and cliff faces shaped by eons of the dance with this watery world. But just remember that no matter how far we travel together in this voyage of discovery or how many leagues of contem-plation you journey on your own, never forget that the truth will al-ways and forever be as plain as the nose on your face. Just follow your nose - it will always guide you back home.

Of course, one could ask and one should ask how it could possibly be that I am pointing out these earthshaking ideas and the official fancy scientists and anthropologists and paleontologists are not? Who the hell am I? Well, it doesn't actually matter who I am, because I am pointing out *self-evident* facts and presenting sound consistent logic.

The dots have been right in front of you for your entire life. I am just connecting them for us all to see.

I could go off and flaunt all of my accreditations like some delicate prat, but it really couldn't matter less who I am or what I do.

Oh, and by the way, it's "couldN'T matter less" or "could NOT matter less" — not "I kuh care less, BoCletus." If you "could" care less, then that means you DO care to some extent. In other words, there IS a lesser degree to which you kuh care. This isn't rocket science, for fuck's sake.

And "literally" means it actually, tangibly, un-metaphorically, un-alle-gorically, un-hyperbolically happened in REALLY REAL life!

You know our world is ending when you have to start doubling up on words like "really real" and "true true."

If you found the mini-skirt at the mall in your exact right size and you "literally died" then we'd be going to your funeral, weeping and shak-ing uncontrollably, and shoveling dirt onto a box with you in it. WE WOULDN'T BE TEXTING ABOUT IT!

WORDS MEAN THINGS!

You know what? I don't even care anymore. I kuh care less.

You know, when I was a wee child going through school as a wee school laddie, they fed us some certified nonsense about how the greenhouse effect worked. Our science book had some complete bull**** about how the greenhouse effect works because every fre-quency of light comes in through the windows but only some frequen-cies get back out and some frequencies get stuck inside. WHAT?!?!

Then why does the greenhouse effect still work inside of sealed rooms and in dark sheds with no windows at all?! *They* don't have any filter windows. Yeah, the windows are generally slowing down the infrared a little, but that's not the main driver of the greenhouse effect.

I realized this as a small wee child!! The greenhouse effect is just the hot air getting trapped because the area is sealed up. THAT'S IT! And they had all of these official diagrams and technical illustrations in our science books about how it all worked and blah blah blah. And then people said I was nuts for daring to call bullshit on our science book. Well, now of course, they've entirely admitted to the revision I just stated. It begs the question. How many other big official scientific theories are still being shown to this day that are just complete and total nonsense?

So, how is it that so many of our most qualified people so often get so many things so incredibly wrong and can still be so magnificently sure of themselves all the while?

Hmmm, geee, I don't know, maybe academic politics, financier agendas, herd mentality, professional rivalry, intellectual fetishes, pet projects, myopic tunnel vision, confirmation bias, ego trips, religious beliefs, Dunning-Kruger effect where people don't even have the amount of intelligence necessary to realize they are not actually as smart as they think they are, and OTHER THINGS I'M OUT OF BREATH!

Or maybe you just haven't heard about it yet because the Illuminati wants to make you stupid so they can use the strongest amongst you to be bred for games — blood sport games in the arena, where life is cheap, and legends are born.

Besides, in order to get an engineering or science or medical degree you only have to be able to store and recall a lot of compartmentalized data in your head. You don't actually have to be that good at connecting the dots between them and noticing broader patterns.

That's why we have the scientific method — because sometimes in history all of our "smartest" people get it wrong, and by sometimes I mean all the time.

That's how any advanced civilization that hopes to continue being advanced makes sure to check itself before it wrecks itself.

The world has no shortage of self-congratulatory pompous windbags patting themselves on the back like some sort of third world leader with all kinds of oversized jumbo shrimp-platter military ribbons and medals on his chest. Like, really dude? You actually did stuff to earn all of those? Is one of them for being a liar?

When my friends and I were younger, but not exactly wee school laddies, we all drove jeeps and buggies and various off-road vehicles, whether the manufacturer intended them for that purpose or not. Sometimes we would play a game when the timing was right and the situation called for it.

When we would all be driving somewhere together and would come to a traffic stop light, the person second back from the intersection would let his vehicle slowly creep up to the one in front until it gently made contact. He would then give it throttle and start pushing the vehicle in front out into the intersection.

The panic instinct of the person in front would trigger to push on their brakes, but this was a futile act to no avail as his wheels were already locked, whining and chirping across the tarmac. The panic would then rapidly escalate into a fumbling anger and into needless shouting and quite frankly inappropriate verbal attacks of a personal nature. The angry party would next throw his vehicle into reverse and hit the accelerator!

The battle was joined! Two titans now locked in an epic game of push-of-war surging back and forth against one another's power with wheels burning and clouds of jet smoked rubber billowing across the battlefield. One noble warrior desperately trying to keep his vehicle out of harm's way, and the other honorable gladiator desperately trying to push him out *into* harm's way.

This primal contest of wills would rage on, engines shrieking into the night, until a police vehicle would scream up beside us with bulged out eyes yelling "HEY!!! YOU BETTER KNOCK IT OFF!!!" or the traffic light would just change. Without taking my eyes off the prize, however, I would shout back over the roar of motors "PIPE down, copper!! REAL men are doing MEN stuff!!"

For the purposes of the legal purposes and other purposes I can neither condone nor deny the experience of these transpirings. The very last statement of this historical narrative may in fact be only loosely based on true original events.

Oh, and when I say "we" always played this game, I mean it was always just me trying to shove them out into intersections.

The reason I share this fine tale is only to prove the point that sometimes in the world there are things that simply need doin', and you just gotta look around for cops, throw it in gear, and get it doin'.

5

V -- MOTHER NATURE WANTS TO SEE YOU NAKED

Yeah. Mother Nature wants to see you naked all over, mmmmm, yeah.

She wants to see all your hair off, mmmm. Yeah, there it is.

She loves it when she can see every inch of your bare skin because that helps you glide through the water better, mmmmm.

Having almost no body hair and developing oily skin cuts way down your friction and drag so you can slide through the waves all day long. Mmmm, uuuuu, I'm almost, I'm almost, I'M ALMOST, HAVING NO BODY HAIR IS A SPECIFIC ADAPTATION FOR AQUATIC MAMMALS AAAAAAAAAAAAAAAAAAAAAAAAAHHHH Hh-huuuuuhhuuu...

Huuu... give me a minute...

Just come get me in an hour...

I can't feel the legs that this idea has grown.

Alright. I don't even want any drama right now. I just want to relax and talk about little mini hippos.

Yep. Did you know they were going to breed dwarf hippos and bring them over to the United States in the South and release them into the wild? Sure enough. It was called the American Hippo Project.

The idea was that the hippos would go around cleaning up all the waterways that get clogged up with all the water plants. I'm not sure why the idea never panned out. Maybe it was because of the possibility that they could breed themselves back up big in the wild and return to being the single most dangerous non-human animal in the world.

I guess it'd put a damper on your fishing trip to get an enormous set of saber teeth through your boat.

But while you were sinking and a metric ton of pissed-off-for-no-reason hippo was roaring at you, you'd probably be thinking to yourself "Hey! Why is it that hippos and humans are both hairless land mammals? What gives?"

Back on shore I would explain out to you over the bullhorn that humans and hippos and elephants and rhinos and pigs are all land animals that either currently spend most of their day in the water or are directly descendent from a recent ancestor that spent most of its day in water. Elephants even still like to shower themselves in water and roll around in mud a little, like the way pigs enjoy doing it to get cooled off.

Hey, what do you get when you breed an elephant with a rhino?

Pfff. Hell-if-I-know!

HAHA. Get it?! Hell-if-I-know?

Ha! You get it. Wait. Where'd you go? Did you make it over to the other shore? HEY! DID YOU MAKE IT OVER TO THE OTHER SIDE? WHERE ARE YOU?

Meh. I'm gonna keep bullhorning the story. I assume you're still listening.

So, assuming you're not a land mammal that spends its entire life underground like the subterranean naked mole rat, if you are a hairless land mammal, then you are an amphibious water dwelling land animal. And your naked human butt is an aquatic swimming butt.

We've already established that you have a wonderfully adapted diving nose for swimming. Now we've established that as a naked land mammal you are in a club in which EVERY OTHER member is a water dwelling land mammal design.

Yes, a few aquatic mammals still have their hair, but the reason they kept their hair and it adapted to be insulating and waterproof is because their fur was already half way there. They were *cold weather* animals with thick multi layered coats to begin with. It was way easier to just develop waterproofing for these advanced coats than it would have been to entirely lose all of that complex cold adapted fur.

Animals, like apes, from warmer and temperate climates, however, have just a basic simple fur that is far easier to lose than it would be to stack on all the complexity needed in order to have a proper aquatic fur coat.

The only "body fur" we didn't lose is the bushes in our armpits and loins. This pubic hair plays many rolls, but it's interesting to consider which roll is the chicken and which is the egg.

The most simple role some of the hair plays is as a physical buffer for skin-on-skin areas that could get chafing.

Beyond that, it's fairly clear that these body crotch regions sweat out pheromones and smells that attract mates — mates for breeding, not like for being Australian friends. The lower scent zone helps guide the plane to the runway in the bush when there's gonna be a rumble in the jungle, and the upper scent zone beneath the shoulder helps spread your sexy aroma around when you're gettin'-down-to-business end is below water.

Now, the spongy hair in these areas does two notable things. First, it wicks out this sweat and pheromones giving it far more surface area to evaporate off of and spread around to all your fellow office workers so they know where you're at sexually, and second it creates a buffer pad to air-out the spots between your skin-on-skin zones so they don't get some nasty skin infections from trapped moisture and adult greases.

"But what about all the hair on my head?"

OH JEEZUS CHRIST!! Where the hell did you come from?! I saw you go down on the lake when the hippo attacked. I thought you were dead. Good god. Never come up from behind on someone like that. Ohh man.

This is good. This is good. I can stop using the bullhorn.

Okay. Jeezus, dude. Alright. The hair remaining on the top of your head is to shade your head from the sun and help keep your brain cool. That baby punches WAY above its weight in energy usage and needs to be cooled like a heavy-duty computer, because that is exactly what it is.

Having long hair not only provides more shade, but it also creates a huge increase in surface area for water to evaporate off of while you are above water and gives your head its very own evaporative cooler.

When we moved out into the water most of the day, we no longer had the protective shade of the jungle canopy, so we needed to make a few changes. The head bush likely would have been one of the first things to develop kind of like the way it looks on macaque monkeys.

It's also of note that our long head hair in the water may help stream-line us and make our head more aerodynamic or hydrodynamic and help smooth the corners between our head and shoulders.

The best human free divers can stay under water for a long long time with practice. The groups that have even been free diving as a contin-uous part of their existence and culture actually have a larger spleen, which allows them to stay underwater a lot longer. But we would still be spending at least a third of our day with our little heads poking above the water surface, and in sunny places it needs its shade.

"Well, why do I have extra man hair on my front head area then?"

Oh! Yeah! I'm really glad you asked that. Your facial hair is likely still there for purposes of shade and evaporative cooling, but it's actually a better question to ask why women DON'T have hair on their faces.

Woe! Easy! I didn't mean "better question" like yours is bad. I meant better like it's more of a scientific imperative. Woe! Why are you clenching your fists? Scientific imperative just means it's more difficult to understand from a cause and effect relationship. Did that hippo wack you in the head? You're not lookin' so hot. WOE! WOE! I mean one of your eyes is pointing the wrong direction now and you seem noticeably disoriented and combative.

I'm gonna continue on. A leading school of thought just believes that women have a lot of physically neotenous traits. WOE! EASY! Relax! Neotenous just means having mutations which cause you to retain some aspect of childhood. WOE! HEY! Big guy! Same team! Childhood just means a state when humans are much younger and smaller.

Can we just establish that nothing I'm saying is meant as an offense to you or a weird passive aggressive slight or something? Let's get you lied down. I'm gonna call an ambulance. WOE! WOE THERE! HEY THERE! Ambulance just means a pickup truck with a cover on the back and people coming to help you. Nice people. There you go.

For Pete's sake. Alright. It seems like women who had mutations to make them look younger for longer were more selected for over time because the alpha males understood that youthfulness correlated with fertility.

So, women lost their sweet mustaches and sexy lumberjack beards and became physically smaller adults and retained higher voices into adulthood to name just a few of the physical neoteny traits that they adopted.

All those traits are a trick by the way. They didn't actually make the women more fertile. They just made them look that way. It seems that this thing where women get all dolled up has actually been goin' on for a veeeeeeeery long time. Keep tellin' them you think they'd look better without makeup. It's never gonna happen. They've been at this game for about

ONE...

MILLION...

YEARS...

But now you know the science. And you know that under all that blush saying "I'm healthy (X Pick me" you're still dealing with the fertility of a lumberjack.

Speaking of sexy beards, one theory which is probably way more scientific is that men retained their facial hair because they need the extra cooling for their extra big and awesome extra intelligent man brains.

My first and only logical proof is such. Men's jobs are obviously real jobs and women's jobs are obviously play jobs as evidenced by the fact that men can't wait to *not* talk about their job, and women can't wait to definitely talk about their jobs for an hour needlessly after they get home.

As support for my theory I present exhibit A.

So, there I was minding my own business in an actual high-speed chase like in the movies because some nut that was going twice the speed limit honked at me for being in his psychotic way. So, I spoke back to him in sign language and told him to go f*** himself. Then this literally insane person in a huge truck brandishing a gun was trying to catch up and get beside me for God knows what purpose and I tried to prevent him from doing so by weaving through traffic at high speeds with my hazard lights on honking for everyone to get out of the way and driving up on shoulders to get around traffic and going through red lights as soon as safely possible, all while calmly on the phone with the police with this actual insane person following me for miles and miles visibly screaming and pointing his gun and going completely mental.

I was in a literal action movie scene. He was racing after me for miles trying his hardest to get his vehicle up next to mine. He eventually realized by where I was leading him that I was actually luring him to the police, and so he veered off to flee.

That's when I got behind his truck and started chasing him back, hanging my arm out the window and pointing my gun at him.

He didn't even realize I was behind him though. You should'a seen the look on his face when I pulled up beside him at the next traffic light and gunned him down.

This whole incident, however, occurred on a drive home from work, which made it tangentially related to work, which meant there was no way in hell I was ever going to talk about it, and I never did until now, simply to enter it into the record.

I never had even the slightest impulse to share that story with anyone. But heavens to Betsy if Stacy said "Thanks for all the help" but she did it with a mildly flat and sarcastic tone. OoooooOOOOH shit just got REAL.

Like I said — play job.

Oh! You like that story?! Haha. Me too. Don't worry. The ambulance is on its way. Hang in there, my big hippo fisher. You almost caught that hippo. Yes, you did!

Anyhoo. Maybe the extra face hair for extra brain cooling theory needs a little work.

Getting back to a hard science. What's another conspicuous connection to the marine environment?

Well, it turns out that human babies and marine mammals such as seals and sea lions are born with a white waxy coating on them called vernix. As such, it is exclusive to aquatic mammals and humans. The vernix coating is waterproof and antimicrobial and protects the babies from germs in the aqueous environment.

They have believed that the vernix coating is there to protect the babies when they are inside of the aqueous environment of the womb, but if this was the case, then why don't *all* animals have this? Why would the only animals that have this feature other than us be aquatic mammals?

It seems pretty obvious that the aqueous environment germs that the newborn babies are being protected from are the germs in the water that they are born into.

What else is special about babies? Well, despite the fact they're born with an obvious aquatic mammal wax layer on them, they can naturally swim for the first six months of their life and beyond if they stay in practice. They instinctively swim and can surface for breaths and can float on the surface because of their perfect body fat ratio. When toddlers go into the shallows they also learn to walk on their own with their little baby feet on the sand and with the water supporting their body weight.

We have way too many aquatic mammal adaptations to list all in this one book, but another big one is our dive response where our body knows it's underwater and knows that we're not just holding our breath in the shower or something.

When we dive our body does things like slow our heart rate in order to make our oxygen last longer.

Our body KNOWS when it is diving. It KNOWS water.

I myself was even born in water and I loved it there. I was perfectly happy until they pulled me out, and I've been pissed ever since.

Even the bases of our fingers have webbing between them unlike any other ape. The *only primates* that have that webbing are ones that swim! Did you hear what I just said?!

For those who may be confused, if you wanna get all technical about it, primates are a broader classification that includes monkeys and apes. Apes are just one type of primate. And humans are just one type of ape.

....primates
...........apes ....monkeys
...humans

But I will reserve the right to refer to any type of primate as a "monkey" whenever and however I see fit. And when I say "human," I am referring to us Homo sapiens, not any other type of hominin "human."

Shoot. That's another thing I have to explain. Hominins (with an "n") are all the human-like apes such as ourselves that walk upright on two legs. When you see the word "hominin" think "human-like." This includes the Neanderthals and Denisovans and us Homo sapiens and all the other Homo-this and Homo-that between us and the basic jungle apes.

The more general category is hominids (with a "d"), and this family includes all the apes. So, that would be us and all the other hominins and the gorillas and orangutans and chimps and bonobos. But the line between hominids and hominins can get a little blurry when discussing the earliest apes to branch off into the upright walking, hominin family. So, if I use the wrong word from time to time I DON'T CARE... freakin' STEVE!

.... hominids
.... hominins
.... Homo sapiens humans
. me not caring

(The broadest category is hominids. Then, within that you have hominins. Then, inside of that you have Homo sapiens humans. Then, within that you have me not caring.)

Getting back. As we are told, however, we are the only single webbed hand primate that has this webbing for some completely different reason.

Suuuure.

Really, how badly do we need to torture logic? Somehow we are the only hairless land mammal that supposedly isn't meant to spend its days in the water like ALL the rest. We are the only species of mammal at all born with vernix on us that doesn't actively live in the water right now. And we are the ONLY PRIMATE with finger webbing that is not there for swimming, as we are led to believe.

Yeah, right.

And this is just a few major things. There are way too many nitty gritty things to cover that would bore you silly, like special organs and bones inside us and all sorts of stuff.

The excuse they give for our finger webbing was that it was for extra grip. Really? Gripping what? Fruit and coconuts? The stuff jungle apes already grab? They have way more need for palm grip than a Savannah ape. Why don't any of them have this grip webbing?

Or was the extra grip for throwing rocks while we persistence hunted on the Savannah? Too bad that's not how you throw a projectile. You hold it in your fingers, not your palm.

This definitely isn't the first bogus theory about human development that has come from people who sit around all day in their heads and never actually go do things like physically throw a baseball, POR-TIUS!

The standard ape style hand is already specifically built for grabbing. Every time I look down at my wonderful, grasping, ape-style hand, I thank a tree. We owe them so much. They were the first ones onto land to pave the way for us, and they started as just simple green moss. Then they became small little weak shrubs that couldn't get it up. But then nature gave them wood and they got hard and we just started grabbing them. We already have these hands which are perfectly de-signed for gripping wood. We don't need any webbing for that.

Not to mention, we also have the webbing on our flipper feat.

Really though. Go look at images of female and young male proboscis monkeys playing in the water.

Go look at Jane Goodall with her chimps. You are looking at small hairy arboreal tree apes sitting next to a slender hairless water ape. That is what you are seeing.

Either that or it's a girl getting hit on by a bunch of Italian guys.

"WOOOOOOE! HEEEEEEEEY! WUDDU YUH TALKIN' ABOUUU'?"

Actually, slicking back our hair is a human thing, because we're the ones meant to be wet. Being in water is entirely natural for us. Even being in colder water. We have a special type of fat called brown fat around our vital organs and concentrated in our upper abdomen. This special brown fat is literally there to generate heat when we are specifically in cold water.

Many biohackers now take advantage of this and take cold showers and sit in tubs of ice water up to our chests for extended periods of time. This may sound horrific at first, and we all know the terrible shock of a sudden dowse of cold water, but our bodies know exactly what this is. They kick on the brown fat like a little nuclear reactor in the core of our body, which starts generating and dispersing heat to the body.

Biohackers use this for things like burning lots of calories while doing nothing and also getting the lymph system in our body to start circulating automatically, which is really good for us.

Cold dips and breathwork actually have some pretty astounding effects on our body. You would be amazed. But it's what our bodies are specifically built for – the breathwork associated with swimming and the thermal regulation needed for diving in water of almost any temperature.

There was a group of people who lived in southern Patagonia at the extreme southern tip of South America and they would famously walk around butt naked in the heart of winter and go swimming in the frigid Antarctic water like it was nothing.

That type of thermal regulation is like a muscle that most of us do not use anymore, but it is astonishingly good for our health to reawaken this sleeping ability.

There have been swimmers that have swam across the English Channel and across from Alaska to Russia and from South America down to Antarctica without wearing wet suits in water that was only a few degrees above zero.

Now, that's an extreme example of cold, and the main woman that set these types of records was pleasantly plump for more insulation and was generating more heat by staying constantly in motion, but people that start swimming in colder water in general are amazed at how quickly their bodies adapt and how readily in time they begin to crave the cool dip.

We're also able to do this because of our subcutaneous white fat layer beneath our skin. None of the other apes have this type of fat layer the way we do and it actually closely resembles the blubber layer of super advanced marine mammals. If you ever called someone a land whale, you were actually more correct than you even knew.

You know what I mean, like the time you saw that gigantic fat land whale woman on one of those mobility scooters in the U.S. riding down the sidewalk.

At least you think it was a woman.

You saw her chuggin' along on her mobility device in a trucker's hat and shooing away small school children trying to climb aboard her to get to school.

You had to wait while "she," you're guessing, slowly crossed the street in front of your car. Then her scooter wheel dropped into a huge pot-hole and she spilled over into the road.

Your friend Jeff was with you in the car and totally freaked out for some reason and said "We gotta help her!" And before you could ask him how he knew it was a "her," he was gone and running over to help. But then he just ran up and stood there like "WUDOO I DO?! WU-DOO I DO?! WUDOO I DO?!" while she was just sloshing around on the ground, waving her arms and gargling and making frog noises.

Your friend Jeff was completely in shock and frozen. So, you yelled out of the window "Jeff! Get out of the way! I'm gonna try to push her up with the car!"

Like that kind of land whale.

You had to get'r out of that high traffic area before a speed boat hit'r.

HEY! It ain't MY fault she ate half the friggin cheesecake in flippin Bwoston, Massachusetts. It's only MY job to point out she's fat. How's else is she gonna realize she's fat if people don't keep constantly reminding her?!

We should remind everyone of everything all the time so they don't forget. Just sit in the airport and point things out. "Hey, buckteeth! Yeah, you. Yes. That's why I'm pointin' at yuh. You've got bucked teeth, sir. You might wanna get that fixed."

"Yoe! Lady wit' duh weird shaped head! ..."

"... look, look, see, this is how I do it. Now, this is how you do it. Yuh see? Fix that. Okay. Who's next?"

Hey, how's my hippo fisher doin'? Hey! You awake? Takin' a little nap over there? Whatever happened to that ambulance we called? Should'a been here by now. I don't see anything. Meh. Ambulances are overrated. What doesn't kill you makes you stronger... or leaves you permanently crippled. Not bad odds.

Yep, yep, yep. Lots of crazy stuff happens in traffic, though.

One of my favorites is when you're at a traffic light waiting to make a turn from the center lane of the road and the other direction of traffic is taking their turn and some of them are making turns across the road, which brings them cutting across right in front of you all nice and slow, and you always think "They better not clip the corner of my car on the way by or Im'a be piiiiiiiiiiiiiissed."

And there's that weird little tension because they're thinking the same thing like "What if I accidentally just clipped the corner of their car? I wonder if I could gun it and get away." and you look back at them reading their mind and wag your head "No... you would not get away. I've got a horse's power fifty eight hundred liter V20 infinity stone in this thing, even though it doesn't look like it, and I would catch you."

But it's this strange little intimate moment where they're crossing directly in front of you nice 'n' slow and almost facing directly towards you. And just as they're going by they look up and you both make eye contact through your windows like "Hello, stranger. I wonder who you are. I wonder where you're going today in your big fancy car."

And then they're gone.

I always wanted to use that moment of eye contact to just throw my chin up while looking them dead in the eye and mouth the words "*fffff-fuuuuuuuuuuuuuuuuuuuu ckyooooooooouuuuuuuuuu*" and follow them with my head while holding eye contact the whole time so they know I'm definitely talking to them.

Leave them to ponder that the rest of their life.

"What did I even DO?!"

6

VI -- BLUFFY THE
VAMPIRE FAKER

There we were in one of the most Godforsaken deserts on Earth in the heart of Arabia with daytime heat so hot it would cook you alive and nighttime delights so charming as to introduce you to things like camel spiders larger than a man's hand and faster than streak lightning.

My friend James and I were bunked up for the night in our large canopy tent with a sheet divider down the middle for privacy.

James was telling me from his side about how we hadn't seen a woman's face in weeks and he was describing what the last women he had seen looked like. I said over to him from my side of the tent "Mmmm. Keep going. This is working for me." So, James replied "Well unzip your sleeping bag so I can do it to the sound of you doing it."

You see, the human mind is an amazing thing. We have such wonderful imaginations. Sometimes we imagine ideas which are extravagant in their eloquence, and other times we conceive of notions that are complete poppycock horse sh**.

One such idea is the wildly popular theory of human evolution known as persistence hunting!

The basic theory of persistence hunting goes that since humans have no claws or fangs or chimp strength and are basically weak little by-atches, we became land...

Sorry. I have to stop to laugh.

The theory goes that we became one of the *land predators* by just running after our gazelle prey until it finally keeled over from overheating.

Just one of the gang -- lions, tigers, bears, naked weak slow fangless clawless not even intelligent yet monkeys. Yep. Just hangin' out with our bros, the other land preds.

I actually initially thought this idea was quite amazing, ironically, since I myself did long distance cross country running, making nationals every year I ran and placing top thirty each time. Not too shabby. I first got into running after some bullies were picking on me and my friend Jenny started calling my name out over and over and yelling at me to run away. So, I started running as best as I could and then all the buckles and straps on my leg braces just started exploding off of me and I nevEr sTopPed rUnnInG."~

The idea, though, is based around the fact that humans cool themselves down by evaporative cooling through sweating out of specialized glands in our skin for secreting an aqueous salty solution -- sweat.

We can jog for extreme distances like ultra marathoners do because of this type of cooling. Most animals on the other hand cool themselves through their breathing and panting, which means they cannot huff and puff *doubly* hard while they're running in order to both get enough extra oxygen AND cool down at the same time.

Thus, we just slowly jog after them and track them until they keel over, or so the theory goes. It turns out, however, with all the people groups around the world still living in essentially their same ancestral Paleolithic ways, the phenomenon of persistence hunting is virtually non-existent.

There is one group of native Indians in Central America called the Tarahumara that have incredible sets of legs on them and regularly run ultra marathon distances between tribes.

The Tarahumara are somewhat similar to the specific tribe within Kenya that are the Spartans of running and who culturally self-selected over time for extreme pain endurance and thin calves and ankles, which provide a major mechanical advantage while running.

The Tarahumara, however, are the only group who have been known on rare occasion to actually run deer to death.

I just wanted an excuse to say Tarahumaaaara a bunch.

Ayahuascaaa!

Central America is one of the most beautiful places on Earth with a rich and vibrant culture of life. If you're lucky enough to go there you can enjoy so many of the enchanting experiences this land has to offer.

You can stay at friends' houses where you have to keep your head cocked to one side the whole time to keep from dragging it across the ceiling.

You can practice your mission impossible last second backwards limbo lunge to keep from going eyeball first into ceiling fans.

You can stay with another friend who said he had a spot for you and then discover that by "spot" he meant the back of his truck outside exposed to the elements and you wake up in the morning with a strange man staring at you, standing in the back of a jeep, wearing a uniform, holding a vehicle mounted machine gun, aiming it directly at you, first thing when you wake up in the morning.

"Uuuuuuuh... Do I put my hands up... or is that a sign of disrespect in your culture?"

You can have some of the most delicious shards of broken glass street tacos you've ever had.

You can get a picture taken with a real live monkey and then laugh and start to look down and then the woman who handed you the monkey panics and says
"DON'T LOOK IN THE EYES!!"
"DON'T LOOK IN THE EYES!!"

Your stupid friend can sleep with a prostitute and then spend the rest of his life gagging every time he smells baby powder for reasons he still refuses to explain.

It's like Las Vegas if it was its own subcontinent.

And this is only a sample of the endless possibilities of ...
~~Centraaaaal Americaaaa ~~

They actually do have some of the most *perfect human food* in the world that combines vitamin C rich fruits like lemons and limes with incredible seafood. Definitely try the ceviche.

The Tarahumara are really cool though. The reason we don't see anyone hunting like this, however, is because it turns out it's actually pretty ret****d. The wear and tear on your joints and knees, EVEN WITH heavy collagen consumption, is debilitating if you do it constantly.

This sustained endurance running is one of the worst things we can do to ourselves for free radical oxidative damage and aging. It's better for us to do impulse interval exercise like how children and puppies play. You don't see kids just running in circles non-stop around the playground for hours without stopping, or if you do you might want to forewarn him about girls like Jenny.

The calorie consumption of the hunters to have to do this and bring the catch all the way back to the tribe is absurd. Or bringing the ENTIRE tribe along for the hunt is even more absurd.

Not to mention, they actually did an incredible experiment and gathered some of the best long-distance cross-country runners from around the world to persistence hunt some antelope in the plains area of the United States Southwest where there's virtually nowhere to hide except over the horizon. Aaaaaaand the antelope handily trounced all of our best runners. It was a total joke.

It was like pitting Sloth from The Goonies in a cage match against a bunch of hungry pit-bulls. There was no chance.

But that didn't dissuade the fans of the run-forever-and-then-back-again-with-something-as-heavy-as-you-are theory.

There's another reason you don't see modern subsistence level Pale-olithic people running around doing ultra marathons, at least not the way you see people do who drive those mullet vehicles that are half-car-business up front and half-SUV-party in the back, and always post inspirational memes like everyone didn't already become immune to them years ago. Yeah, yeah, yeah. It's not the depth of the journey, it's the depth of your soul or some shit.

It's because indigenous people don't have fancy water bladder back-packs and a convenient way to carry around a ton of hydration.

There also aren't many checkpoints set up on the Kalahari with little plastic cups of electrolyte sports drink that you can just whip onto your face and then throw on the ground as you shuffle along toward that shining city on the hill when all your countless hours of training and sacrifice will finally pay off and you'll cross that golden finish line -- this thing you've been completely making your identity about and never stopped talking about at social gatherings for a year, and no one else has ever started caring about for more than five seconds.

"What? You ran a long ways? What, like was it through enemy gun fire and snipers? No?! Oh... Okay. Well, I got a better one! You guys know how I told you I like to check my own ass for..."

Humans can run like crazy but if we don't seriously rehydrate and re-place our salt and electrolytes we will start to cramp up quite badly the worse it gets and we'll eventually get incredibly sick.

If you notice, when we get sick they give us salty soup with soluble electrolytes and minerals, and almost always when we end up in hos-pital they give us an IV with a saline solution, which is effectively the same type salty solution as our blood and as sea water. They've even used sea water in IVs before during desperate times of war.

Getting low on salt and electrolytes makes you very ill, which is why we've learned to administer it to those who are infirm.

The idea that primitive pre-human running apes were hoofing it around the scorching Savannah for miles and miles and miles losing liters of water and entire grams of precious body salt and electrolytes without any timely ability to replenish it all is absolutely mental.

Maybe a few runner apes could replenish their electrolytes by FI-NALLY catching an antelope after a brutal marathon of tracking pursuit and then biting into its neck with their wicked chimp teeth and sucking the blood out like some crazy demon monkeys, but what if they don't catch the antelope? They're fuckin' dead. Their success rate has to be one hundred percent. Nothing is 100%. No predator is that good.

Imagine stumbling in to the finish line of a brutal marathon and being told "Sorry, mate. We forgot all the water and sports drinks. Gotta run back."

We lose one and a half liters of water an hour under these hot or strenuous conditions. Where are we getting all the water to replenish this when we've ran and ran and ran so far that we've gone all the way to the horizon *numerous* times from the location we started?

And what about every other member of their pack that have been waiting around back home sweating their asses off in the sweltering Savannah heat, losing their own water and minerals?

The way humans excrete their valuable salt out of their skin is absolutely remarkable. We take it for granted, but it's NOT normal. It's a very unique thing in the animal world.

Most land mammals can cool themselves through panting and the moisture that evaporates out of their breath without taking their essential salt with it. They're bodies go way out of their way to preserve their body salt and electrolytes, and the animals themselves will go to extremes to get salt in their diets.

Moose and elephants will dive down into fresh water to get to plants with salt. Goats and sheep will climb massive cliffs and ruminants and other animals will travel miles and miles and go into deep caves to access areas where the geology has exposed salt in the rocks.

The rain and hydrologic cycle of Earth wash free salt and minerals out of the land and back to the sea, one way or another.

Our bodies act like they're spoiled for salt. What in the holy hell is an animal like that doing landlocked so far inland?

The fact that humans just gush tons of salt away directly out of their skin is like an amateur rapper who lights cigars with burning rolls of cash and throws stacks of bills into the air at clubs makin' it rain on them... when he hasn't even produced a single album yet and still lives at home with his mom on government assistance.

"BOY! Why you takin' pichuz for the internet burnin' my RENT MONEY ON FIRE... ?!"

It's completely ludicrous.

So far we've watched and seen how bonkers it is to assert that we were running such insane amounts constantly just to survive, and then we also observed how humans blow their whole salt paycheck in the first five minutes and are in a world of hurt on land without it. There's just one more angle to this we need to look at.

As touched upon in a previous chapter, some human populations ended up in cold environments AFTER becoming big brained and intelligent. They used their intelligence to make the technology we call clothing, which protected them from the cold air and buffered them from the selective forces which would have minimized their body's surface area heat loss and which would have reduced them down to short little fat round people with no necks.

Some people like the Inuit and Eskimo actually have started shifting back toward lower surface area bodies from living in environments so brutally cold that even thick clothing aren't enough to compensate for the icy death grip of winter.

Wearing clothing, however, did nothing to shield their faces from the cold and thus did not shield them from the selective pressures that reduced nose sizes to protect them from frost bite.

Clothing also couldn't change the fact that they needed to be light skinned in order to soak up as much coveted diffuse sunlight as possible. We are not cave dwelling naked mole rats. No matter who we are, we gotta get our tan on.

Maybe try putting your TV right up against the window and running the video game controller out onto the lawn so you can play outside every once in a while.

If your parents still get mad about your screen time, don't tell them that nothing you ever do is ever good enough in your life and nothing you ever do ever makes them happy no matter what. Just tell them you're doing a paleo lifestyle experiment for science.

If that's still not good enough for them and they threaten not to buy Evil Redemption VII for your birthday, don't threaten them back by saying that then maybe you'll just kill yourself and see how they like that. No, just tell them that you have multiple storage backups of documented digital evidence that they actually feel like they're not truly smart enough to do their jobs at work.

They'll interrupt you at this point with an exasperated scoff and a chuckle of false confidence with involuntary micro tells of nervousness and fear. Just continue on unabated that you also have evidence that they feel like they are seriously sexually inadequate, and they are harboring violent sexual rape fantasies, and they also have feelings of ecstasy about having bowel movements, and if anything ever happens to you it'll all get leaked to the press.

You won't be having any more problems from those jive turkeys, muh man. Enjoy Evil Redemption VII, brotha!
HIGH FIVE!

So, why in world bring this up? Well, these are not the only examples of humans starting to take on small new adaptations after leaving the beach front property of the garden of Eden.

One of the only other notable tweaks to the basic human starter pack is caused by extreme drought periods, monsoon seasons, and cold winters where land food becomes nearly impossible to find during part of the year. This creates selective pressure for a more cautious, planned out, and fore-thinking temperament.

This adaptation appears to have manifested in much the same way that some wolves became dogs. A dog is essentially a wolf puppy in its mind. That's why dogs would rather curl up next to you while napping on the couch instead of, like a wolf, consider ripping your throat out.

This is referred to as mental neoteny, which is the retention of childhood mental traits into adulthood. In humans this may have been selected for in harsh environments for the imagination aspect of childhood which would aid in mentally simulating future conditions in the harsh times and thus aid in planning accordingly. Quite fascinating if you ask me. This may also have some unintended side effects such as literally developing a more playful eccentric puppy-like personality.

Again, why bring this up? These are a couple examples of fairly simple alterations to the basic template of a human to give you an idea of the relatively small amount of time it has been since the modern Homo sapiens emerged into their completed form. Certainly, it has been enough time for some light customization, definitely, but that amount of time is *nothing* compared to all the time it took to make us into us.

Now, there's one more honorable mention to make here, which is the whole reason for this discussion about post development adaptations.

Sports scientists in their endless quest to gain every micro shred of advantage possible accidentally stumbled onto another specialized human modification.

It turns out that certain groups with specifically West African ancestry have higher ratios of fast twitch muscle fibers to slow twitch muscle fibers. What this means is FAR more powerful sprinting and jumping. The best performers from all other groups of humans are barely overlapping with the worst performers within this fast twitch muscle fiber group.

That super speed would have come in handy once. When I was little

I had these friends Stewart and Andrew who were several years younger, and so they were even more littler. They lived about several clicks away and they were even poorer than I was, so they couldn't even afford used bicycle parts to make their own bikes.

One day we were coming back from their family shack over to my family shack. They were walking and I was on my bike. There was this caravan mobile trailer house thing about half way between our "houses" and we had to cross by it, and these stupid cunts had about fifty German Shepherd dogs for some reason. Why do you need so many goddamn German Shepherd dogs?

But on this day as we crossed by they had both sets of their gates open and this horde of German Shepherds came pouring out onto the street barking like the hounds of hell. I just remember looking back in abject horror and saying "RUUUUUN!!! I CAN'T SAVE YOU!!!" and then pedaling my mix-and-match banana seat girl's bike ass off for dear life.

I literally never saw them again. I have no idea what even happened to them. I never saw them at school again. I have no fuckin' idea what even happened, and I don't think I want to know.

By the way, if you ever want to feel like superman, just go into a shack house or a trailer mobile "home" with wheels under it and just start punching around wildly. Your fists will go through anything they contact -- the doors, the floors, whatever. You can shove your hand through the outer wall and run down the entire length of the house keeping your arm out shredding a hole through the whole side, just jumping through every interior wall in your way. Just make sure to get the hell out of there before the home owner gets back.

They get maaaaaaad.

Oh, and it turns out that we definitely weren't the only ones that developed this type of enhanced sprinting ability. The Neanderthals as well had acquired this incredible trait of being able to dash in explosive bursts. Far from being the stereotypical, clunky, oaf-like cavemen, the Neanderthals themselves were very powerful sprinters. They had a much more robust musculature than we do and an equally impressive bone structure to match.

So, what is the point of all this sprinting talk?

Well, the entire reason that this type of sprinting ability even exists in the first place is specifically because it enables us to engage in the predation style know as ambush hunting.

Ambush hunting involves stealthily stalking a prey animal and very carefully sneaking up on it until you are right up near to it and sufficiently within range to break out of your cover and start sprinting after it. Ideally, this then enables you to take the prey by surprise and to get up close enough to impale it with a spear. This is the basic idea behind ambushing hunting, and this is precisely why this type of intense burst power has evolved.

So, why then is it worth mentioning that many of us now have the ability to be these effective ambush hunters? Well, if we clearly have always had the ability to develop into these super sprinters, and if we originally started out as jungle apes that already possessed 500% more muscle strength than we do now, and if we also were initially equipped with an entire array of ravenous gnarly chimp teeth, then this means we already had ALL of the ingredients necessary to become perfectly viable ambush predators. We had everything we needed to become functionally just like lions. We were perfectly capable of developing the speed, the strength, and the biting power we needed. And we could EASILY have developed all of this lion-like ability way back when we were still those smaller brained jungle apes. This was ages before we ever even dreamed of acquiring a larger calculating hominin brain.

I'll say it again because this is important. We had ALL the ingredients necessary to become a proper certified land predator. The process of becoming a faster runner and getting sharper claws and bigger teeth is actually a very simple and straight forward progression, as far as the mechanisms of evolution are concerned. Basic features like this are very intuitive and elementary, and they can easily develop at an exponentially faster rate than the tedious trial and error of forming an advanced analytical human intellect.

So, the current model being taught right now says that we were running around on the open grassland plains of the Savannah for many hundreds of thousands and even millions of years with our simple ape brains, just scavenging about for rotting leftovers until we were finally able to start murdering things with the deadly arts of fast walking and slow jogging. Alrighty then. We can work with that. Let's go ahead and just say that this was actually the case for a moment. Okey dokey. Well, why then, might I ask, did we never even once start to undergo the process of ever developing any grasping claws or sharper teeth?

If your whole biological existence over ages upon ages traversing epochs of geologic time had entirely consisted of just sitting around all day grinding away on an endless succession of raw land meat, then it is completely beyond question that you and your entire genetic group would be developing one serious set of carnivore teeth. There's no way around it. Under these environmental conditions there is absolutely no conceivable way that you and your entire extended gene pool would ever be doing anything other than perpetually adapting an ever more reinforced jaw and commensurate set of large slicing molars. Every tiny little mutational increment toward having an ever more efficient set of claws and teeth would directly translate into increased strength, better health and improved mating success. Every ounce of this progression would be rewarded by the environment. Thus, all of these small little changes *would* necessarily proliferate. In other words, you couldn't NOT develop these traits while existing through such staggering amounts of time in this manner.

Not to mention, having a nice set of your very own personal shredding claws would definitely be a marked improvement over having to carry around a solid chunk of serrated cutting stone with you all the time. And regardless of when it was, exactly, that you were finally able to reach this milestone of at last being able to fabricate your very own effective cutting tools, by the time this event ultimately did occur, you would *already* have long since been in the process of becoming a paleo Edward scissor hands. It makes no sense at all that we would've spent so much time as such a prolific butcher, and yet against all odds, would have only ever retained just our basic original template of finger nails for peeling off things like fruit skins, rather than actually getting any real claws for peeling off things like faces, Clarice.

What you definitely would NOT be doing is perpetually spending your days out there just mashing away on a bunch of raw muscle tissue using only our flat little frugivore teeth. That's like trying to eat bubble gum. It's a joke. Actually try it. And try to do it without gagging. This is why actual big game hunters are already born fully equipped with a complete arsenal of pointy-stabby-cutty things all over them. Our little light-duty human teeth are just not built for this type of heavy-duty shearing and slicing activity - and they never were.

And it's not like we originally *did* get some of these extra carnivore utensils before ultimately just losing them all again once we finally started developing larger creative brains for making tools and cooking food. We never even acquired any of these advanced attributes to begin with. And not only did we not *gain* any of these extra features, but we even LOST all of the initial strengths that we originally DID possess.

Again, this makes no sense. If we were indeed an indigenous animal from this type of hypercarnivorous environment, then without a doubt, there would have been an enormous evolutionary pressure on us to develop larger claws, sharper fangs and shearing molars, in addition to adapting a turbo force sprinting speed and stronger digestive system. We undoubtedly would have become the ape version of a lion, and this all would have happened long, long, LONG before we ever started becoming exceptionally intelligent.

The fact that we never actually became some sort of predatory lion-ape in this type of harsh and brutal environment just completely defies the laws of nature. And the idea that we would instead have turned into some sort of weak, sweaty, naked, vulnerable, persistence hunters in this dry arid climate, surrounded by super predators and serial killers, makes absolutely no sense whatsoever.

We could easily have become a serious primate predator, and yet we did not.

And again, it's not just that we COULD have become a chimp-cheetah, but by all rights and by all lines of coherent and rational thought, we absolutely SHOULD have.

No, no, no. The assertion that we bizarrely did *not* take the path of least evolutionary resistance makes about as much sense as saying that we stood in front of the ocean and threw a rock at it and still somehow missed. It's an insult to reason.

And if the whole idea of evolution in general still isn't quite your cup of tea and you're not really that interested in delving all the way into it on a textbook level, then I would just say this. The organic selection process from naturally occurring mutations is actually quite a wonderful gift because it means that there is always an opportunity for things to get better. Things can change, conditions can improve and there will always be a reason to hold out hope for a better future, as the saying goes. But without the benefit of the mechanisms of adaptive evolution we would just be stranded forever on this hamster wheel of perpetual nonsense stuck exactly where we are. Lucky for us though that's not the case, and life on this planet still has enormous potential.

One fun and very interesting way to observe the mechanisms of adaptive selection in action is to just look at dogs. Dogs have only been adapting away from their wolf counterparts for a relatively short amount of time, and yet they have already made an entire suite of significant changes. One of the many conspicuous differences is that the jaws and teeth of a dog are already appreciably less sharp and under developed, and so they are a little more appropriately suited to grinding on vegetation now and thus somewhat less effective at slicing through meat.

And all of this has happened with dogs in only a miniscule fraction of the amount of time that we hominins have been moving away from our jungle ape counterparts. There is just no possible way in a million years that a burgeoning Savannah ape was out there shimmying around on the prairies, gobbling up bloody giblets of meat for two entire orders of magnitude more time than what separates dogs from wolves and yet somehow never made even the slightest morphological shift toward adapting a more phenotypically carnivorous mouth. No. Sorry. There's just no way that happened.

Let's get real though. The developing ancestors of the human animal never spent any significant amount of time successfully thriving as a serious plains predator of the open grasslands... and we've got the mouth to prove it. We only began mimicking the behaviors of land carnivores well AFTER we became human, but that is not the lifestyle that made us! This is not where we are from.

All of the hominin groups around the world that have ever possessed these enhanced, jet pack, sprinting abilities have each and every one of them acquired these capabilities long after the point of becoming the basic forms of what we are. These features were aftermarket upgrades that were added on later. They are the newest features, not the oldest ones.

And besides, it is a well known fact that human men usually want to be very muscular. They love to work out and lift weights so that they can get as big as possible. And this of course tends to be very desirable to human women. But in nature, this is actually what herbivorous and plant eating animals do. It's only animals like oxen and gorillas that get physically massive like this. They build enormous amounts of muscle in order to become as structurally imposing as possible. Huge muscles are for animals that are built to stand and fight – not ones built to chase things. That's not a feature of wild carnivores. Predators need to be quick and stealthy, and they must strive to be as lean and efficient as possible.

The attraction within our species to the archetype of the hulked out, yoked up, beast mode male is actually an affinity which dates all the way back to the jungles - ages before any thoughts of ever jogging around in the sun or heading out for the hunt. That's one of the most ancient and primal instincts we have. Go check out hairless chimpanzees and you'll see exactly what I'm talking about. Getting bulked out and dieseled up is an archaic throwback to the days of swinging around on tree vines like Tarzan – not from eras spent running around with spears.

And speaking of attraction to muscular males, what about pregnant women? Can you imagine a prego trying to hoof it around all day across the scorching grasslands with everyone else, just jogging around incessantly so that the tribal group can continue chasing after animals on the persistence hunts? Can you picture them actually following along to all of the far flung places that the hunts would constantly be dragging them off to? Does this sound very realistic - expecting a pregnant woman to be running around and doing actual marathons every week? Yeah right. We aren't African wild dogs that can run as fast as a motor vehicle and thoroughly exhaust an antelope within one to two kilometers and then go inhale all the meat super fast before a bigger predator comes along and then return the short distance back to their den so that they can regurgitate it all up for the pregnant women and children to eat.

As slow as we are as humans, it takes extreme marathon distances for us to finally exhaust and overheat an ambulatory ruminant to death – that's if we don't just bore them to death first by how slow we are.

I weirdly happen to actually know a lot about this form of brutal tracking pursuit from actual first hand experience. It just so happens that I myself was in fact a national league, cross country and long distance runner when I was a wee youth. On numerous occasions I personally tested this persistence theory myself. I tried to pit myself as man against beasts to see if I could successfully run a bunch of fat cows to the point of exhaustion. I wasn't trying to persistence hunt them, per se. I just wanted to see if I could actually tire them out to the point where they would let me approach them and I could earn their trust and they would accept me into their tribe as one of them, like Tarzan but with cows. Hey, I was a kid, shut up. I ended up chasing them though for what felt like half way across the continent, running all day across the terrain and nearly into the night. And for all of my trouble I only ever succeeded in annoying them. They would just trot off into the distance and then sit around peacefully eating grass and regaining their energy while waiting for me to catch up again, and then the cycle would start all over. It was a joke. And they obviously were not sharing my vision of becoming Tarzan of the cows.

I even tried this with a baby calf once because she got out into some range that she wasn't supposed to be in. I thought I could easily catch her out there with my super runner awesomeness, but she just kept leaving me in her dust for kilometer after kilometer after kilometer. I was honestly getting pretty demoralized that this new born baby girl was just totally trouncing me. It wasn't until my friends finally caught up on the horses that we were ultimately able to catch her and gett'er back to her mom. Persistence hunting is something that sounds fun to say with your mouth but is actual insanity to literally do.

Anyway, I don't really care what anyone says, there were never any simple apes out there running around like that and then trying to run the entire way back at our blazingly slow pace, jogging through predator infested terrain for marathon distances carrying huge chunks of bloody meat. You might as well just attach a bunch of dinner bells to your butt. Never mind trying to do this whole entire, double marathon, round trip to get there and back all in one day before the darkness comes and you suddenly find yourself at a complete disadvantage in the domain of things much more deadly and powerful than you are.

The only way to make this idea work, where the hunters go and retrieve the meat for everyone else and bring it back to the tribe, is to give them some serious edge weapons so that they can actually defend their kills. But if you say that the hunters were already in possession of these types of weapons, then that would mean that they had already achieved the highest levels of intelligence thus far obtained in the animal kingdom. So, giving them artisan, hand crafted weapons just defeats the whole purpose of saying that they had used this strategy to become intelligent in the first place. You may like to ponder that one for a moment or draw a diagram.

However, if you would rather go back and say that they instead had actually achieved their intelligence first by simply scavenging around the Savannah for road-kill leftovers, then you have to acknowledge that this would've taken an extremely large amount of time. If they had actually spent over one million years literally living like some kind of flightless monkey vulture then they most certainly would have gotten their Nosferatu claws and Dracula fangs by then. So, there again, the Savannah theory fails. And if this whole logic trail of the last paragraph just made your head spin, then don't worry about.

Moving on.

Of course, you could easily just avoid all of these issues to begin with by simply gulping down all of the meat into your stomach for safe keeping so that you can discretely smuggle it back home to your tribe while staying under the radar of all the super predators. Then once you make it back home to all your friends and family you can say "sorry, guys, we weren't able to catch anything" and then say "JUST KIDDING" as you start to laugh and regurgitate up all the food.

Now, this is clearly a really good idea, but how exactly are you supposed to keep from actually digesting all the food before you ever make it home? It takes less than two hours to fully digest a meal, and you're gonna need a whole heck of a lot more time than just two hours to make it all the way back home. We don't have anything like a bird gullet that would enable us to securely hold all the food in our neck so that we can transport it across long distances without digesting it. And if all of that wasn't enough, for some strange reason I've never been able to get past this lingering suspicion that none your pregnant wives are ever going to be very interested in having a little taste of anything you've ever vomited up or regurgitated. It's just this weird feeling I have.

Well, then I guess there's only one thing left for us to do. We're just gonna have to bring the whole entire tribe along for the ultra-marathon adventure. Oh, wait. I forgot that women get pregnant sometimes. Never mind. That's not gonna work. This persistence idea is not going to pan out, guys. Well, it was a great theory while it... wait... waaaaaait a minute... wait, wait, wait, wait, wait... hang on a minute... what if... what if we've been going about this whole thing all wrong? What if we've been going way too easy on pregnant women and toddlers? Maybe they need to get off their butts and stop complaining and start doing some marathons and competitions. Come on, ladies! You got this! Don't worry, we have the unique ability to sweat. That fixes everything.

Seriously though, any theory about human origins that cannot fully account for the ever-present challenges of accommodating pregnant women and small children is just not a serious theory. Women tended to be pregnant a lot. It sort of helped us with that whole keeping our species alive thing.

We would've needed a whole lot more than just the ability to sweat in order for a much simpler brained version of ourselves to even be caught dead anywhere near the gladiatorial events of the Savannah. Not to mention the fact that landlocked environments are inherently hostile to all forms of life such as ourselves which are so oddly in-efficient at managing their salt and water. In fact, whenever you see an organism that is so conspicuously wasteful of a particular nutrient like this, then this is a clear and unambiguous sign that this animal was originally custom built for a place and environment where that resource was in ample supply and relative abundance. It's not "wasteful" if you're in an environment where the resource is everywhere.

The whole reason we hominins expend our water and sweat in this manner instead of just using our breath like the other land animals is specifically because this odd ability is precisely what enables us to lower our body's core temperature while we are underwater, and this is how we are still able to cool ourselves even when we are holding our breath.

Experienced free divers can stay deeply submerged for a very long time. If they were unable to cool themselves through their skin then they would start to boil alive. Burn victims who have lost most of their sweat glands across certain areas of their body often report how incredibly overheated they get and how quickly it occurs. This is no trivial point.

We are endowed with this wonderful ability to perspire across the entire surface of our body so that we may go and spend our days out swimming and foraging alongside the ocean, not so that we may be spending our lives wholly preoccupied with legging it across the prairie.

I am very grateful, however, to everyone who has ever been involved in helping with the development of these thought provoking theories about persistence hunting. They have captured our imaginations and inspired us to dream. These great visionaries have invited us to join them on a journey of discovery and ushered us into a strange new world where we have seen ourselves bounding in great strides across the open plains of the untamed Savannah. We have imagined ourselves running like the wind, wild and free, with our entire tribe and everyone we know steadily pacing by our side completely naked and everybody's stuff smacking around.

Well, it was a beautiful vision... while it lasted.

Anyway, there's nothing wrong with a little jogging here and there, especially when you have all of your clothes on, but too much of it over the long run is definitely not a good thing. This type of sustained monotonous impact is cumulative murder on our joints, and this is true no matter how well we supplement.

The human leg is actually very poorly designed for absorbing impulse shocks. Virtually every other type of fast locomoting land animal has legs that are zig-zag shaped like springs, while we instead have these weird L-shaped legs like golf putters. And it turns out as well that this kind of regular endurance running is unfortunately one of the worst possible things we can do for causing free radical oxidative damage and aging. It's not an intermittent hermetic stress that causes your body's adaptive systems to respond by getting stronger - it's a sustained chronic assault that just grinds away your systems and degrades you. So, if you're still a big running enthusiast like I am, then you really need to keep making sure that you are definitely going big on the antioxidants... among other things.

Believe it or not, however, it's actually a million times better for us to just get our exercise in the form of swimming - to whatever extent it's practical for you. It is very well understood at this point that swimming is without a doubt the absolute best form of exercise for the human body. This is a very important point to understand. Swimming is the only exercise that we can do from the very first day of our lives until the very last and every day in between - purely a coincidence, I'm sure.

I was never really much of a swimmer myself. We didn't have many places to go swimming where I grew up, way out in the desert. But for some strange reason I always had this freakish ability to run forever, and this was despite growing up dirt poor and not having any professional guidance or coaching. I always did everything that the professional running coaches will tell you not to do, and yet I would still somehow manage to dominate competitions. And the fancy kids on the big city teams would always see what I was doing and make fun of me for it, and I never knew why until years later. Oh, and by "fancy," I just mean that they were regular poor people. I was extra poor people.

It all started when I was a wee youth laddie. We had just moved out to this weird little town in the middle of nowhere with only about a few hundred people in it. The townsfolk started to notice how I would run all day almost without stopping, way out deep into the wild. So, they then proceeded to inform me that what I was doing was actually a sport. Yes indeed, it turned out that running a lot was an actual sport. So, they created an entire fake track team from scratch just to take me to the national competitions every year and try to gain some notoriety and prestige for their tiny little micro town.

We couldn't afford proper uniforms, so we had severely mismatched shades of black and cutoff sweater pants for our track shorts. We also couldn't afford a proper advisor, so our running coach was our football coach was our basketball coach was our ping pong coach was our... You get the idea.

We'd go to big races and I'd come in somewhere in the lead with my donated clothing uniform, and then some very long time later long after the last of the real runners had come in, the next fastest of my team would arrive -- good ol' Nick, the fast one.

Then the others would start coming in. The last was usually Steve. Steve was shorter and basically looked like a human lego man with his proportions.

His lego head had freckles and a big gap between his front teeth like Sponge Bob. He initially complained about being on the team and said his legs weren't long enough, but our "coach" just said "Well, they're long enough to reach the ground, aren't they?"

The length of his feet was barely longer than his thick stump legs. So, it basically looked like he had no feet and his legs just went straight down to the ground with caps on the end instead of shoes. He would pretty much just walk the whole way and take snack breaks with stuff in his pockets. He wouldn't even jog off the starting line; he would just start his walk.

Then for the very last stretch he would give us a healthy jog for the crowd and smile and wave, which was weird too because it didn't seem like his body had any joints, like some invisible hand was holding him up and doll walking his block body across the floor.

That was my team, and they were great. We got ironic awards for slowest team ever to complete the races.

But we steady always knew to keep it on the real,
and we never blinked once when teams were talkin' trash,
cuz real gangsta ass thugs don't run from shit,
cuz real gangsta ass thugs cain't run fast."~

WHAT?!

7

VII -- BRAINGHIS KHAN

I was starting my own terrorist group. I needed only the most talented actors who looked the part. We had to look legit.

I actually had a plan to start a hyper realistic corporate training where we would run into your bank in paramilitary balaclavas and check-ered keffiyeh scarfs yelling in Soviet bloc accents then run back to the manager's office and throw him on the ground with a gun in his face, "WHERE KEYS TO SAFE?!" until he pulls a real gun and starts firing on us and we break character back to theater divas.

"Wait wait WAIT! Ouch ouch OUCH! GARY, STOP! WE'RE WITH KEEP IT REAL TRAINI'EEEEEEEEEEEEEHH!!!"

Running away with our wrists limp again, still carrying our play guns.

"HE'S GOT A GUUUUN!!"

"EEEEEEEH!"

Something has been happening to us, though, as we've been moving more into modern times. We aren't the same dieseled up cave commandos we used to be.

Our average body strength and intelligence has been shrinking for quite some time. We've been getting frailer and weaker and dumber and more mentally ill.

What in the world is going on?

I know of a particular community that moved out to the middle of nowhere in the 70's when all kinds of new cults were forming. They thought they were going to go invent some new higher evolved way of living or something, but they just ended up going insane and literally eating each other. Oooooooookay. What the hell was that all about?

This is hardly an isolated incident. This has been happening to humans since the dawn of humans, on small scales and on large scales and on everything in between.

Entire civilizations have gone mad and started eating each other and collapsed when they still had plenty of normal food like vegetables and fruit and crops and livestock. They would just mysteriously lose their minds and implode with no obvious or clear catalyst - no major wars, no major toxic exposure, no reason to be starving.

So what the hell is going on here?

Well, some have blamed aliens. And still others like to even try to claim that the ideas of human evolution I have been discussing in this book belong with Bigfoot and aliens.

Well, smarty pantses, it just so happens I actually know a thing or two about aliens. Something made me curious once about the subject and I ended up going out into the field "with" a group of people that supposedly could get aliens to come fly around, and they had interesting alleged video as well of these things.

And by "with" them I don't mean there in their group with them; I mean there observing them from a distance like animals. I and my brother were a distance well off monitoring both them and the sky to see if they had the goods. It turns out they did not have the goods. But I gave them a fair shake.

We also stayed well off because we realized that there was a distinct possibility that a large saucer craft could show up and a Pleiadian general with blue skin could step out and say "Listen up, you maggots! You all signed up for this program to be ambassadors to the universe, so you are now property of the Galactic Federation of Life. War has broken out in the galaxy and now we need all the fighting power we can get! Grab your gear, buttercups, and let's MOVE! LET'S GO! LET'S GO! LET'S GO!"

Then the leader of the group that organized the whole thing (and is secretly on the Galactic Federation payroll) would tell them "Don't worry, ambassadors! This is what we trained for. This is your chance to spread love to the universe."

But by this point the Pleiadian general would be amongst them in the crowd right in their faces "THIS AIN'T GONNA BE NO LOVE MISSION! THIS IS GONNA BE A GALACTIC BLOOD BATH! NOW MOOOVE! MOVE IT, MOVE IT, MOVE IT! I DON'T HAVE ALL DARZALCLAY!"

Noticing the crowd's reluctance, the saucer would most likely then raise up above them and open a round port in the bottom like a camera shutter and shoot a tractor beam down on the crowd. The only logical thing for the tractor beam to do at this point would be to create a strong horizontal force to rapidly pull everyone to the center with all their lawn chairs and drinks and stuff and compress them into a big smashed ball and start raising them up to the port beneath the craft.

Once the ball of ambassadors and their backpacks and food and stuff got to the port it would become apparent that they were still too big to fit through. So, the only scientifically reasonable approach to this would be for the tractor beam to successively lower them down and then slam them back up into the port until it did it hard enough to get them all through.

Needless to say, we did not want to be any part of this, so we stood off. What I *can* say though is that it definitely was not these aliens that caused those civilizations to start eating each other and collapse. That was ALL on us.

And it was probably for the best that I didn't see any aliens that night. That would've completely blown my freakin' mind out the back of my head, and I would'a made my whole life about it and had like fifty bumper stickers on the back of my car about how we're being lied to.

So, for the LOVE of everything HOLY, can we FINALLY get to the ANSWER of what HAPPENED to these CIVILIZATIONS?

Yes, we can... right after I tell you this next story.

Actually. Forget it. Let's just get to the answer. When we scrutinize the details of these societies there is actually one profound correlation that emerges from the mist of data extremely strong. No matter how well-nourished some of them may seem to have been, there is one specific type of food that is conspicuously absent.

Seafood.

They didn't have any seafood in their diet.

Why does that matter? Well, for one thing a full *HALF* of your brain is made out of omega-3 fats alone! Where do you get these omega-3s to build your brain from? You get them from seafood. Every single *major* source is seafood.

Let me say that again. Every single *major* source of the omega-3s you need to build HALF your brain are ALL SEAFOOD.

Aside from elephants, which are also semi-aquatic, all the biggest brained mammals on the planet are seafood eaters -- whales, dolphins, humans! It's no coincidence! Nothing else on the planet builds brains like seafood does. It is the undisputed heavyweight champion of BRAIN BUILDING! No other nutrient source can TOUCH the number of brain children of the mighty seafood.

THE OCEAN'S BOUNTY AND THE FOOD OF THE SEA... IS

BRAINGHIS

F***IN

KHAN!!!

Alright... I'm calm.

It's okay, guys. I'm calm now. You can get off.

I said I'm CALM! I said get OFF ME!
arrrahrrraaah GET OOOOOFF!! I SWEAR TO FFF-
FUAAAAHHRRAAHH!

OKAY!

Fine. I'll just lay here all night. I don't care.

I can keep doing it right from here. This doesn't even bother me.
OUCH, you're on my ANKLE!

Alright. So, there are a few *minor* sources of omega-3 in some specific
land plants, but they are in a form that your body has to convert over
to the ideal form it wants, and the conversion rate is terrible. If you're
trying get by on sources from land plants, you are basically trying to
catch flies with chopsticks. Yeah, you'll get a tiny little amount, but
that's not enough to build and repair and nourish your brain to full
capacity.

A quick note on the details. Resist the urge for your eyes to roll back
in your head for just one second here. The specific forms of omega-3
your body needs are DHA & EPA. This is the kind in seafood. However, a handful of land plants have a *different* form of omega 3 called

ALA, which your body can convert into DHA & EPA, but the conversion rate is very poor (It tops out at around ~25% in some and is as low as 1% in others). Land plant sources of omega-3 are simply not viable; it's basic math.

I may as well mention this little beauty secret here too. It's also an anti-aging tip. The colors of reds and pinks you see in the marine food web, like in flamingos and lobsters and shrimp and salmon and krill and things, come from a pigment which is the *single most powerful antioxidant* you can get from food or supplements. It's called astaxanthin if you'd like to know.

You don't have to quit being a wine alcoholic and pretending you're doing it for the resveratrol, but just for your own knowledge, that stuff isn't even in the same universe as the antioxidant power of the red and pink colored seafoods. Those of us that get a lot of it notice cool things like increased endurance stamina and being able to stay in the sun a lot longer without burning. Fun stuff.

Back to brain nutrition. Notice we haven't mentioned anything about land meats. We haven't said anything about beef or pork or chicken or turkey or wooly mammoth. That's because land meat *isn't* even on the list. It simply *doesn't* have the omega-3s in it that you need to build your full-size human brain.

About half or more of what will become your adult brain is built while you are breast feeding. In that time your mother's body will take whatever it must from her in order to build you. It works a lot better if omega-3 is in her diet, but it will take it from her either way. Mom should be getting omega-3 to protect herself and to maximize you. If she isn't getting enough in her diet, then her brain will literally shrink in size. When this goes way too far it starts to cause postpartum depression.

As soon as you're done breast feeding though, your development will seriously struggle if you are not getting omega-3 in your own baby diet from seafood or seafood supplementation like clean fish oil. Pay attention, babies. Your little body will be grasping at the trace amounts of omega-3 it can get from some of the land plants you are likely eating. But this is like trying to run your engine on heavily watered down fuel.

And this is where I'll say it. This isn't a mystery or a controversial topic or just a theory. This isn't up for debate or some edgy research we're still not 100% on. This is just a plain and simple nutritional reality that you likely have never been educated on. If you are not getting these sea nutrients, your development will be stunted and you will have lower cognition than you should have had and experience higher instances of things like depression and mental illness and even accrue major effects to your eye health.

In other words, without seafood nutrients you are dumber and more mentally ill and even a little blinder.

Not to mention all the other crazy problems like thyroid diseases and a laundry list of unfortunate things.

The longer this goes on and the worse it gets, the closer you and your population will get to insanity. Your body will be telling you to eat, eat, eat because it knows it's not getting what it needs. So, in your stupor and unstable psychology you'll eventually turn to the last thing you haven't tried yet -- human flesh. You want a zombie apocalypse? Then cut off the seafood supply.

This is why people are afraid to have their cars break down in the middle of nowhere and start hearing banjo music.

Smaller populations got landlocked away from critical sea nutrition needed for brain health and then went all zombo and lost their shit, and their family tree stopped branching and compounded to make things a million times worse, and now they want to eat you and your friends, but none of you knows the first thing about self defense or karate for some damn reason.

Can you imagine if Texas Chainsaw Massacre came after Tony Jaa or Gina Carano?! They would just feed that chainsaw to him in the first five minutes! BOOM! Movie's over!

Besides, one of the most famous observations in archeology was that across the entire continent of Africa, early humans had moved inland into the interior and weren't doing so hot. They only had the most simple set of just a few basic tools.

These interior populations could have been descendant from overflow groups that were pushed away from the coasts due to simple overpopulation, just as we have been doing throughout our entire history.

As it happened, Earth's climate shifted and affected the environments of the interior and forced these inland populations back out to the coasts again and back onto a seafood diet.

Maybe by now you can guess what happened next. Yes indeed. Their brains came roaring back and brimming with vibrant life again. They started making art and jewelry and new tools and inventions and technology and living in caves by the sea.

You want to know what a real cave man looks like? Well, there you go.

"No. That ain't true. I gotta friend who hates seafood and he ain't never ate it and he's bout smart as they gyit."

Really?! You mean your friend Dusty with the dead eyes? You mean the one who still hasn't figured out there are other ways to breathe besides JUST through your mouth all the time?

Same guy?

Is this the same Dusty that lives and works on a farm but also owns samurai swords and nunchucks for some reason?

I just want to make sure we're talking about the same Dusty. I'm thinking of the Dusty that carries around Chinese ninja stars with him everywhere just in case some shit pops off and suddenly that ancient form of mystical weaponry is uniquely suited for f***ing some sh** up.

Even if he actually had enough muscle memory from practice to somehow not freeze up in a life or death situation like that he would just end up nailing the robber in the back with his rage inducing stabby stars and then get himself shot to hell. "AAAAAHH! WHICH ONE OF YOU MOTHER FFFU...?"

Okay, yeah, THAT Dusty. Right right right. Yeah. Same guy.

"But I don't like the smell of fish. Myeh. Seafood smells gross. Wya wya wya."

Well, too bad so sad. That's the smell of life, son. Get over it.

"Myeh, but seafood looks weird, like alien food. Myeeeeeh."

Have you looked in the mirror, you freakin' long legged, flat faced, butt naked, proboscis ape? You are one of the weirdest damn things out there.

And when I say "seafood" I do mean seafood. Fresh water food is excellent and still contains the profoundly critical omega-3s, but they do not have all of the critical *trace* electrolyte minerals that the ocean has, like iodine, selenium, zinc, magnesium, etc.

These ENORMOUSLY important trace minerals are simply not consistently available in ANY land food or fresh water source for long. Salts and minerals and electrolytes get washed away from the land by the rain and rivers and aquifers and the hydrologic cycle, and make their way back to the sea. One example is in the rain forest where the soil is extremely poor and is little more than sand, because the rain washes everything away.

The government in the United States even famously mandated that the trace mineral iodine be added to their table salt because they had such a large intra-continental landlocked population living on corn and beef and pork and wheat, and they ended up with severe cognitive impairment and thyroid diseases and health problems from not getting this and other essential sea minerals, and this was even though many of them still had access to fresh water river and lake foods.

This is also where many of the archetypes originated of the backwoods hillbilly rednecks that were dumber than a box of hammers.

We've got to have those ocean omega-3s, and we absolutely must get all the trace sea minerals. No ifs, ands, or buts... unless you actually WANT to be literally mentally retarded, in which case go ahead and knock yourself out.

As for me, I'm going to keep stating the obvious and acknowledging that the sky is blue.

I'm never going to stop seeking the truth and you can't make me -- not even if the Illuminati tries to blackmail me again with recordings of me baby-talking my dogs in helium balloon voices...

"Hello, hello, hello. How are you? How are you? How are you?"

... or speaking to them like a slow wit, simple farmhand petting a little mouse he found somewhere.

"I'm'a take ceh you, Mistuh Bingus. Mean Billy not goen' hurt us no moe'."

8

VIII -- MAMA'S GOOD
INTELLIGENCE...

... AND DADDY'S LOTS OF MONKEY

I got really good at hand-to-hand combat when I was a little kid because I was always fighting this one girl.

Now, you should never ever hit a girl for any reason whatsoever. There is no justification under any circumstances for ever striking a woman. There is, however, an exemption written into the law for girls capable of putting your head through a plaster wall.

Her name was Becky, and she was built like a brick shit house. She was bigger than any boy... nay... bigger than fifty boys.

She wasn't fat or thin -- just a big ol' girl -- with generations of farm and sadness flowing through her veins, and she could put a solid set of biological brass knuckles across your face.

And for some reason, in any weather, she always was wearing these weird pioneering settler clothes and long sleeve full length dresses with poofy shoulders, and things that adults thought were really cute, but little boys thought were making them suffocate from laughter.

Now, I don't know why it was exactly that Tyrannosaurus Bex always had such a huge giant chip on her shoulder, but what I do know is that there was a high price to be paid for every joke about wagons and dysentery.

The Brecking Ball Beckhemoth had an arsenal of moves. There was the hold you by the face at arm's length as she dropped a meteor of burning fist down through the atmosphere into your head. Then the classic wait 'til you try to throw a kick and then grab you by the leg and start spinning you around in the air while walking you over to something metallic to throw you into.

She was quite creative. Too bad our parents never showed us how to just flirt like normal people.

There are times when the world certainly needs the shameless ridiculousness of men, and there are moments when it uniquely requires the singular touch of a woman.

As nature herself has directed it, there is a division of the useful talents, providing specialization's advantages, such that the roles of the sexes create balance.

The males have their own distinct duties and their specific contributions to make, such as providing their families with resources, while shielding their groups from invasion. They must vigilantly guard against predators and exploit sources of food with inherent risk while modeling assertive behaviors and performing tasks which are difficult and laborious.

Nature of course balanced all of this with the complement of women, who themselves have tended more so toward ceaseless detailed attention for protecting the vulnerable young ones and a close watchful eye for danger and gathering important resources and passing wisdom to the next generation.

These principles are just as true now as they have been for millions of years. That's why nature continues these patterns with the wisdom of an old master painter. Some things just simply work. They are time tested to be quite effective. Each is given special strengths and talents for the mutual benefit of kin folk without them.

For many an age of the past, in a time well before our own species, our predecessors were living down beneath the sprawl of great soaring jungle canopies. These ecosystems were vibrantly brimming with squawking birds and little chattering monkeys. They were built for this primeval living since the late chapters of remotest antiquity.

These labyrinthine forests were to shape their biology, and to spawn the whole diverse primate family. Prismatic fruit was the driver for color vision and thus produced our enhanced optical clarity. Our opposable thumbs were thus selected for climbing and vines and tree branches. The enveloping jungles are what gave us these tools to go explore so many disparate new habitats.

Throughout the natural world, there have always been such landscape vistas where the emerald green shrouds of jungles, fervently press up against their limits, like the winding and rocky shorelines of the vast unyielding oceans with their smooth soft sandy beaches formed by the endlessly crashing wave motion.

These are the borderland regions where blended mergers intertwine, and whole universes intermingle as far-flung worlds impact and collide. This is where tree dwelling species will encounter the deep blue waters and they can live in between the two extremes of towering heights and coral reef bottoms.

The roving bands of primate species will surely experience this convergence. Their paths will cross like destiny when they finally behold the vast ocean surface. Given a sufficient amount of time, these fateful meetings are unavoidably inevitable, like the slow erosion of a steep mountainside under Earth's grand hydrological cycle.

In the epic quest for better sustenance and the perpetually migrating travels, foraging animals like jungle primates will often follow the rivers and channels. These routes will eventually meet ocean, if they continue on their downhill journey, just as the aquatic sea mammals before them whose last visit to the beach became enduring.

From the very first breathtaking exposure to this strange new watery realm, these new things would have seemed quite alien from the sights and sounds and smells. Small sand crabs were strategically scurrying along the sunny and open seaside past surveying flocks of shrill seabirds out searching the pools during low tide. These exotic new foreign curiosities would have been good cause for careful precautions, but in time their observational studies would push them beyond these reluctant emotions.

These daring first jungle explorers started gathering mollusks from along the seashores, finding plenty of coastal food sources like plentiful clams and small crabs and rich oysters. If clearly the seabirds were eating them, then certainly they would at least have been worth try-ing. But this then armed them with a nutritional super power, and their abilities were about to start igniting.

These distant descendants of small mammals from earlier reptiles and emerging amphibians were finally returning to the oceanic cradle and rediscovering their primordial life origins. They had again found the genesis wellspring and now returned with their new special features like dexterous fingers for articulated grasping and color vision for closely studying sea creatures.

When they finally began incorporating these aquatic brain foods into their diet, they would not have needed to wait long for the first stun-ning effects on their quality of mindfulness. The sharp boost to their memory and focus would have been so conspicuously present. No other food web has anything like this. They had uncovered the foun-tain of intelligence.

Generations of mothers became empowered. They could now nourish themselves and their babies, while teaching the precocious wee young ones the life skills needed to feed their new abilities. By supplying the essential sea nutrients to their growing children and their own life-giving bodies, they were thus giving their offspring the ingredients for the next level of development now dawning.

It was thanks to these first enterprising mothers, that we all got this great chance to transcend. This is where we parted ways from our ape cousins and struck it out on our unique quest towards human.

I can say I personally do have a remote sense of what it must have felt like. I experienced a similar sudden input of critical omega-3s.

I had actually gone for many years from my teens into young adulthood without getting sufficient sea nutrition until I rediscovered it again.

The first thing I did was start taking fish oil and I was amazed. I was amazed at how quick the results were.

So, it made me curious.

I looked up the safe daily limit, which equated to one full bottle of what I had, so I personally felt totally safe to start taking large amounts such as several full grams, which of course were nowhere near the amount of an entire bottle.

This is when things really got interesting! I'll spare you the full story, but in the first couple of weeks alone I started remembering my memories and my dreams from my entire life and even seeing how they connect to one another in incredible multifaceted ways. Be forewarned though if you want to see this for yourself, all of your dreams and many of your memories have strong emotional components to them, so to be suddenly flooded with hundreds of them can be QUITE intense.

Don't just back up your files and photos that are in the digital format. Back up the ones between your ears too.

And this is the beauty of the power of sea nutrition. You can prove it to yourself!

You

Can

Prove

It

To

Yourself

You don't have to trust me or science or anybody! You can just do it yourself and get your socks knocked off. IT'S FUGGIN' AWESOME!

And it's not just fish oil that is a supplemental way to get omega-3s. Krill oil is another good way.

I started giving a high quality krill oil supplement to my dogs in their food and gave them a safe double dose. Within a week you could already tell how they had become more thoughtful and contemplative and more interested in studying things. It was hilarious. Even people who meet my dogs always notice how they have a piercing gaze, like they are working things out. Haha. "Why do I feel like your dog is getting in my head?"

Because he is.

Krill oil even comes with an added bonus, but you'll need a friend like the one you occasionally park really close to on the driver's side of his car so that you can't even squeeze a tennis ball between your vehicles.

Krill oil has a natural super antioxidant in it that has a deep crimson pigment. Sometimes the capsules will leak a little and get on the buffer inside the bottle. If that happens you can pull the buffer out with a little krill oil on it and leave it on your friend's desk and just watch him sit there and stare at it and try to figure out why there's a big white cotton pad with bright red stains all over it sitting on his keyboard.

As mentioned before, the omega-3 types our brain and body want are DHA & EPA. DHA is critical for high-speed processing, signaling and thinking. DHA is also critical for eye health, which isn't too surprising since our visual input is a direct extension of our brain and needs work very fast!

It's like when you walk in on your older brother and for a fraction of a second he's still head banging and shredding out on an electric air-guitar to the music, watching himself in the mirror, and he suddenly consciously realizes you're there but it's still gonna take him another 1.5 seconds to physically stop his body from doing what he's doing.

He's just living out his deepest rockstar fantasy right there in front of you and powerless to stop his body for another couple of head bangs, but his eyes are cocked to the side looking straight at you, and he sees you seeing him... with a totally dead serious hardcore expression on his face.

"You want a piece of me?!
YOUUU want a piece OF MEEE?! Well, you can SUCK ON THIS!!
BJJJ BJJJ BJJJ BJJJJ

AIR-GUITAAAAR!!"

I hope he reminds the crowd...
... to get their omega-3 DHA, *mothuh fff***uUuUuS!*

|m|/ > < \|m|
    U

(•) (•)

D:

Alas! Alas, the sea nutrients themselves are not the end of the story. The nutrients are surely the necessary building blocks, but now you need a builder.

It's not enough to have the raw materials for building a bigger brain -- something needs to select it. There needs to be an advantage to being ever so slightly smarter, in order that this mutation will have more than the average number of children and be passed on and continue to dominate in this manner.

So why aren't otters and grizzly bears as big brained as us? They get tons of amazing seafood.

Well, the basic reason is that long long ago there was The Great War between Otter and Bear and Man and Elf.

Wait. No. That's wrong. That was a different thing that happened. The simple reason that otters and bears don't have quite our level of intelligence is because their environments simply are not selecting for it despite the fact that they have all the raw materials needed to build big juicy brains.

Bears have already adapted to be clever enough to know how, when & where to get more calories and nutrients than they can shake a paw at from sources like clams and salmon. There's no significant advantage left to be gained from any small increment in intelligence. They're basically maxed out in their environment.

So, a slight increase in brain size would only be a calorically expensive burden that is more likely to disadvantage the bear who bears this trait rather than help them. You see what I did there?

Now, let's say the bears were to accidentally dig out a Native American style channel trap for catching fish in a river or just happen upon one that had formed naturally. They would definitely notice all the fish trapped in it, but it would require more than just a small increase in intelligence for them to understand the full significance of what they were looking at. Only then would they deliberately seek to recreate it.

And even a fish trap like that is not likely to confer an appreciable advantage over what they are already doing. They have certainly saturated the available niches of their environment.

It's an even simpler story with otters. There isn't really anywhere to go, intelligence-wise, at least not with anything less than a quantum leap in cognition. But of course nature doesn't work like that. It only gives you small little bits at a time to see what works, and that's just the way it otter be.

teehee

So, what in tarnations made nature keep making a beach monkey smarter and smarter and smarter until the damn thing could read a book about evolution?

Well, people can build entire careers on this one question, but we can get to the jist of it right in this moment.

The estuary bay environments have always been a favorite place for humans because they have everything we need. It has the fresh water from the river; it has the incredible food from the sea; and it has some land foods like seabird eggs and coconut and fruit which contains vitamin C.

Some people don't even think we came from primates, but the vitamin C is one of the dead giveaways. Only a handful of species on the planet cannot synthesize their own vitamin C and thus *must eat fruit*, and we are one of them. It was an ability that was much easier to break than it would be to put back together. This vitamin C dependency helped tether us to the land and could have been central to keeping us from going full dolphin mermaid.

Oh, and by the way. If we went full mermaid we wouldn't look like an actual fish with scales and gills and shit. That's a completely different branch of the animal kingdom. If we continued on that path our skin would get more like a dolphin's. I know there is a popularized rendition of mermaids which illustrates them like actual fish, but that is silliness.

Anyway, with these apes enjoying these perfect garden of Eden conditions, it wouldn't be long before populations would approach the maximum carrying capacity of the environment.

THIS is when competition would arise and THIS is what starts to confer advantages on the SMARTER monkeys.

So what's all the competition over? And why would that suddenly start to benefit the smarter ones?

The vast majority of the time there isn't much advantage in Eden for being slightly faster or a little bit stronger. You don't need those attributes to gather your fruits and nuts and mollusks and crabs. Your food isn't running away from you, at least not *that* fast.

Those bigger muscles could only get so big before they were just a caloric burden that would require you to eat more food. Aww, poor muscular guy. :'(

The only major advantages here in this new environment are to find clever little ways to get more calories and nutrients and find little niches to exploit, and BOY are there some niches to exploit in a jungle estuary ocean world. You've got all kinds of things going on. You could make a song about it. You've got the fresh water river and the salt water ocean and the brackish water bay and the jungle, hey, hey.

And you have unique types of plants and animals that live NEAR these different types of water and others that live IN them. There are LOTS of things going on in this area.

The big show, however, is the ocean.

In most places if you were born slightly weaker but also slightly smarter like a weak nerd, your intelligence wouldn't matter and you would just get out-competed on account of your weakness. You would get bullied away from the easier food sources and you wouldn't get chosen first to mate or breed with. Most habitats benefit the jocks, not the nerds.

Oh, and sometimes the theater kids get picked first, like with peacocks. If you get born as a jock in their society they consider you a sucker.
"Hiihiiii... peacock sucker!"

Hey! Don't shoot the messenger. Their words, not mine.

But the ocean, of course, is a total game changer. Mother nature was just waiting for a tree monkey to come along with her opposable thumbs and start fidgeting around in the water. Or maybe God set it up that way to finally give nerds a chance.

There are actually numerous examples of species where they carry genes such that every so often they have a male child that is less masculine and more effeminate and a little tricky and basically sneaks by the stereotypical males and gets resources or gets "access." Nature does play these sorts of games where it's not always one size fits all. If it fits, it ships.

In seaside apes, if you are born a little weaker, you are likely to be sidelined because of your size and as such you would then have a lot more time on your hands, like the proboscis monkey males that sit around all day doing absolutely nothing and touching themselves while the one alpha does ALL the women by himself.

You're not even worried about benching weights on the beach and trying to challenge the alphas, because you know you're weaker and there's just no chance. And you don't have to rear children or breast feed or anything like that. So, now you have all the time in the world to go explore. And if the fates did make you a little more clever, you have the potential and the special opportunity to go and discover new things that no one else has ever seen.

These nerds, however, were never going to genetically survive unless they got a little crazy. Lucky for them, when you have ocean front property the world is your oyster.

If you figure out how to catch the crabs without getting pinched, you can save yourself an hour of playing with the thing until you tire it out. You might figure out how to slap the surface of the water with a switch stick and stun the little fish that are impossible to catch. You could get better at finding the little nooks and crannies where tiny morsels like to hide in the tidal pools and sea cliffs. The list goes on and on.

The ultimate frontier for you, though, is the water. The braver and smarter you get, the more opportunity awaits you in the reefs under the sea. There are so many things to figure out how to eat and plenty of hazards to figure out how to avoid. Incredible opportunity awaits you with both your bare hands and with simple tools, such as using a primitive wooden hook to stick in the clam's mouths to dislodge them much easier. Or you might figure out you can scare lobster and other things out of their hiding spots using a stick and then find clever ways to catch them.

The underwater environment itself is like a womb for the mind and intellectual habits. It suddenly silences all the monkey chatter and the world around you and turns you back into an individual, alone with your thoughts.

There are parrots and birds that can maintain incredible mental maps of their surroundings and retain astounding recollection of their region's topography. Our early ape ancestors would have continuously benefited from being able to possess richer and richer multidimensional versions of these mental maps, along with knowledge of how to apply themselves in many diverse and varying ways throughout these environments.

So, you may not have been able to hang out all day squaring off with the buffest dudes in front of the women, and you may not have what it takes to fight with them over the more obvious and simple food sources. If in this case it happens that you are also not especially innovative then you are pretty much screwed. BUT if you've got a little trick up your brain sleeve you now live in a world with the opportunity to play it... and you've got all the time in the world.

The women may not really need you as a provider since they can fend for themselves in these jungle beach habitats, but if you were the clever monkey finding new tricks, you would actually be more well-nourished and more healthy than many of the jocks, and you can provide a greater abundance of food for her and her children in addition to what she already gets for herself.

Here again, there was a first woman who had the opportunity to step us out into the light of a completely new paradigm. She could choose the usual chest thumping hot boy, or *now* she also could choose this other guy who's obviously kind of skinny and not the biggest stud around but he's definitely well fed and he shows up with a lot of extra food, walking around on two legs a lot so he can carry his excess surplus in his hands. You know, I think I'll give this a chance.

First these women started filling their baby's tums with these wonderful brain foods gathered from the sea, and then they began to select from unorthodox providers who were certainly creative and a little bit witty.

This is the recipe for intelligence. Supply the brain food, then start selecting the provider nerds, and of course live somewhere that this is all possible.

BOOM!

Fast forward this process through staggering amounts of time and you've eventually got yourself a nerd fish monkey that is quite bright indeed.

Got a problem with humanity existing? Don't like all the things we've done?

Blame women.

Much of the psychological resistance to this origin of man, however, comes from the fact that machismo and being the badass shark face-punching provider now has to scoot over and share the stage with more patient and quite things that we cannot force into being and that simply happen in their own way and on their own time.

Bravery and vigor and having the nerve to stare down a predator are still necessary and needed and vital, but they are not the only games in town.

Going back to the value of the clever loner nerd, that right there is the key. You now live in a world where brains can compete with bronze. In your new reality there is alpha male *intelligence* in addition to alpha male *muscle*.

For these early hominids, the alpha males would no longer be just the biggest jocks, now they would also be the brightest nerds.

This exchange of the provider males bringing food and resources as a one-time gift to win her over for the moment and then get some action in return, before a physically large alpha finds out, is a similar method to how certain cuttlefish literally sneak in for some access. Or it's like that super prancy hair dresser guy that is actually getting more lady action than anyone else. This is one viable and very simple means by which genes for higher intelligence could successfully propagate.

You know, there's a name for that sort of exchange though.

In Latin American regions there's a delicious rice drink called horchata, but my little brother and I preferred not to slang it up like the locals and instead just be very respectable and polite tourists and refer to the drink simply by its formal title.

"Yeah, Can I puedo get a number tres ceviche with a large prostitute'chata and no ice. Yeah, no ice, thank you. I'm poor. I need all the drink I can get. Muchas gracias. Which one? Oh, this one. Yeah, that one. Yep. Exactly. The lady-of-the-night-chata. Yes. Thank you gracias."

Still brings it with two thirds ice.

Sheeeee understoooooood. Don't take her side.

Hey! YOU can drink your turnin-tricks-chata whatever way YOU want and I'll drink my money-for-sex-chata the way *I* want!

Anyway. This type of one-time gift giving during mating occurs in a few different other species as well, like a chubby little sparrow like species known as Great Grey Shrikes. Bonobo apes even do gift giving simply to make new friends.

I don't know how many potential nerd suitors would have been propositioning your many greats grandmother with food for sex, but that may have had a little something to do with why the most ancient known human statuette carvings are of fat ladies. The hottest chicks were getting the most offers. So, what are you supposed to do? You gonna let all that food go to waste?

Maybe that's why money was invented. She was like "I can't keep eating all this."

Then some nerd figured out how to turn sea shells into Mardi Gras beads. Those things have been around for a long long time... and basically for the same reason.

Alas, courtship selection may have started as a horchata relationship in the early days, but in time it evolved into something more. It evolved into a contributor provider relationship where the males would actually bring her *extra* food during her pregnancy and then into early child development. This enhanced the chances that his nerd offspring would make it and be healthy. Haha. Nerds! Freakin' nerd offspring.

So, with time the women started selecting provider partners based on their quality as a provider in both quantity and quality. This evaluation was and still is multifaceted and based on a myriad of features like physical fitness and coordination, but it also has a lot to do with what quality of gifts he brings her. Is he bringing her small amounts of the same old played out stuff? Or has he figured out something cool and new or even found a way to bring larger portions than most others can?

This is the secret sauce. This is where intelligence gets selected. Just take a moment to soak that in. OKAY, moment's over!

So, do we see these hypothetical sorts of gift giving providers anywhere else in nature? Well, as a matter of fact we do. And the animals that do it wear skinny jeans too.

Which animals am I talkin' about? Obviously, I'm talking about crows and ravens. What else would I be talking about?

Male crows have to impress the ladies with gymnastics and singing and dancing -- yes, literally doing all of those things. Some males will also do crazy dangerous acrobatic stunts to try to blow the minds of the females. And boy, does it. They watch the death-defying feats on the edge of their seats. Silly animals. Glad we don't do stuff like that.

But there's something else the boy does to try to impress the girls, and it's something so subtle and so simple it could very easily be overlooked. And what IS that thing? Well, he brings her some gifts. It could be one type of food or some other type of snack, or it could be some fancy little bright thing like a shiny Mardis Gras bead off of a necklace to give to her, basically for the same purpose.

By doing this gift giving they demonstrate their quality as a potential provider. They're saying "Look at me! I'm good at finding stuff!" And this is it! This is the magic moment! This is where intelligence gets selected!

But it even goes further than that. She is selecting him on his fitness as a provider because he will bring her extra food while she is pregnant and while they are raising their young babies together. He will help her feed them as a team and they will even both teach them about the world. This is where his abilities as a provider and even an intelligent mentor are really put to the test. If you are a good provider, then your babies are better fed and healthier and stronger. If you are an intelligent mentor and a wise teacher then you give them the survival advantage of quality knowledge.

This is the same provider instinct that grandmas from the old country are programmed with to try to fatten you up.

So, they figured out the power of teamwork for raising children and that 1 + 1 doesn't just make 2, but instead 1 + 1 makes 11.

We don't know yet, though, whether they are aware of the fact that there are only 10 types of people in the world -- ones that know binary and ones that don't.

"So, crows and ravens give gifts? Whoopity doo! Why are you telling me about an animal with a bird brain?"

Well, my heavens. That is a bizarrely adept question coming from someone with such a fresh mouth. I bet you don't talk to your parents that way. Do you? Well, please don't being that gutter mouth into my classroom, or we'll have to have a little talk about your behavior after class. And I do mean an actual talk, not like a setup for a porn scene or something.

Alright. Where were we? So, are these birds even really that smart, or is this just some misery loves company thing hyped up by bird watchers to try to get more of us normals sucked into their degenerate lifestyle?

Well, I'm not sure actually. I'll try to think of a few things and we'll see if any of it sounds like it's exceptional by animal standards.

Let's see. They have impeccable long-term memory. More like im'peck'able. Am'uh right? They have enough visual acuity to quickly learn faces and pick you out of a crowd and remember your *specific face* for years. "Hey, there's that Frank guy that left a bunch of fries under the table a couple of years ago. Long time no see, Frank. KAWWW!"

Isn't that a trip? They can remember SPECIFICALLY you! You're not just some random human. They know YOU!

They aren't just simply one of the few animals on the planet that *use* tools, some of them even MAKE tools from scratch like skewers and fishing style hooks. They freakin' understand what the concept of a hook is. They even comprehend concepts like water displacement. I know humans that don't even understand that.

What else? Oh! They demonstrate key aspect of self-awareness. Did you hear what I just said? Yeah. Self-awareness. They engage in behavior that is indicative of them being able to think in their own mind about what YOU may be thinking about in YOUR mind. WHAT?! "Hey, Frank. You look like you lost something. Yuh need help? We're good at finding stuff. REAL good."

Our ancient ancestors even revered and respected them for their wisdom and intelligence before we went all full ret*** again and turned them into a bad omen or whatever nonsense. They even have distinct funeral rituals when one of their friends or family has died.

They communicate with each other and have vocabulary and language. They can put multiple word concepts together to form basic sentences. They've exhibited behavior of teaching their own learned information to their children and passing on knowledge intergenerationally. That is the very definition of *culture*. Whaaaat? They can teach their kids about you if you suck or if you're a nice lady that will come bring snacks when they ring your doorbell.

BOOF! Mind BLOWN!

"Look out for that guy, Jimmy. He threw something at us once. ... So, don't you forget that. ... You see that face? ... Take a goo'oo'ood, haaaard, loo'oo'ook... at that big, flat, monkey face."

They even like to play like puppies and like to do things that have nothing to do with eating or mating, like using flat pieces of bark as surfboards to ride warm air updrafts. I'm not making that up. Some of them literally wind surf. Are you fluffin' kiddin' me?! \uuu/ Surfin' crows?!?

Of course, it's always important to respect animals for what they are and to not project all of our human values onto them, but the act of playing is by no means an exclusively human attribute.

Even laughter occurs in a large number of species where the same parts of their brains light up in response to the same types of stimuli, like being tickled, and they enjoy it and have convulsive breaths just like us. It's just that they don't have our vocal cords. So, we don't recognize the sound of their laughter, but they are definitely laughing.

Crows, however, are more intelligent than parrots which have larger brains than them, and they are as smart, if not smarter, than apes. And this is despite having such physically small brains. It really makes you wonder about the correlation between brain size and intelligence.

They can even learn our human words and not just simply parrot them but actually put them to use. They've been known to play other animals off against each other and play clever tricks on them and do things like gather up a bunch of dogs and then get those dogs to stampede through a bunch of people so the people might drop something valuable. In that instance with the stampeding dogs the crow actually got the dogs to follow him by saying the human words "Here, boy! Let's go!"

You could go on and on about things they do, like how they know the right height to drop a nut or a *clam* from based on its mass in order to only get it cracked just enough and to not splat it everywhere. Or how they take years to develop as babies, which is another hallmark of high intelligence.

Hell, they might've even been one of the birds that taught us how to start foraging for seafood like this. They do live everywhere, even native to Hawaii.

What would be *really* insane is if one of our nerd ancestors even learned gift giving from them by noticing them giving little presents to each other, or some sad nerd shared a little bit of his extra food with a crow and then the crow brought him a pretty little pebble as a thank you. If you share food with them or otherwise help them they *will* actually thank you by bringing you shiny little presents. And they *do* remember you and your face.

Sure, that's a bit speculative, but I could see that. Giving things as gifts that have absolutely no utility is not something that comes naturally to males. I could see a nerd offering some of his lunch to a chimp chick he had a crush on, but offering a pointless little rock just because it looks nice? Fat chance.

I'm just saying. We take things for granted now, but in a survival scenario and a primitive state I could see our ape'cestors, male and female alike, finding things and thinking "ooo neat" and then just discarding them when they realized there was no food or tool value.

Mothers would be inclined to keep random things away from their children unless it's a "toy" which helps teach them about something they will be doing as adults. In fact, fanciful color on animal organisms in the jungle and in the water tend to be a sign that they are toxic and dangerous. You probably wouldn't want your children developing an affinity for things that look like that. Besides, it just creates an extra chore of figuring out how and where to store the thing. It's not like they had pockets.

So, maybe it wasn't until the act of gift giving from a little friend that the lightbulb went off about the sentimental value that could be imbued onto an object simply by a gesture performed with it. "It was just a speckled little rock, but now it's special because my little buddy gave it to me."

It wouldn't be the only inter-species relationship we've had. Of course, dogs have been our constant companions on every continent. But there's another species that seems far more familiar with us than would seem to make sense. It's another animal that treats us like an old friend.

The species I am referring to is dolphins. They've been known to help humans in trouble and to save people's lives. That's in the water, to be clear, not like if you're stuck in a burning building.

They also help human mothers give birth in the water. How insane is that? We've probably lived with them and possibly crows and ravens too for hundreds of thousands of years. They might remember an old bond between us even though we ourselves forgot it, and they are just waiting for us to remember again. Maybe all of these extraordinary creatures have been our partners on the land and sea and air.

Who knows.

We'll always have some of each of them in us though -- the big seafood brain of dolphin with the problem solver mind of crow and the childish imagination of dogs. Woe.

So, I don't know. Does that answer your question? Does that sound like an intelligent BIRD BRAIN to you? What's uh matter?! No snarky comments now?! THAT'S RIGHT! You think I would bring that hish up for NOTHING?!

KING KONG... ain't got SHIT... on certain MEMBERS... of the corvidae BIIIIRD FAAAAMILY!!!

So, there you have it. That's the recipe for the smartest animals on the planet -- seafood diet and courtship-caretaker selection for being an omnivorous forager provider.

And yet, how would any of this ever have happened on the Savannah? The grasslands are a pretty simple and one-size-fits-all type of place. What does being ever so slightly smarter on the open plains get you, especially if you are physically smaller? It gets you kicked out of the gene pool club is what it gets you.

Not to mention, Savannah survival is a team sport. Going off to do something on your own is a great way to get your ass eaten. You know what I mean.

It's a similar story with dolphins. They are getting massive amounts of wonderful sea nutrients and they have huge brains, but they also tend to live in relatively uniform environments and developed successful hunting strategies in packs using teamwork. Plus, they have those stupid flipper hands.

These pack hunting strategies don't create nearly as much opportunity for slightly more clever individuals to stand out. And these hunting party strategies are inclusive and don't force some individuals to branch off and do their own thing.

But in our complex, hide and go seek, scavenger hunt environment with rivers and jungles and tidal zones and cliffs and reefs and open ocean, there was more than enough opportunity to branch off and niche split.

The emphasis really WAS on the physically diminutive and more modest nerds, though. A male who was born with both jock strength AND nerd intellect would still be inclined like the others toward simply taking the path of least cognitive resistance and just using his strength to physically compete for the more obvious food choices. In this classic competition format his slight cerebral enhancement would tend to only have marginal benefits, if even that.

It would still be the classic nerds that would be carrying the brain ball farther down the field. Chimpanity keeps producing both nerds and jocks for a reason. We need both of them.

Thus, this dualistic alphas phenomenon gave rise to the eternal tug of war burning deeply within women's hearts between the gravity and the pull of two worlds.

There's the sexy hunk who's a little bit thick and definitely more of a broke ace beeyatch, and then there's the other guy with a serious dad bod situation but also has a lot more cheddar and a provider game that's a hell of a lot better.

And no one can tell women what they should choose. They have to find it within themselves -- they've gotten us this far.

Of course, men in these gene pools will end up being a mix of both extremes, but most of them will still tend to lean to one side or another.

It wasn't just the women, though, who inherited a struggle. As is often the case with jungle apes, they will occupy a region where there is no major threat to them from an outside predator or starvation. The threat, instead, can come from their own success and their own expanding population which then must compete for resources.

The dark reality is that this type of overpopulation has been leading to war since we were swinging from branches and limbs -- vicious ravenous wars that do not cease until one entire group of males is completely wiped off the map.

Paleos and Savannists see our killer instinct and often attribute it to chasing prancing little antelope across the plains. But, alas, the instinct comes from long before any thoughts of galavanting around on two legs. The instinct is from slaughtering ourselves. Our predator is each other. Our prey is us.

There are many facets and faces on the jewel of humanity, some are surely breathtaking and others are quite ugly. But all of these things are intimately linked, and life itself is always ebbing and flowing.

The formula for creating us demanded payment. Our intelligence came at a cost. The population success that leads to competition causes both terrible war and innovation.

We're all in this boat together now, thanks to the women. And we're going to have to put our heads together to invent our way out of this so we can eventually find our way back home and get back to what really matters -- hanging out on the beach around a campfire, looking up in the night with wide wondering obsidian eyes, spending days swimmin' and chillin', and checkin' out all the beach bodies struttin' around.

Anyway. We need to get back to the good ol' days when we got to watch women walking around on their two hind legs, wading around in the surf, flashin' T & A all day. Or, in those days, I guess it was more of an A and flapjacks scenario.

Ah, the old days when we just let it all hang out. Being naked isn't always pretty though. I've seen it myself.

I had a job bagging groceries once -- yep, the servant boy on the end that you never make eye contact with cuz you've already blown your dealing-with-strangers, fake enthusiasm all over the cashier and you've got nothing left for the bag beeyatch.

Am'a right?

One time my manager tried to get me to throw a guy out of the bathroom but conveniently didn't tell me why. So, I went over and opened the door to the men's bathroom and THERE was a man obviously from very very far away with very very different customs and traditions.

There he was in the stall with the door open, laying between the wall and the toilet with one arm around the bowl of the toilet and the other hand beating the living sh** out of his hard d***.

It was an instant "AAAAH MY EYES! Oh, you f****n' sick f***!"

It was worse than that game where your friend is reading a book and then asks you what you think of the book and lowers the book down and it's just his balls between the pages of the book. "AAAAAH MY EYES! NOOOO!" Then you try to karate kick the book while keeping your eyes protected with your arms.

You know... THAT game?

You guys played that, right?

Then my manager tried to pull rank and get me to go back in there. I was like "The only way I'm opening that door again is if you give me a handful of flash bang grenades to throw in there."

I ain't tryin'uh get no aids!

You could pretty much get away with any crime if you just did it with your whole junk hanging out, running out of the bank with bags of cash and your scenario flopped out in front of the world. You could just run straight at the cops, flappin' and slappin', and they would dive to get out of your way. "UUUGH SICK!" Scrambling over each other to escape.

Yep. It's not just for doin' crime either. Does some organization keep sending grown ass men to your neighborhood and your house, hoping to tell you all about the good news of something irrational? Just act like you're really interested and keep nodding and smiling, and without breaking eye contact, just...

Actually, no. ... I changed my mind. ... Don't do that. ... That's not what we're all about in this book. This is a family book. ... But what I will say is this though.

If some organization out there just keeps sending grown ass men to your house... so they can share with you the good news or something, you might wanna have some questions lined up for 'm.

Like...

How many wives do I get if I sign up for your thing?

Is this gonna require me to commit suicide at any point?

How 'bout standoffs with the government? You guys ever been in any shootouts?

And finally. Is there any way *I* can interest you in joining *mY* religion instead? ... I'll *double* whatever these guys are promising you in the afterlife. And it's basically like what you're doing now. ... except you'll be selling cookies in a micro bikini.

9

***

IX -- WE STILL DON'T KNOW WHY HUMANS HAVE CHINS

Can you believe they still don't know why we humans have our chins?

I'm serious. They don't know.

This might seem like a weird thing to dedicate an entire chapter to, but I don't really care if something seems like something.

Trigger Warning: This chapter is a wee bit technical, but I've made it as intuitive as possible.

Humans have these big jutting, square jaw, superhero chins and no one knows why. Many clever theories have been proposed and investigated, but nothing has ever quite seemed to hit the sweet spot, and I'm not even going to bore you with all those loser theories.

So, it's still this open mystery, really, why no other hominins like Neanderthal or Denisovans have our type of chinny chin chin. They have these sort of sloped back chinless jaws, almost like a snout, and more sloped back foreheads than humans.

Our chin is so unique that they automatically know they've found one of us in an archeological dig site as soon as they see that chin.

It's still a mystery though. Luckily, for some of us, mysteries are like catnip.

I don't know if what I'm about to say should be announced while riding in on a Roman chariot while a servant behind me whispers in my ear the words memento mori to remind me that I am but a mortal man, but I'm just going to open up my trench coat and say it. Plus, I don't want some servant class person that close to me. Haha. Eww. Can you imagine?

First, however, there is a little bit of a backstory that helps everything make perfect sense.

Well, living on the beach and wading around in the water on two feet, and carrying gathered food around in our hands, and even treading water in the upright position would all have led the way in making bipedal walking one of the very fist adaptations to occur in the coastal environment.

If you watch chimps when they have to cross water, they do it standing upright on two legs and their increased buoyancy from being in water helps them stay stabilized. Water is the perfect place to learn how to walk upright.

They do walk kind of funny though like when you have to get up too early and you're over the counter trying to remember how to build a cup of coffee.

Can someone just bring my bed into the kitchen and hold me up there so I can do it laying down? There's really no reason for me to have to do this standing up.

Or can I just get one of those sassy rectangle robots from the Interstellar movie to carry me around in the morning?

If you want to get a good sense of what these early upright walking apes would have looked like, go check out Oliver the humanzee. He was a real-life sort of missing link, somewhere around half way between chimp and human.

Oliver wasn't really a humanzee, which is technically half human and half chimp. There actually is no such thing. Humans and chimps are not close enough to breed together, so don't even THINK about it. He was genetically full-on chimp, just a very very different sort.

Oliver walked upright all the time, and he did it in a very natural human way with his legs locked -- not like the way a trained chimp would with his knees still bent. He had a very human-like face that freaked a lot of people out and a much more human-like *hooded nose* and even male pattern baldness like a human man.

Oliver much preferred the company of humans to that of chimps. He also came from the Congo in the heart of Africa, which has been a preserved habitat since extremely ancient times.

It's quite implied that Oliver was something of a throwback recessive set of genetics from the time our ancestors were about midway. This could very well have been a transfer over into the chimp population from our midway ancestors when they were still close enough to breed -- similar to later humans mating with Neanderthal. And this midway point was likely just before the next chapter of our hominin development where we starting fighting off the other ape competition for our new coastal spot and "deselecting" for other jungle apes with any amount of psychological attraction to our coveted new beach front property. They might even still be avoiding the beaches to this very day because of the legacy of our distant ancestors.

But you can go and see old pictures and videos of Oliver and you can look back through time many hundreds of thousands or even millions of years at a living breathing fossil.

Spoo'oo'oo'oo'oo'ooky.

Anyway, water also encourages an upright posture if you are often snorkeling around in it face down, because doing so puts you in the same body position as you would be in an upright standing position.

Humans can actually see underwater *without goggles*. The trick is to get the pupils to shrink to their smallest size. Studies have observed any child being able to learn this trick. This ability surely must have been being selected for during this early snorkeling phase of our development.

Apes that hadn't fully adapted their diving noses yet would have had to keep their faces straight down in the water like a snorkeler to make their head like the upside down cup in the water with the pocket of trapped air in it, as we discussed before.

Upright posture made us even more adapted to the bipedal head angle where our face points the same direction as our chest.

Then, as this development was well underway we slowly started getting better at actual swimming and diving.

Because of our head orientation we could look all around us in very convenient ways while exploring reefs underwater, rather than having to always point our face in the direction we are going like a dolphin.

It was also very important for us to have our head on a swivel since we are not the swimmin'est thing around with the biggest teeth. We have to be able to look around quickly without having to move our whole body every time.

Because the head posture for upright walking was already so established and convenient for surveying the world around us both above and below the water, when it came to adapting to swimming we favored a forehead first swim alignment, rather than a nose first orientation like all the other underwater swimming mammals.

The unique forehead first alignment was a good compromise between the two extremes of head tilt. This forehead first rather than face first swim style helped shrink our snout and muzzle down and flatten our face. This helped round off our head to be hydrodynamically balanced and *more compact* and symmetrical like the tip of a torpedo, thus giving us more speed.

But here's the issue. If we stayed like other hominins and never developed a chin then our head would be nearly a round ball up on top of our neck like a joystick ball on a spring or like a spherical lollipop on a bendy stick.

The problem then occurs when we want to tilt our joystick head back. We've now placed our entire head off center towards the rear and the drag resistance force of the water is pushing our joystick head more upward and to the rear and thus pushing our entire body to veer off course in the upward direction.

We needed something to stabilize us when we wanted to tilt our head back. Instead of just a round ball head, we want a head shaped more like a Spartan helmet where the top is still round but the face area is long and flat and extends downward.

And that's what we got -- these longer straight faces that project down into our protruding chins that extend the line of the face. When you tilt your head *backward*, the only thing that actually goes *forward* is your chin.

Now with your head back and long straight face and forward extending chin you have this great surface from about your nose downward which has a drag force on it wanting to push your head back forward. This force is countering the force wanting to push your head more backward.

And we get great stability from having two opposing forces pushing against each other, like biceps and triceps or like the balance scales for weighing things having weights on both sides instead of just on one side with the whole scale flopped over.

The main benefit though is now your head can be tilted back to see where you are going and the center point of force of the hydrodynamic drag resistance is aligned over the base of your neck.

What that means in plain speak is that the resistance forces pushing back on your head and face are balanced. So, your head isn't being pushed off to one side or the other and causing your whole body to be steered off course, as was the case with the round joystick head.

Another way to understand it is your face and chin are a scoop style stabilizer fin.

The diving nose and rounded head and flattened face and protruding chin were all the logical conclusion of a natural progression, and now humanity has more chin than a Chinese phone book.

There is a principle that is well beyond the scope of this book and this chapter, but it's worth recognizing that certain things in nature just make sense, which is why they will find a way over and over. They don't need one lucky break as is often claimed.

Eyes are an example of something that make so much sense that they have developed from different sources independently, multiple times.

Our eyes are windows to the soul and still quite magical no matter how you look at it. This has inspired claims that evolution cannot produce them. It's a fun idea, but it's not correct.

It starts with photosensitive cells on the body that can sense changes in light. Cells like this on the face near the putting-food-in-hole prove most useful -- the area where the sensory organ for smelling is already located. These skin cells occurring in specific little patches like polka dots give a more accurate detection of direction of light and shadows. Once these polka dot spots start to form into dimples and get deeper and deeper, they get the ever more accurate ability to determine the direction light is coming from, and this dimpling continues until the cavity is a fully inverted bulb, like a round pore in the skin for maximum ability to perceive direction. The pore gets sealed off with a transparent organic film which protects the internal environment of the pore and continues to allow light in. Further shaping in this film causes a lensing effect to make for an ever clearer and higher resolution image. The rest is history.

Every single tiny step towards the ultimate creation of our beautiful eyes makes perfect sense, just as the formula for us as a species makes total sense. I just want people to know that every part of you is perfectly elegant and exactly how it is supposed to be when you understand what we are.

I just want people to appreciate what they have. If I was a doctor I'd tell every parent randomly that their child will never walk again or never be able do some other thing so they would all feel a sense of accomplishment when they "beat the odds."

"Excuse me. We've been getting complaints that you've been telling parents their children will never be able to do things again."

"Did you tell a kid he would never be able to go to the bathroom again and have to hold it the rest of his life and then wink at his parents?"

Uuuuuuuuuuuuuuu'u'u'u'uuuuuuuuuuu'u'u'u'u'uhhh... no, they're liars.

The reason for mentioning all this here, though, is that this is a natural place to point out that, just like eyes developed independently multiple times, the development of human-like species likely happened more than just once and in more than just one place. It then appears that the best form of us won out over all the rest and here we are.

A last interesting note about all that is since we can see from the data and numerical probabilities that our galaxy must have plenty of intelligent life in it, much of this life is likely to be in a very similar template to our own human form, including much of it mammalian since we mammals have a special ability to develop large brains under the right circumstances.

And I know you don't care to know for yourself personally about the alien girls but later on I know you're going to ask me for a friend of yours, and the answer is YES, they will be hot.

I'm not answering your other question.

His other question.

Back to the point! Get it? The point? Of our face? On the bottom?

No, I'm not ashamed.

So, why is the chin so unique to humans? Why don't the other archaic humanoids and hominins have it?

If you look at people's skulls you'll notice that there tends to be some flattening on the upper area of the back of their head, basically back a little ways from the crown. It's most obvious if you look at their silhouette from the side. If you draw a straight line across that flat spot, that line is the direction their body is going through the water when their head is tilted back.

What is also quite fascinating is that the chin actually extends out from the face *perpendicular* to that line. It extends out in the exact direction that would generate the most amount of drag resistance per length of chin, like sticking your arm out of the car window directly sideways, which is perpendicular to the direction the car is going. And at that angle of head tilt, the chin is the *furthest point away* on the head from the center axis of the body in the direction of travel. Being out at this furthest point gives it the most bang for its buck for introducing drag. In other words, it's the most efficient location for this purpose, and we certainly do know how mother nature likes her efficiency.

Now, if you look at Neanderthal and Denisovans skulls, they also had these flattened streamlined areas on the top back of their skulls too. Theirs were just at a considerably more horizontal angle. And the tips of their jaws (their version of a chin) also extend out *perpendicular* to that optimal direction line for them to swim. That same perpendicular relationship holds true for them as well, which is what you'd expect since they are hominin fish monkeys too. Their optimal angle of tilt was just different than ours and at a more horizontal angle. Thus, their jaw tip extends at a much more steep vertical direction downward, whereas ours is much more forward. So, they would have swam with their heads tilted considerably farther back and closer to a face first angle than us.

The difference in head tilt may have been because we swim much faster than they did. They were more squat and cold-adapted than us, with wide barrel shaped rib cages compared to our sleeker human design. It all translates into higher swim speeds for us.

This head tilt angle in us and our fellow hominins is also right about at the perfect balance of allowing us to tilt our head back a little to see forward while still having our eyes largely protected by our brow ridges from the flow of oncoming water. And at this tilted back angle our heads are hydrodynamically balanced over the fulcrum of our spine and neck. Our ape cousins don't have this perfect balance. There's isn't even close.

Also, you can see this perpendicular relationship between the flattened area on the upper back of the head and the direction of chin protrusion in humans as you walk around and observe people from everywhere in the world. If you're having trouble seeing someone's head then just ask them if they can remove their hat or wet down their hair. They will be thrilled that you are studying the shape of their skull and gladly accommodate you.

Now, our jutting hero chin might not seem like it would be big enough to make all that much of a difference, but in dense water a little bit of chin can actually go a very long way.

Airplanes have small horizontal flaps on the very back end of the tail. Pilots give a slight offset to these flaps to keep the plane going straight forward and not veering up or down on its own. Any pilot will tell you that a very little bit of trim and adjustment to the horizontal elevator flap goes a long way. Just ask any one of your many pilot friends.

And like the elevator flap, the chin doesn't just go a long way toward balancing the head, but it also helps keep the head from veering the whole body off course. The more balanced and neutral the drag and lift forces on the body are, the less energy is wasted just trying to stay on course and the more energy can be put into flying forward. It may add a slight amount of overall drag, but it translates to big gains in overall speed. A little bit of chin goes a long long way.

People with a little extra body fat can actually swim faster and longer than a lean counterpart simply because they expend less energy to stay up and more energy going forward.

And conveniently, our chin fins are on the tip of our jaw, which of course moves, so we can customize our viewing angle and drag profile for intense sprints or for just cruising along.

You can get a sense of the importance of these forces by just going out and swimming, yourself. Just tilt your head back a little too far past the equilibrium angle for looking forward, and as you tilt back beyond that you'll feel how your head is being pushed back and it's diverting you off course. That's basically what would happen if you didn't have a human chin and you tried to tilt your head back even a little bit to see where you are going.

Then you can also put a padded chin strap on to extend your chin and see what a dramatic difference it makes in drag -- drag as in the force in physics, not like... the other... you know.

The bottom line is that your chin enables you to have your head tilted back a little and see forward while staying balanced in the drag forces and going nice and straight.

And whether or not it's all just a happy accident, it's still a stunning coincidence indeed that we hominins and humans have all these perfect relationships

Maybe our collective unconscious has been trying to tell us something in these archetypes we are drawn to of the flying superheroes with the big square chins.

At least now our torpedo heads are big enough to figure out how to make things like artificial flippers and outsmart their own drag resistance. We are entered into a completely new paradigm where we consciously affect our own evolution. We don't have to evolve big pointy cone heads to get smarter and stay fast.

While on the topic of mysteries and head drag, it's kind of hard not to mention male pattern baldness.

Male pattern baldness has been a bit of a mystery too but, like a professional swimmer cap, it may have been a move towards decreased drag for the head while in phases of temperate climate when there was less emphasis on shielding the head from the sun. It would potentially make sense since small marginal things like that can make big differences in performance, and that performance was critical for competing males all up out here strugglin' to be the most impressive to the ladies.

As noted in an earlier chapter, having long hair can contribute to reducing overall drag by helping to fill in the gaps between the head and shoulders and make everything more streamlined. This would make sense of why the baldness is specifically in a pattern and not the whole entire head. Maybe Olympic swimming will get a sweet new look soon. You thought birthday clowns were bad before. Wait 'til they can swim like a mother fucker.

Go put *that* on a romance novel.

Ancient ladies voted though; they were pretty big fans of the chrome dome and pony tail combo. Maybe it didn't really matter since everyone's hair was usually wet and plastered to their heads anyway. You know, mullets have been popular as far back in history as we can see. Maybe that was the guys with full heads of hair trying not to be such losers and fake it like they were starting to get genes for going bald.

"Hey, baby. Don't worry. The men in my family are DEFINITELY going bald. Look, I can barely grow the top."

"Okay :)" she says with pretty flowers in her hair so you don't go running your fingers through it and then smell your hand. Hey! Wuddu yuh want? No one invented soap yet.

You can only smooth talk a woman and seduce her with a sweet mullet, though, if you have the power of speech.

We'll bring it on home down the country road with this. While we were on the train to brain town with our ever growing heads and increasing breath control for diving, it was a bit of a foregone conclusion that we would eventually develop the ability to speak. But there are driving forces for auditory communication which are specific and unique to a water world.

When we are underwater we need to use all of our appendages for propulsion and position control and keep our head moving around and scanning our environment. That only leaves us with one single option for communication -- sound.

Even more importantly, when we are between dives and treading water, we often need to keep our hands in motion to keep our heads poking above the surface, especially if we are denser males with lower body fat. And even then our heads still need to be pivoting around and monitoring our surroundings. While treading water we are definitely limited down to only one consistent way to stay in contact, especially if there are big rolling waves blocking our view of one another.

A swim life has an enormous and unparalleled demand for audible language. Our first words as a species may have been things which were just as obvious to the ear both above water and below...

SHARK!

You know, my friends laughed right in my face when they saw me pull out my phrenology kit and say I had to check their skull shapes first before I could be friends with them. They laughed even harder right in my face without even thinking of my feelings when I measured Tony's head and saw he had an underdeveloped lobe for love of home and country, and from his patterns it was obvious he was going to be an unreliable husband.

Well, maybe now they'll agree, phrenology has its benefits.

Maybe now they'll agree."~

10

$$9 \, {}^{3}\!/_{4} \, 1 \text{ -- WERE HUMANS} \\ \text{REALLY CREATED...}$$

... BY A MAGICAL SPACE WIZARD?

Don't wear your freakin' shoes in my house, bitch!

You think I'm cleaning this place all the time so you can just wear your gross parking lot oil shoes in here where I'm laying on my carpet?! You freakin' maniac. Now you just contaminated that whole area over there by the door that you just walked all over with your world-filth and grease shoes. You mother fffffffffffff'freakin' son of a...

Man, your lucky I'm not... mmm...

Oh hey. I didn't see you guys over there. Thanks for stoppin' by the chapter. This is gonna be another sweet one. We've got all kinds of stuff to talk about. Sorry about that guy though. I just had to deal with him first. He's been acting like bitch ever since I safety tested some old rotten food on him.

Hey, don't judge me. I needed to see if the food was bad or not. What else was I supposed to do?

I kept getting sick from something I was eating and I wasn't really sure what it was, but I had a feeling it was coming from this weird old trail mix I had. So, I invited Zack over and offered him some, and he wasn't even suspicious by the way I offered it. He just gobbled it all down without even thinking... like the greedy animal he is.

Then about twenty minutes later he was doubled over, clutching his guts.

As soon as I saw that, in my mind I was like "YES! I freakin' KNEW IT! It was that DAMN trail mix." And he was all pale and groaning and saying "ughhhhhh, my stomach... I don't feel good. I think I ate something..." And I was like, "yeah, you're damn right you ate something. That trail mix was toxic."

"You've got a belly full of poison, my friend."

Don't worry about Zack, aka kaZackstan, aka Zackajawea, aka Burnt Anakin when he shaved his head. I gave'm some charcoal for his tum-tum. He was fine.

That's my older brother, by the way. He made my life hell growing up. So, I can do whatever I want now.

He was always trying to beat me up when we were kids because I kept growing taller than him, and he kept getting all insecure about it, as if I was willing myself to grow taller. Then somewhere along the line I just outgrew him once and for all and there was nothing he could do about it. And I was also really tough by then too, so there was double nothing he could do about it.

Sucker.

I just told him to go get one of those surgeries where they break your legs and stretch the bones longer.

But that just makes him even more mad.

I'll tell you what though, there's still a lot that we can do to heal those old wounds. We just have to learn how to show some kindness.

And what better way to do that, than with some free old trail mix.

We humans are not so great at eating stuff that's going bad. We couldn't even dream of eating something like rancid old rotting road-kill or some festering carcass, even though a true-blue obligate carnivore would just gobble that stuff right up.

A legitimate dyed-in-the-wool scavenger is specifically evolved to eat rotting leftovers. It's an extremely useful ability... if you can evolve it. You would never just give that up. Even if you recently figured out fire like five minutes ago in your evolution, you still wouldn't just lose that incredible ability. It's a survival super power. My dogs and I come across things all the time, like day old dead birds on the ground. It would be an extremely useful ability from a survivalist standpoint if I could just pick the thing up and start munching on it and get all those bird calories for free like my dogs try to do.

If we had ever possessed that ability in the first place, then we would still possess it. This isn't like losing the ability to synthesis vitamin C when you're living in a jungle full of vitamin C foods. This is losing the ability to scavenge things when you're living in a world full of opportunities to scavenge and get perfectly good calories for free.

Again, if we had actually had that ability to begin with, then we would still be like our dogs who can eat our fully cooked food as well as some rotting old leftovers just as easily. ... ... Not only can they digest our newfangled processed foods, they can still eat things like salmonella that would make us sick as hell.

And we supposedly were a bunch of contemporary contenders to the vultures and hyena. ... Well... their stomach acid is ONE HUNDRED TIMES STRONGER than hours!!! ... They can eat rabies and cholera and anthrax. ... ... THAT, ... my friend, ... is a mothuh effen sCAV-ENGer!! ... ... not us.

And, no, fire and cooking don't explain this. Humans have been dropping like flies from foodborne illnesses left and right. Our species is vulnerable as hell outside of a nice, salty, sanitary environment. ... ... Haven't you ever played Oregon Trail. Humans used to die at the drop of a hat in the old west. We ain't no land critter, partner. We're from Margaritaville.

There's a basic rule which states that survival favors the generalists. This means survival tends to favor the animals who can do a variety of different things. So, again, it would be totally absurd for us to just lose some perfectly good natural ability, especially one that would still be so incredibly useful to us.

Okay, well, then what would happen if the rule was different? What would it look like, instead, if the rule said that animals will just ultimately adapt to doing whatever is the most efficient thing for them and then subsequently lose all of their other abilities? Would that explain how we could lose an ability like scavenging? Well, yes. Yes it would. But if that was the rule, then you'd also see things like entire nations of people just living off of raw dairy milk and little else. That would be a far more efficient way to live while still getting nearly all of your essential nutrition. And these people would necessarily have a severely impaired ability to digest anything other than dairy products. ... But we don't actually see things like that, of course, because that's not how things work. It's not merely about what's most efficient - it's about what's most versatile.

And not only that, but scavenging and roadkill are not even the least bit appealing to our species. Little kids will taste test just about anything, but even a little kid doesn't look at a dead bird the same way our dogs do and think "mmm, that smells delicious." It's just not in our nature. And if it really was in our nature, then you can bet your bottom dollar it would take a whole lot more than a few hundred thousand years of fire to erase it.

But let's say it did. Let's just say, hypothetically speaking, that we really did lose some primal scavenging ability from becoming too obsessed with fire and cooking. Okay. Well, then why are billions of people around the world to this very day still eating raw aquatic foods like sushi and shellfish? I thought we became so obsessed with "predigesting" our food using fire that we just completely lost our taste for raw meats altogether. Hmm. That's interesting. I guess that trend only applied to the land meats. Funny how that works.

No. We never became so obsessed with cooking things that we just lost our taste for raw meat and then subsequently lost our scavenging ability. No, that's obviously not what happened. And if it was, well, then you'd never see people eating stuff like sushi and clams and oysters and tons of other raw seafoods. We would be too "obsessed" with cooking after all.

No, the fact of the matter is this, we eat things like raw seafood because we *can*, and we cook things like coarse land meats because we *have to*. That's the only way that we, as a pescatarian animal, can actually digest that stuff. Otherwise you're just having to bite off big chunks of land meat and then swallow it whole. You can't actually chew that stuff. It's like trying to chew bubble gum. That's why carnivores just shear it off and then swallow it.

But true carnivores can get away with that because they have a digestion system that works very differently from ours. They hold onto their food inside of their stomachs for much much longer than we do - anywhere from eight to sixteen times longer. This gives them the ability to fully break down those bigger chunks of land meat. If we tried that though, we'd just end up wasting a lot of it and not fully digesting those solid chunks. I mean, you could survive like that if you really had to, but this would be one of the single worst ways for a hominin to get some calories.

So, chew your food, sonny Jim. There's a reason why we have these mashing style teeth. It's because our digestive system expects a mashed-up purée, not a bunch of solid chunks of bubble gum.

And once again, the whole entire reason why we're always cooking these land meats like this is because we have to. That's the only way the that we can actually break these things down enough to then thoroughly digest them.

Simply put, there are some things we *can* do, and there are some things we cannot do - eating old trail mix being chief amongst them.

As a matter of fact, there's actually a LOT we can't do. And there's even a whole huge list of things inside of us that are either flawed or weak or vestigial or just plain nonsensical.

Biologists can clearly see at this point that our human anatomy is a freakin' mess. It's just a total disaster. If we really were the product of intelligent design, then there is a very real possibility that our Magical Space Wizard was a little tipsy when it happened. Either that or he was a total bookworm who just wanted to watch natural evolution unfold.

Our human anatomy is *seriously* jacked up though. And we're not even talking about pathologies. Even the most "elite" specimens amongst us are a biological dumpster fire.

Let's talk about our eyes for a moment. The inside of our eye is like a dark movie theater. Light comes in through the little pupil hole in the front, like a how a camera shutter receives light through its aperture hole. The light then shines straight back into the dark inner space of the eye, just like how a film projector shines its image into a dark movie theater. This then casts a color image directly back onto the rear surface of the eye, again just like how it does on a big movie screen. However, the main difference here is that the large inner surface of the eye is actually much more like an active TV screen than like a passive blank theater canvas. This hi-def screen at the rear of our eye is completely alive and it's filled with millions of tiny little sensors. These millions of little sensors are continuously absorbing the light that they receive and then sending it off to the brain.

Now, each of these little photoreceptor pixels has its very own little wire to carry the signal away. In some types of animals, like octopuses and cuttlefish, these little wires are all coming out of the *back* of the screen, in the rear, as you would expect. But in our eyes these little pixel wires are actually sticking straight out of the *front* of the screen. What?!

Yep. The little wires in our eye screens are all sticking straight out of the front. So, then they have to stretch all the way across the remaining length of the screen until they can reach a big round hole in the middle. And since these millions of little wires are actually blocking the whole front of the screen, they also have to be transparent like a bunch of little fiber optic cables.

Needless to say, if you had TV or a computer or a smart device and it was built like this, with tons of little wires sticking straight out of the front of it and then running through a big gaping hole in the middle, it would be a pretty bizarre sight. But that is precisely how your eyes are built. And those big gaping holes in the middle give you a blind spot. But then your brain just photoshops it in so you don't notice it.

..........O..X.........

You can see this for yourself by using this visual aid. Just hold it up directly in front of you, and then look at the right-hand side where the X is. And now cover up your right eye so that you're only looking at it with your left eye.

Now, without looking away, start moving the book closer to you and then further away. At some point you'll start to notice that the symbol on the left-hand side is completely vanishing. It's disappearing from your peripheral vision entirely. This is because of that blind spot.

The whole reason, though, that we ended up with this carnival attraction deformity is because one of our extremely ancient ancestors, from one the most archaic chapters of life, accidentally got their photoreceptor cells wired in backwards. But since it wasn't really causing any catastrophic problems, this mutation just continued and kept on going.

This oddity in our eyes, however, is quite far from being our sole exclusive design flaw. There are many many mANY more where that came from. And I'm not even gonna list 'm all, because it gets pretty boring pretty fast. But suffice it to say, that almost all of 'm are pretty technical in nature. ... like this one for example. There's this nerve up in our head that has to go all the way down into our torso and then all the way back up again. It gets wrapped around this completely unrelated anatomical feature while we're still in development. And so, this nerve just ends up having to make this huge round-trip journey for absolutely no reason whatsoever, other than the fact that it develops out of sequence within the fetus.

Awwwwwwwww... poor little fetus.

He's trying his best, okay?! How's he supposed to know what a *sequence* is?! He's barely even zERO!

Awwwwwwwwwwwwwwwwww...

What if we figured out a little IQ test for fetuses? ... just like a little test he can grab with his hands.

Never mind. But like I said though, a lot of these little glitches aren't exactly the most riveting anatomical subject matter.

Some of these little defects in our manufacturing were never really under a lot of pressure. And that's because none of 'm were really causing any problems. And nature doesn't care if something is "perfect" or not - she only really cares about what gets the job done.

These design flaws that we've been discussing so far have been inconsistencies which are essentially innocuous. They're just hanging out and minding their own business. But we're also riddled with a host of other problems which are anything but innocuous.

So, what sorts of flaws are we talking about? Well, the two most famous examples are our knees and our back. These two specific areas are absolutely infamous for their tendency to sustain injuries. Even the strongest human beings alive are always just one little angular tweak away from a major debilitating injury.

Then the next two biggest troublemakers are our shoulders and our ankles. These two major joint complexes are just constantly getting sprained and injured, especially during sports and exercise.

Each of these four specific anatomical regions will continue to faithfully serve you and perform its roles, just as long as you regard it with precaution. They all have their limits. And regardless of who you are or how invincible you might feel at the moment, these limits are really quite considerable for all of us. What's also pretty interesting to note here, is that all four of these weaknesses actually derive from the same original catalyst. ... and it's a rather simple one at that.

But before we can get to that we have to discuss the exception. There is a sole member of this group which stands out amongst the rest. One of these four major body parts is special. That body part is our shoulders. This is the one we really wanna focus on.

Our shoulders are one of the most injury prone areas of the body. They are constantly being torn and damaged and sprained. This can happen during sports or exercise or during just about any other type of strenuous activity. So, this has led many experts to assume that our shoulders are just more one of our design flaws.

The main reason that it's so easy for us to injure our shoulders is simply because our shoulder bones aren't really attached to any other bones. Our shoulder bones are just floating there, for the most part. They're only really being held in place by the surrounding soft tissues. And that's why we can wiggle them all around and move them every which way. So, whenever we exert a force against our arms, the strain is almost entirely being borne by the muscles and tendons around it. Unlike basically every other part of our body, the shoulder doesn't really transfer its forces onto other bones through direct contact.

The only exception to this isolation is the collarbone. It connects the shoulder bones to our sternum, right in the middle of our chest. But the collarbone is mostly just a pivot. It's mainly just there to keep our shoulders and arms from getting pulled off. The main reason it's there actually is to enable us to hang from branches. The skeleton hangs from the collarbone which itself is then hanging from the shoulder bones. In other words, when we are hanging by our hands from a branch, the weight of our body is being held up by these very secure bone to bone connections from the chest to the collarbone to the shoulder. Or, if you'd like to put it another way, you could say that our collarbone is there for bearing the tensile forces of hanging.

So, it's clear to see at this point just exactly what makes our shoulders so vulnerable. But what do the mainstream schools of thought have to say about this?

Well, it just so happens that our shoulder's incredible agility has one very unique side effect. It enables us to throw stuff. A practiced human child can throw a ball with about FIVE TIMES the velocity a fully grown chimp. And that's despite being *five times weaker* than them.

So, here's where it gets interesting. If you take all of that incredible throwing power and you put it directly behind a well crafted dart spear, it can finally be enough to down an animal.

Hunting with a throwing spear is extremely difficult. That's why you don't really see anyone doing it unless they're fishing. Modern tribes prefer blow darts or bows and arrows. The spear itself has to have just the right flexibility and weight distribution; you have to throw it using an atlatl; it has to have a very hard bone or stone tip on the end; and finally, it has to be thrown with a great deal of power and precision with a very large range of motion. And that's precisely what our shoulders finally gave us. Their fluidity, flexibility, and agility is a pretty big deal.

So, anthropologists just assume that our shoulders have evolved for throwing stuff. But then they further assume we haven't had enough time yet to become injury resistant. So, their basic thought is that we spent millions of years developing a throwing shoulder but not developing it well enough to resist injury.

They just assume this same basic idea about a lot of our vulnerabilities. The presumption is that our species simply hasn't been around long enough to become genetically stabilized, so we're still this sort of weak, fragile, unfinished product.

But as far as our shoulders are concerned, the only thing that really makes any sense from their perspective is to believe that we developed them from a need to be throwing stuff constantly.

Well, I just have to be direct about this. This idea doesn't actually make any sense. As we well know, adaptive evolution works in small little itty bitty steps. When a small new mutation occurs, it can be beneficial, detrimental, or indifferent. If the mutation is advantageous, then it becomes reinforced and selected for. This process will then repeat itself, over and over and over, until those small little itty bitty steps have finally added up into something huge.

Now, when it comes to throwing a projectile for hunting, it either works or it doesn't. You either *can* throw a spear effectively, or you cannot. There's no middle ground. There's no trophy for second place, much less a trophy for fifty eighth place. If you can only throw a spear with about half as much power and accuracy as what's required, well, then this does absolutely nothing for you.

This is no trivial point here. A person's throwing ability is either good enough or it is not. In order to hit a smaller target like a fish, you will need speed to track its movements, power to deploy the projectile, and then accuracy for the small target. So, you will need speed, power, and accuracy. Alternatively, in order to down a larger target, like a ruminant, you will also need an explosive burst of raw power. All of these things require agility and fine motor control in no small measure.

So, the ability to throw a spear well cannot proceed from the act of actually throwing a spear. It has to come from some place else. In other words, it has to occur by accident. It has to be a side effect of some other process. It's not just something that would simply develop of its own accord.

So, I'm just gonna go 'head and give this theory a little help, since it obviously needs it. A much better supposition would be that throwing stuff was originally selected for through rock throwing.

You could, for instance, speculate that rock throwing helped hominins fight each other. And maybe it could even help scare off some bigger predators. ... Maybe.

Alright, so let's look at that idea of throwing rocks to intimidate bigger predators. ... Oh wait, they don't care. They kuh care less about our stupid rocks. You throw a rock at a lion, it'll do precisely *nothing*. ... Okay, never mind. Then let's talk about throwing rocks at our fellow hominins.

Oh wait! ... I almost forgot. ... Before we talk about throwing rocks at each other's heads, we first have to talk about jackals.

There's another theory that rock throwing could've been used in order to scare off some of the smaller types of predators like jackals. The idea goes something like this. First a bunch of apes go out and try to track down a group of jackals that've just killed a small antelope. Then they start making all kinds of noises and shrieking and screaming and weakly throwing their rocks at them and trying to scare them away. Then they go over and finish eating whatever's left of the little antelope.

Alrighty. Let's try to flesh this idea out so we can make sure that we're understanding it correctly.

So, these loud, clumsy, stinky apes are just planning to sneak up on some jackals apparently - animals that have a phenomenal ability to detect sounds and odors from a very large distance away. Okay. Not lookin' so good. But okay.

And these jackals are also quite familiar with these weird ape animals by now. They know how they'll try to steal their catch if they're able to. Thus, they are *acutely* aware at this point that they need to hunt as *far* away as possible from these ape creatures.

But then, apparently, these ape creatures are *still* able to spot these little jackals, from an extremely far distance away, so they can just start running like hell and trying to make it out to them in time. ... And, of course, this can only work if the jackals have actually hunted something smaller – something which is just a little too big for them to eat really quickly or run away with. Otherwise, these apes will only succeed in burning up an enormous amount of calories. They'll be running all the way out there for nothing. It's extremely critical that our apes are getting out there quickly, because there's just no way, in a million years, that we can support any number of individuals off of just a few old nasty scraps and some left over bone marrow.

So, according to our theory, these not-so-fast-running apes are inexplicably able to get all the way out there before the jackals can finish eating everything.

Ooooooookey dokey.

Well, this is gonna be a little bit tricky, though, since these smaller predators have long since developed the ability to consume their catch extremely fast, for *precisely* this reason - so that some larger, opportunistic predator doesn't just roll up and steal their catch from them. So, this means that our prairie apes are going to have to do something that not even species like lions and hyena can do. ... Wow. ... That is... ammmmmmmmmmazing. None of the super predators can do it, but these tree-dwelling fruit animals can? Hmmmmmm.

Is it getting hot in here? Do you guys feel that? I feel like it's getting hot in here. Alright. I think we're alright. I think we can make this work still.

So, what happens next? Well, according to our theory, our apes have now managed to catch up to the jackal hunting party. And they have done this so astronomically efficiently that the jackals have barely managed to begin eating the carcass. Our apes must be pretty exhausted at this point. Traveling at half of light speed tends to take a lot out of yuh.

But this is no time to rest. They've gotta scare off the jackals still. They've gotta start screaming bloody murder, and flinging sticks around, and jumping up and down, and waving their hands in the air. Apparently that's how you do it.

So, now this fairly large group of apes can *finally* go over and start dividing up what's left of this little antelope carcass. However, this only leaves them with about one quarter of a serving a piece, if even that. And their group has dozens of hungry members in it. So, there's just not that much to go around. And you wouldn't be getting lucky like this every day either. Predators like jackals tend to go days without hunting and eating. So, this little quarter of a portion that you get is only gonna happen about once a week. And that's if you're lucky.

Either that or your apes are gonna have to go running around, all over the savannah, traveling through *multiple* clan territories, sneaking up on jackals, with a stealth, speed, and odor concealment superior to any other predator on the savannah. *That's* what it would take to be consistently eating like this. So, now they're just burning up like eight *thousand* calories a day attempting to acquire about two or three hundred.

Allllllllllllrighty then...

If my calculations are correct, eight thousand calories is just *a little* bit more than two or three hundred. ... Realistically though, they would never be running around all day like this. It's one thing to be running really far in pursuit of a target that you've already identified. It's another thing entirely to just be running around *hoping* to find something. ... And if you even go a day without actually finding something, then you just wasted thousands and thousands of calories for absolutely nothing. ... So, no, they would not be marathoning around every day like this - not even in a smaller group. They would either be meandering around like nomads, or they would be remaining in a single region like a wolf pack.

But regardless of what they did though, they would never be getting a whole new antelope every day. This means that, normally, they would still be foraging throughout the forests and relying on random occasional luck to spot some jackals.

Now, these apes would have to be living in a somewhat sparsely wooded area along the outskirts of the savannah. This is a place where there weren't as many super predators lurking around. They were usually out hunting in the grasslands – the places where the herds are always grazing... ... usually.

Well, what happens when even just a *single* individual from one of these ultra predator groups finally detects their aroma from far away? Or what happens when one of them finally hears the faint murmur of those insanely loud attempts to scare the jackals? Sound can actually travel pretty far, especially under the right atmospheric conditions. And so can smell. Or what happens if even just a just *single* individual from one of these roving, nomadic, mega killer species just happens across their little hideout?

There was a species of baboon back in the day called Dinopithecus. It was the size of a werewolf. And it had teeth like daggers. What happens when of those guys just stumbles across your camp at night?

I'll tell you what happens. They walk in there like a boss and just start slaughtering everyone. ... And now they know you exist; and they know you're delicious; and they know you're really really slow and defenseless; and they know you're nearly blind in the darkness. But more importantly, they know what you *smell* like now. So, if you keep hanging around the savannah, they're gonna hunt you down relentlessly. They're gonna track you by your smell now. And the only place they're not gonna track you is in the jungle. It's too far out of their element. So, they'll hunt you until the very last person, or until you go back to the forest where you came from.

Oh, and hiding in the trees would definitely not help you. Those predators can climb as good as you can – even the hyenas back then. So, tree climbing was not an option.

Needless to say though, listening to your buddy get snatched up in the darkness by a nightmare-inducing werewolf baboon would not be pleasant. So, your group would just attempt to relocate. And you would think you got away, until they came and snatched your *other* friend.

This pattern would just continue until you finally returned to the jungle. But by then though, this entire saga would be an absolute horror story. Everyone would know it. It would be communicated down through the generations. Yes, apes communicate to their children. They have culture. They hand down information. They would absolutely be teaching their children about this terrifying, blood-curdling danger where the forest ended.

And to think, it only takes *one* single mega killer to do this. Just one. That's all it takes. And then it's game on.

Or, I guess I should say it's game *over*.

Sure, they could fling some sticks and make some noises and stuff, but all of those theatrics will do precisely *nothing* to dissuade an apex predator, especially when it's nighttime and you're sleeping.

And, no, these apes were not intimidating anyone with sharpened sticks. We're gonna have a lot of fun with that topic in some later chapters, but for now, let's just establish that these early hominins still lacked the intelligence and the coordination necessary to use a thrusting spear. They couldn't make a spear yet that was viable, and they definitely couldn't wield one, even if you gave it to them. ... Hollywood makes it seem like you can just file down the end of a tree branch and you're good to go. ... Yeaaaaaah, no. ... Maybe for cavalry... that'll just run right into it at a really high velocity, but that's about it.

But we'll be talking about all that later. Right now we're still talking about jackal stuff and super predators. These early hominins would have been utterly defenseless against any of these serious land carnivores. If they would've even *tried* to go out and be proximally adjacent to the territories of the big plain's predators, they would have found out *quite* quickly that this endeavor was not sustainable. ... They'd all be living on borrowed time.

Predictably however, some of them actually *did* go and try this. Aaaaaaaaaand they got chomped.

This whole theory of trying to steal stuff from jackals is just a complete festival of fantasy. It's such a whimsical notion that even a fully modern human cannot demonstrate that this works on the same group of animals over and over. Usually when we go out in nature, we notice that everything around us seems pretty empty, but that's mostly because we stick out like a sore thumb to the other animals. With all of our smells and our sounds and our giant clown feet, we make it pretty freakin' easy for the other animals to avoid us.

It's like when you wanted to be a super stealthy ninja warrior as a kid. But then you found out that you couldn't actually sneak up on anyone, with your guts always gargling and your knees cracking. So, then you just gave up and started fighting in the open with honor.

Not sneaking around like some ninja COWARD!

If you see an animal off in the distance it's usually because they're letting you see them. They're assuming that you're not a threat at that distance. The only other way to get *somewhat* close to them is to remain perfectly still, in some sort of camouflage cover downwind, or to simply move about extremely slowly through the forest. But neither of these tactics will ever help you with jackals. Neither of them will get you very close to a middle-size predator who is already *keenly* aware of what a threat you are.

One of my buddies has a wolf he rescued as a puppy. Unfortunately though, he could never be re-assimilated back into the wild. So, now Flynn just has this massive wolf Sonny who rides around with him everywhere hanging out the side of his big ass truck, lookin' like some kinda Super Bowl commercial. Sometimes when they're out hiking in the wilderness, Sonny will just stop all the sudden and freeze and become super focused on something way off in the distance. So, then Flynn just pulls out his binoculars to take a look, and sure enough, he can see some other predator out there, way out past the range of normal eyesight.

These animals are acutely aware of one another, even when the other one is a fellow ghostly stealth lord. Just because we humans can deceive *one another* doesn't mean we can also trick *them*. They can hear and smell and sense the world like they've got super powers compared to us. They know when there's some big-footed ape out there trying to ninja around their environment.

What's pretty amazing, though, is to watch a wolf like Sonny up close and in person. He's a *massive* animal. Most larger breed dogs just stand about where my waistline is, but ol' Sonny's head is up at my chest level. And even with all of that mass, he still glides around in silence like he's made out of vapor. His feet are so light and springy it makes him look like his body is filled with helium. There's really nothing more sobering than watching an animal that weighs more than you do just bouncing all around like he weighs almost nothing. It's enough to make your hair stand up.

Yeah, ol' Sonny and Flynn and I are a spectacle. I actually inherited a little Pomeranian girl who is *obsessed* with being carried around places. She always demands that I pick her up so that she can see the world like we do. ... So, you'll just see Flynn over there with his massive *literal* wolf, on a steel chain, lookin' all *badass*, and then me over there right next to him with my little micro wolf.

... lookin' all *badass*.

Sometimes I try be clever and sneak up behind Sonny. But I swear, it doesn't matter if I'm completely downwind, walking through soft grass with my stealth shoes on, I can't even get within fifty paces before he turns around and looks right at me. I have no idea how he does it. Maybe he's like a submarine where he can just tell where you are by the ambient, white noise fluctuations... like when you're blocking or reflecting it. I'm not sure exactly. But he's definitely doing something.

And that is *exactly* why no ape will *ever* be sneaking up on *any* professional predator.

In all these years, no one has ever observed an ape attempting to stalk some wild predator, much less trying to steal their kill from them. It's a total fantasy.

And there's another reason why you don't see this happening. An existence of trying to fight off an entire group of medium size predators all the time would be extremely stressful. Animals don't do stuff like this. They try to AVOID stress, not seek it out. That's why it's called stress.

But let's just say that the stars align and then these apes are somehow able to overcome this. Okay. So, what happens next? Well, nothing. It still doesn't work. Even if these apes could start stealing a ton of food from these jackals, then all these little jackals would do is move away. Yeah. They would do exactly what they would do if any other animal started stealing from them - they would just leave. Because here's the thing, these medium size predators like jackals can easily go and relocate into the areas where the bigger predators are. They do it all the time. They have the stealth and the speed and the perception to remain enormously aware of their surroundings. They know exactly how to avoid these bigger super predators. But guess who *doesn't* know how to avoid these bigger super predators. Yeah... you...

Who knows though, maybe these apes figured out some sort of rudimentary voodoo magic and it enabled them to start stealing from the jackals. Well, it still wouldn't matter. These jackals wouldn't just roll over like idiots and let you do that to them. They'd simply move on to greener pastures filled with serial killers. And then good luck trying to follow them.

Jackals are essentially generalists. They're survivors. They've got options. They could even just adjust their normal hunting strategy so they were only hunting smaller game that they could run away with. These animals are not just gonna give up and let some weird group of slower, less capable, jungle animals show up and start parasitizing off of them. Are you kidding me? No way. Nothing about this idea is realistic.

This whole jackal phase is one of the most critical linchpins for the savannah theory, and yet nothing about it is even remotely consistent. It would make about a hundred times more sense to say that the apes were on a tightrope, peddling back and forth on a little unicycle... while juggling several handfuls of wet spaghetti.

That would be about a hundred times more believable.

I guess that would make it a flying spaghetti theory.

... sort of like the flying spaghetti monster Christmas decorations my sister puts up during the holidays to troll her neighbors.

And then some of her neighbors try to tear 'm down. So, then she has to put up security cameras to protect her flying spaghetti monsters.

I can neither condone nor deny the actions of either party.

But this whole jackal thing is *sooo* far out there. I mean, honestly. It's so far out there it reminds me of when you're trying to read a book and you're getting really drowsy, but you're still reading it because it's really important to read, but also your eyes are getting really heavy, but it's okay because you're not sleeping - you're just resting your eyes for second because you gotta go deal with this box of little animal babies that someone just dropped off at your house like a box of little orphans, and now you're just trying to take care of these little animal babies while you're going to school and working full time and raising a family of dogs, and you're just waiting for the zoo to come and get them, but they're growing really fast, and you can see 'm getting larger by the minute, and one of them is even turning into a massive timber wolf, and now you're just trying to get the damn thing out of your house, and you're trying to cram it in the trunk of your car and do that thing where you push it in really fast and slam the trunk down without slamming it on part of the wolf, but it's really hard because you're trying to shove the wolf in really fast while it's trying to rip your hands off, and now you've just given up on the whole cuteness thing and you've changed your mind about the wolf being a cute animal baby, and now you just wanna kill it, but you don't have anything to kill it with, so you just go to the grocery store to see if anyone there has a gun you can borrow. Then later you get back home with no success and there's a bunch of police in your neighborhood having a shoot for some reason, and then you try to get down low for safety, and you're crawling across the ground trying to sneak up behind the police guys so you can ask 'm if any of them can help you, because you've got a bunch of dangerous animals in your house that need to be detained. So, then one of the cops just starts shooting into your house at random and you're like "WAIT!!! STOP!! MY BROTHER'S IN THERE!!!" And the cop just starts laughing and shooting more. And you can still hear the jaguar inside attacking your brother.

Then you suddenly realize that you're still reading this book and everything is safe and your brother is safe and your house is safe and there aren't any wild animals in your house, but now you might wanna get a gun at this point, just in case.

You never know. It might make sense. It'll definitely make more sense than the jackal thing.

This jackal thing is wack, son. And even if we *could* make it work somehow, magically, then it still wouldn't last long. ... because I have a joker card the actually trounces all of this jackal stuff. It's called hyena. They would find out about our little jackal trick and just come steal it from us. They would double steal it.

And you're not gonna be scaring off any hyena with your little hand-waving trick. When hyenas roll up in a squadron, even the lions run away. And they're not gonna be leaving anything behind for you either. They eat everything - the bones, the hooves, the teeth, all dat. Yes... literally all of it. And they're not gonna be leaving YOU behind either. They actually hunt things way more often than they scavenge them. Hyena can run as fast as a motor car. Your little jackal trick is gonna end up costing you everything. ... assuming you could even do it... ... which you couldn't.

And that's not even to say anything about the buzzards and the vultures. Your very existence would represent a serious competition to them. So, there's a very real possibility that they would just learn how to call in the lions after you. This is what the ravens will do with wolves quite often. They call in the wolves so that they can come in and chase off their coyote competition.

So, not only would you have to contend with every other limiting parameter, but you would also have an orbiting satellite always ready to call in a ground strike after you.

We've just been sitting here trying to make this whole scavenger ape thing work, and it's like we're trying to shove a square peg in a round hole. And it's just not working, but we just keep shoving it in anyway. And now our face is turning purple but we refuse to relinquish it. And there's an older teacher who sees us struggling, and now she's asking us if we need any help with it, sweety, but we're just recoiling from her and hunching over it to protect our little project. And then she reaches out very gently to touch our shoulder and we start screaming.

But then there's this perfectly good idea over there, just hanging out on the beach, being completely ignored. And this other idea is so simple and so natural and so eloquent and so obvious that there are numerous primate groups around the world at this exact specific moment doing this exact specific thing. There are funny little macaques out there, right now, snatching up their clams and having breakfast.

I mean, the answer is just staring us right in the face.

Maybe that's the problem though. We think that everything has to be difficult. When something seems too obvious we just ignore it. It's like that one time when you were trying to find that one, pipe-smoking, British, detective guy and his trusty sidekick Watson, but then the two of them just drove directly in front of you in that really loud, conspicuous motor car, and you didn't even notice because it was so over-the-top.

It's kinda like that.

Things don't always have to be so difficult though. ...

It's elementary, my dear boy.

And even if the jackal thing did work, it still would not explain how we got so good at throwing stuff.

If flinging stuff at jackals and screaming bloody murder actually works to scare them off, then your ability to throw a projectile is obviously sufficient. You can clearly get the job done with your current abilities. There's no need for you to get better at it. ... And if it suddenly stops working because the jackals have gotten bolder, well, then you just go over there and swat'm with a stick.

And, let's just be generous, and say that even if you did get a mutation that gave you a whopping five percent increase in your throwing ability, what is that tiny little improvement going to change in this scenario? It won't change anything. ... What, so now you can throw a rock five percent better than every other guy throwing a rock at the exact same time? ... So what. There's no advantage there. There's no selection process.

So, throwing rocks at predators is not a thing. It doesn't matter what size they are. If they're a larger predator, then it'll just accelerate the rate at which they eat you. And if they're a smaller predator, then it'll just condition them to avoid you.

And not to get too off track here, but this whole entire argument is a moot point anyway, because consuming a bunch of land meat would not change anything. Chimps already do that, and it obviously hasn't boosted their intelligence. When chimps come across a little animal, they form a massive circle and they essentially trap it. But this tactic only works within the jungle. ... And like I said, it obviously hasn't helped them expand their brain size.

So, now, finally, that brings us back around to the original question. Is it ever a good idea to throw a rock at your fellow hominins?

Well, with all of what we know about human history, including all of the traditional tribal peoples still living in their exact same ancestral ways, we essentially never see any kind of serious rock throwing. It's not a very useful tactic. Most of the time your adversary will simply dodge them, or they'll just block them with a shield while they're running up on you. Or they'll just wait until you're finished before they rush your position. But even if a rock actually does hit you, it's only very *rarely* that this blow will be decisive.

Let's keep in mind though, we've been talking about this as it pertains to us fully modern humans. Our early ape ancestors only had a tiny little fraction of our current throwing power. And they were a lot more resilient to physical punishment. So, not only was it enormously improbably that they would actually hit something, even if they did, it would almost invariably have no impact. ... so to speak.

The other chimp would be like "Woe! What was that? Did a fly just land on me? ... Hey, Dave... ... If I wanted a kiss I would'a called your..."

Chimps can't throw for sh**. In their world, throwing something just equals, "get it away from me." There's no accuracy to it. And the velocity is minor. And their bodies are extremely resilient. And they don't line up in a phalanx on a battlefield – they fight in the trees and hide in the bushes and spread themselves out a little. ... This is not a recipe for throwing stuff.

Besides, if you're just heaving a bunch of rocks at an approaching enemy, then all you're gonna be doing is exhausting your fightin' arm... before the fight has even started. You might even throw your shoulder out. Good luck fighting under those conditions. ... You never wanna waste your precious energy like that. You wanna save it for the stuff that's actually most effective.

And I'm not really gonna get into it here, but I grew up in a pretty Madd Maxx hellscape myself. One of our favorite things to do was to have these big stone-age rock battles with each other. And I can definitely say from firsthand experience, that rock throwing, in and of itself, is not a terribly effective strategy. It was a dramatic thing to do, that, in reality, was actually extremely unlikely to ever injure someone. ... Like I said, we would just dodge'm or block'm or catch'm.

... I'm not recommending that, by the way. I'm just saying... we did it. Kids these days don't have the same reflexes. ... We're getting *SOFT!!!*

But one way to demonstrate this dodging effect is to play a little baseball with your buddies. The baseball is almost always being flung around at an incredibly high velocity. And yet, it's really not that hard to just reach up and catch it. Sometimes the batter even cracks it right back at the pitcher - at like half the freakin' sound barrier. But even then, the pitcher just reacts and catches it quickly. ... usually. Sometimes he doesn't though. But even then, you almost never see an injury that would be effective in a battlefield scenario.

Humans alone have had about fifty billion battles throughout our history. If throwing rocks by hand was actually helpful, we would've seen it by now. And yet, we have not.

The only time that rock projecting actually *is* helpful is when your position is on the high ground and you're getting that gravity assist. In this scenario, you can easily drop a boulder that is physically massive. This is the only strategy that actually does work. But again, this is just heaving, not throwing. This will never make you better at specifically throwing stuff. ... Oh... and I personally don't recommend this form of entertainment either. This'll just get your friend Chris sent to the emergency treatment, and then his little sister Tay will have to explain to their parents why she did that.

So, to almost conclude this idea here, this type of hominin-on-hominin rock violence is a waste of energy. There is just no evolutionary advantage here. There is nothing that would actually select this. And if there was, then we would definitely be seeing more of it in nature. There are species all over planet, and not a single one is employing rocks ballistics to effectively battle one another. The whole thing falls apart when you just look at it.

And our shoulders didn't come from throwing sand either. Never mind all of the logistical problems of attempting to have a grip of eye-assaulting sand at the ready, even if some sneaky bastards did try this, then the others would just adapt and change their fighting strategy. They would just turn their heads to the side or learn to block it or something... or evolve some really long eye lashes. They wouldn't be getting into these O.K. Corral style standoffs – just threatening to throw their sand at each other and aggressively head faking. ... "You better get bACk! ... You'*better*'*get*'*BACK!!*"

And one final note on the idea of rock projectiles. The only time that these actually *were* utilized was when people were deploying them with sling weapons.

A sling weapon provides a major mechanical advantage for firing projectiles. In a professional's hands, a sling weapon is like a nine millimeter hand gun. But they also have a catch, because with great power... comes great loss of accuracy. ... Or was it responsibility? ... No, no, it's a great loss of accuracy. ... Slings are notoriously inaccurate. This is why their use was exclusively for animal shepherding and large-scale warfare.

The Romans would employ these professional sheep herder guys. They would pay to them to come out and fight for them and launch bullets at their opponents. And they would inscribe them with little puns like "catch this."

Yikes.

But the only reason that this application of sling weapons was even viable was precisely because there was no need for accuracy. They just blasted them out there at literally an entire army.

And, of course, the only other place where you can do this is when you're shepherding. ... because all you need to do is spook a wolf or something. You don't even have to injure it. If you just smack it with a rock or make a percussion sound, then that can be enough to freak it out a little. And the actual sling itself will make a whip cracking sound. So, it's sort of like using a whip to confuse a lion. You're producing these extremely loud noises and they don't understand where it's coming from.

Animals naturally fear these loud noises. That's why movie makers like to put jump scares in a lot of suspenseful films. It works on all of us. It just taps into that ancient primal hardwiring.

Well, actually, I should be more specific. It doesn't really work on those hardcore action movie guys – the ones that don't even flinch when something explodes behind them. You know the guys I'm talkin' about - the ones with a dark past that continues to haunt them... no matter how many asses they kick to numb the pain. ... And then little boys see that and can't wait to grow up and have their own dark past.

But the moral of the story is this, the only time that rocks are somewhat useful is when you're propelling them with a sling, and even then they're not that useful.

Alright, that's enough of rock throwing.

The closest thing to rock throwing that actually *does* work is a simple "throwing-stick." Many people are already familiar with the aboriginal boomerang. Well, this is just a specialized type of throwing-stick.

A throwing-stick is exactly what it sounds like. It's a stick that you use for throwing at stuff. A well made throwing-stick is a very effective tool for hunting smaller things. But this is yet another utensil that requires mastery. Its usefulness will be absolutely nothing if you're only half good or just a quarter good. ... So, again, it's just like with a spear. This is something you only get to use when you are sufficiently evolved for using it – not WHILE you're evolving for it. So, again-again, something else has to select for it.

Alright. Well, then where did our shoulders really come from? We've eliminated everything. What else is left?

Well, I'm sure you can already guess where this is going by now. Yep. Our shoulders are from swimming.

One of the most critical revolutions for our early ancestors was snorkeling. Before that, we were just wading around in the water trying to spot stuff from a standing position – the visibility is less distorted at these higher vantage points. But our standing could only take us to about our waistlines. We needed to start diving and snorkeling.

And dive we did, but snorkeling underwater necessitates agility. When you're scanning along the sea floor you need vector thrusting. However, this is a lot more inefficient with a regular primate shoulder. So, every little ounce of flexibility that you can acquire will immediately make your movements a lot more efficient. ... In other words, you can eat a little less while also gathering more. The effect is twofold.

It's not the least bit difficult to see why a life of constant swimming will select for elegance.

We take it all for granted now, but the gulf which divides our flexibility from the rigid jerkiness of a jungle ape is a vast one. If you watch them while they're swimming in a swimming pool, they just batter and assault the water like a death metal drummer. It's exhausting just watching it. ... But then you see their trainer right behind them and she's just swimming along nice and smoothly like an aquatic ballerina. ...

Hmmmmmmmmmmmmmmmmmmmmmmmmmmmmmmmmmmmmmmmmmm...

... mmmmmmmm*MMMMMMMMM!!!!*

But then several million years ago, after *millions* spent adapting to this environment, we finally at last completed our incredible swimming shoulders. And then some*where*, some*how*, in some long-lost, archaically fateful moment, some hominin would finally discover that he could actually throw stuff.

And so it began... The Age of Throwing Stuff. ... Thus beckoned the reign of our midway ancestor - Halfway Guy - halfway between jungle ape and fully modern human. This was the time of The Halfway People - or as anthropologists like to call them... Homo erectus.

These halfway people were pretty incredible. Their instincts for competitive foraging were now fully engrained in them. And this is why our species is so inquisitive. This is why we yearn to seek discovery.

And so it was, our ancestors were finally ready. They were ready to leave the cradle of our inception. So, they followed the ancient shore-lines to the ends of the world. They went from nearly the Southern Ocean to the Atlantic to the Pacific. And they so nearly reached the continent of Australia that they could almost look out and see it on their southernmost horizon.

And yet, the greatest thing they ever did from our perspective is that they left us with their fossils and their stories. So, now we have these windows into their world. And we can see from how they changed and how they adapted that they were evolving to match these different cli-mate ecosystems.

I actually grew up in a place with a bunch of fossils like this – not these ones specifically, but ones that were like that. My grandma was a geologist and she taught me how to spot them. We used to find 'm all over. And my gran' mammy really liked archaeology, but she also liked her friends that she went to church with. Her old lady friends were generally the types of people who are really fond of saying things like "well, there's no missing link between us and apes."

As if we're not apes.

But I just thought it was a really funny thing to say, since we've already uncovered like hundreds of links. It's like, how many links do you want, Phyllis?

I mean, we all get that mutations happen, right? That's how dogs have different breeds and things. ... I mean, we do understand this, correct?

So, if one of these natural mutations just conveniently gives you an advantage, then it obviously gets reinforced. Right? And then these changes just continue to multiply. Still good? And then if we fast forward this process over a very large period of time, then these changes will eventually become massive. Still savvy? I mean, it's a pretty straight forward process, honestly. So, I don't know what all the fuss is about.

And I don't understand why this is so controversial. The universe is vast. And I'm sure our magical space wizard himself is even a much bigger fan of this stuff than we are.

Wait. What was that? ... You don't like me using the term "space wizard." You don't think he's like some old guy... with like a big hat and a beard and stuff? ... Well, don't look at me - you're the ones who keep saying that we're "made in his image." So, that means he looks like us then, right? ... - according to that logic? ... So, if he's already magical... and then he looks like us also, then that means he's gotta be, like, some old sorcerer, right? ... like with a wand and a pet owl and stuff?

I wonder which one of us he looks like though, because we all look super different. I mean, there's a *huge* difference between us. So, which one of us does he look like?! ... ... Hmmm. I see all kinds of people. I see people that look like all sorts of shapes. And how do we know what health condition he would be manifesting in? Maybe he's got like a limp or a wheezy old voice or something.

Or maybe he's sneaky and he looks the ONE person we would least expect. There's this one old dude that keeps coming to my favorite dog park. He's really fat and he's like always wearing these little short shorts with his huge disgusting diabetes legs all exposed to the world.

It's fuckin' revolting.

He's just always wearing these little short shorts for some reason with his huge bloated elephant legs all fat and swollen and green and purple with fuckin' open fucking wounds and shit. Like anyone wants to see that shit. It's like, dude, cover up your fucking legs.

I'm always looking around to see if anyone else is noticing this shit. I'm just waiting to see if anyone else is gonna look over at me... with a look on their face like, "Yeah! I know! What the fuck?!"

One of my friends is like, "Well, he probably can't wear pants because his legs are so big," and I'm like "Uuuuuuh, dude, it's called freakin' stretch pants............ or a dress."

I'de rather see a dude in a freakin' daffodil dress than have to look at some graphic medical shit.

Dude could just wear some karate pants.

We need robots that'll just drive around all day and protect the public decency. ... - just go around looking for people doing stupid shit – stuff that's technically not illegal but it's still disgusting. ... And they could be like that one pizza-pan-head robot from the Rocky movie – the one they trotted all out like it was all technical. But it was just standing there not doing anything. Even Rocky had to lift the cake out of its hands and set it on the table. But they had all this digital music like it was legit.

... bonka'dinka'deedle, duh'deedle, duh'deedle'deedle'doop.

We can have robots like that just driving around all day shaming people. ... like how in Japan if you don't pay your debts on time, then
someone in like a bright neon jumpsuit might just start following you
around everywhere to embarrass you.

We need that robot to just go up to the diabetes guy and be like
[m'm'm'what][the] [f'f'f'FUCK] [do] [you] [think] [you] [are] [d'o'o'oing]
[you] [fucking][whore]

But God bless that guy though. He's always taking those dogs out and
fostering them. ... God bless the hell outta that guy. ... ... That's the
freakin' master race right there, son. ... I don't care what you look
like.

But put a fuckin' dress on for fuck's sake.

11

9 ¾ 2 -- ANATOMY THAT SOME WOULD CONSIDER TO BE...

... UNNATURAL

Remember that time a secret government research facility raised you from a baby because you were super intelligent and the gubmint wanted to turn you into a living weapon, and then that one old creepy scientist guy told you he was your Pahpah and you believed him because you didn't know any better because they raised you from a baby but then you grew up and you figured out everything because you're a genus and you killed everybody except Pahpah because you still felt a special connection to Pahpah?

Remember that?

That one time?

Well... that was a very magical moment.

You realized right then and there that you truly cared deeply about someone else, and you also killed a bunch of people. It's actually a lot like when our great great great great, great great great great, (five hours later) great great great great grandparents finally developed good weapons and throwing techniques and started sending people out into the wider wilderness to start killing everything.

This was undoubtably a very magical moment, as far as science is concerned. This was the moment when a group of hominin proto-humans finally realized that they could not only fling a club through the air and hit something at a distance but that they could also fabricate spearheads out of animal bones. This combination of throwing-sticks and bone spearheads was about to launch the ancient world into a completely new paradigm and dramatically alter the course of pre-history forever.

Surely, every generation of hominins before them had already realized that they could generally throw things through the air in a very crude and sloppy manner, but this was the first generation to finally do it well. This epiphany must have played out independently throughout the ages in many different places across many different eras, but somehow, somewhere, at some specific point in pre-history there was a *first* group of people to finally do this, and from that very first moment it would have spread like wildfire.

Now, when we said this was about to kick off an enormous revolution, we were of course talking about throwing-sticks shaped like clubs. We are not talking about throwing rocks. Not only is rock throwing virtually pointless for all of the reasons that we've already discussed, but it's also effectively useless when it comes to hunting. If you tried to survive like this by throwing rocks at stuff, you would very quickly starve to death.

Trying to hunt with rocks is an astronomically inefficient endeavor. First off, it will take you a considerable amount of time in order to locate just the right prey species and then to get yourself in range without scaring it off. Also, a rock will only work for getting small game, so your target will have to be fairly small and fairly far away, thus making it an extremely difficult mark to hit. Your striking rate will be utterly abysmal. Hitting within a few finger widths at twenty paces away just isn't good enough - it has to be perfect. And every single time you miss, it represents an entire portion of your day spent locating that animal which is now completely wasted, especially since that animal will definitely not be dumb enough to give you a second chance. Nor will any other animal within earshot. Like I said, you will starve, and you will die. And that's why no one does this.

You need something better.

Enter...

the stick...

If you happen to find yourself a really nice clubbin' stick and you also have a really nice shoulder joint, then sooner or later you're gonna throw the thing. And once you do, it should be obvious rather quickly that this is actually a very devastating weapon.

A good throwing-stick will have about the same basic dimensions as a club. It'll be around two to three finger widths in thickness and have about the same approximate shaft length as your arm.

A good throwing-stick is a very different thing from a good spell casting stick. When it's time for you to decide on your very first spell casting stick, you'll want something that's around the width of your index finger. And of course, it also must contain some sort of curiously enchanted artifact. Furthermore, when you're out shopping for your new spell casting stick and you're looking for the one that's right for you, just make sure that you're testing them inside the lobby of an upscale stick shop. That way, every time you test one that's incompatible with you, you'll be sure to completely obliterate half this guy's inventory. /*~~~BANG!

Happy late welcome to Chapter Nine and Three Quarters, by the way. This is a very special chapter. It can only be seen by us magic folk. So, congratulations to you, and many thanks for coming. And many more thanks for not being one of those mugg families that had the audacity to be born wrong. Haha. The nerve of some people.

Well, I hope you all enjoy.

Alright. As we were saying, these club-like throwing weapons have been aptly entitled "throwing-sticks." The basic idea of a throwing-stick is that you just fling this whole thing with a side arm throw and then it twirls through the air like a helicopter. This dramatically increases its effective striking area. Or, to be slightly more technical, if you twirl this thing at a target, then it'll spin like a disc and have an oval shaped cross section. This dramatically increases its overall striking rate. ... And this... changes... everything.

The moment that a hominin can reach out and strike something, it is no longer purely limited to its garden of Eden diet - clams, oysters, mussels, fish, fruits, nuts, eggs, etc. It can now branch out and start taking smaller game like certain rodents and fowl from the land. The irony, though, is that this land game had considerably inferior nutrition to what they were already eating. But it got added to the menu nonetheless.

The slippery apes had finally learned how to throw stuff. Albeit, they were only throwing clubs with some very basic aerodynamics carved into them, but this was still a revolution for general hunting.

They weren't quite throwing spears yet though. That was still a long way off. And that's a whole different level of sophistication.

However, it's not as likely though that these earlier Homo erectuses... uhh erectuses? Erecti? What's the plural? Whatever. It's not as likely though that these guys would've just been sitting around doing nothing before they had throwing spears. They were already quite proficient with a throwing-stick. As a general rule, whenever life is able to start doing something new, it basically just starts doing it immediately. From the moment they discovered they could use throwing-sticks, it was surely put to use that very day.

This new skill would've made life a little less difficult for those in exile. Every once in a while, there would be waves of dispossessed people who were forced to leave the coastal areas due to overpopulation. Normally, they would just return to a life in the jungles and survive on things like fruits and nuts and bird eggs, but NOW they had the advantage of using a throwing-stick, and now they could pursue a larger variety of options.

Keep in mind as well however that this was yet a million years before the use of fire. So, they weren't exactly cooking or roasting these land animals. They would just be cutting them up and eating them totally raw. And that's not a terribly efficient diet for a hominin. You have to swallow the chunks whole since it's like trying to chew on bubble gum. And our digestive tracts don't really break these bigger pieces down as well, so we wouldn't be getting quite the full benefit of a land carnivore. And it also wouldn't be terribly appetizing, but it is better than nothing from a calorie standpoint.

So, our ancestors never stole anything from jackals, but when they finally had their throwing-sticks they didn't need to. At this point they could fend for themselves out there. They could finally hunt on land with modest efficiency.

But there was still that pesky problem of all those super predators. How are you supposed to chase those little squirrels without becoming a monkey dinner for something else?

Enter... the bone spears.

Not only was this new hominin proficient with throwing-sticks, but they also had created the world's first thrusting spears. They made points by carving sharp edges into bone-like substrates, and it could penetrate through the hides of any animal.

They finally had the ability to defend themselves. For the first time ever, our genus could safely move beyond the jungles. They could live along the edge of the greater savannah. They could even go and live on the open grasslands. But as far as we can tell they would tend to avoid this, because this is where the predators were most highly concentrated.

Even modern humans in Africa with sophisticated iron-age spears will still get mauled by deadly lions because of how quickly they can move. It takes incredible precision and lightning fast reflexes not to get turned into a human shish kebab. This is not something you would wanna play around with, especially for a midway Homo erectus. So, because of this, they just remained around the periphery of the grasslands.

The edge of the savannah was not a fun place, but it held a lot of promise compared to the jungle. If you had to be expelled from a coastal region, then at least now you would have some new options with your throwing-sticks and bone spears.

They would also eventually discover that these bone spears could be useful for hunting megafauna. They would chase after an elephant or a rhino and then just ram it with their spears until they killed it. We know that this species of hominin was very fond of going after these massive animals like this. And it's obvious why they felt this way. They could catch these giant animals on foot and then just ram them with their thrusting spears without any need for accuracy.

And this might be why the megafauna are only really extant within the continent of Africa now. If our ancestors started hunting them so primitively, then it would've given them much more time to get used to us. So, by the time that we got a heck of a lot smarter, they were already thus evolved to avoid us. But if we showed up in other places much later, and we were thereupon equipped with better weapons, then these megafauna would simply not have had the time to adapt to us. They would be vulnerable to us over-hunting them and this could very well lead to their extinction.

Either way, the use of bone spears was a total game changer. You could now pierce right through the hide of a ruminant or even penetrate passed the armor of a massive rhino. And that's something you could never do with a wood spear. Wood is softer and it deforms on impact.

If you thrust a wooden spear against a ruminant, the odds that you will harm it are pretty small. You're bound to give it a bruise or even possibly a puncture, but that's unlikely to take it down, notwithstanding infections. And that's if you tried this with a ruminant. So, forget about actually trying this with a rhino. There's a reason why the cavemen were using stone for their spear points. It's because the wood that you find out in nature just simply won't cut it.

You can't just file down some random piece of wood and expect it to penetrate. It won't. You need an extremely dense form of wood that's been meticulously selected and processed. This would be an example of an "advanced" wooden spear, not a "simple" one. And we don't know that erectus had these highly dense and durable wood materials. And that would be an awful lot to assume about their woodworking resources. ... - especially since you need fire and glazing to give it that extra bit of hardness, and erectus wasn't using fire. ... But we do know that they were making bone spears, so that's what we're gonna talk about.

... And any time I'm talking about wooden spears in this book, I'm talking about SIMPLE wooden spears - not these highly sophisticated fabrications that a much more advanced hominin species such as ourselves would be making, with of course a range of much more advanced techniques for solidifying. ... So, KEEP that in mind... ... Steve!! ... ... And don't get it twisted!! ... Don't try to conflate it and equivocate it... ... STEVE!!

Keep in mind too though that we're talking about *thick* spears for thrusting. There's no way that a shaft of this thickness can impale something without a spear point made from a harder substrate.

Smaller weapons like dart spears are different though. They're thinner, and they can be fired with enormous power. That means the pressures at the point of impact are exponentially larger, and so they can pene-trate irrespective of what their tip is. But these weapons are a lot more sophisticated than what Homo erectus had, and they were yet to be invented for a million years or so. Homo erectus just had hand spears for thrusting, and they were taking down things like rhinos by simply running up to them.

Erectus wasn't dumb though. Sure, they weren't as smart as a modern human, but they also weren't as simple as a basic jungle ape. They were several steps removed from the days of the jungle. It was Australop-ithecus, and then Homo habilis, and then them.

The first step was Australopithecus. They looked more like a typical jungle chimp, but they also walked upright, and they foraged around in the water for very simple things like shellfish. And then they went back up into their tree nests as soon as night fell.

The second step in this process was Homo habilis. Their name just means "handyman." They broke rocks up to expose their sharp cutting edges. And their face was about halfway to that of a modern human's.

The third step, around halftime, was Homo erectus. They resembled a modern human with a sloping forehead. Well, technically, it's us that look like them. Our face was essentially finished around this midway point. So, the second half of our evolution, starting at around Homo erectus, was just us building up our foreheads. And the only other dif-ference between us and them is that our chins have now become a bit more prominent.

But like I said, Homo erectus weren't exactly a bunch of dummies. We know they had their throwing-sticks and bone spears, but they also had their famously useful hand axes. They were about the size of a modern hatchet, and they were sharp enough to fell a massive tree.

The head was a piece of stone that was shaped like a tear drop. The pointy side was the sharp end and it was used for cutting stuff – as sharp things are. Then the backside was firmly wedged within a slot at the far end of the handle. And the handle was about the length of a person's forearm.

These axes were seriously amazing. You can do about a billion different things with them. That's part of what makes them so famous. They can do things like construction and even food preparation. Erectus was surely using them to build cabanas, and that's better than just a tree nest like an Australopithecus. The hatchets are also helpful for making sushi, and you can fillet things fairly quickly just using flake blades.

The existence of these axes has also established that Homo erectus was good at flint knapping. We know they had the skills to fabricate stone edges. So, it was well within their ability to make a point out of stone for a spearhead, and you could use this as another alternative to making points out of sharpened bone.

Alright, so what's the tally? They had throwing-sticks, bone spears, and hand axes, and they likely had some spearheads they made from stone. We haven't actually recovered any of these stone spearheads yet, but we know they had the knowledge to easily make them. If erectus never took the time to make these, then it's only because the bone was already sufficient. It was easy to quickly find and to custom shape it, and they were likely accustomed to the method from using shark teeth and stingray barbs.

Either way, we had a pretty big arsenal. And that's what finally allowed our brethren to survive on the grasslands. Not only were there tubers and root vegetables, but they also could get the largest and smallest animals.

The medium size animals were not within their reach yet though. Those species were still too fast and aware of us. And their technology would not be sufficient for many thousands of generations. So, in the meantime, they would simply make do on eating rhinos and squirrels. Sure, they could catch stuff with persistence hunting, but they very rarely did this, because it was so inefficient. And they always had to carry those bulky spears around for protection. So, they mainly just stuck to the really large stuff.

This lack of technology didn't slow them down though. They definitely got around to at least three major continents. Their groups were always expanding along the coastlines, and they were hugging the shores and oceans wherever they went. Then slowly their numbers would increase and tensions would escalate until eventually a break-away group would just move into the continent. And this now was a viable option all because of their toolkits. They could go after things that were massive and things that were tiny.

What's also very fascinating about these midway erectus people is that they lived apart for many hundreds of thousands of years. They lived in different climates and environments. So, they had much more divergence than us modern humans.

That being said, they still had our basic morphology. As we discussed earlier, they had our faces and our noses and our head hair. And of course they had our stork legs and our big floppy duck feet. And they also had our skin tones and our general lack of body hair. They were essentially just naked like we are.

Yeah, yeah, yeah, we know they weren't "naked," Steve. Our hair just got turned into peach fuzz. Yeah, we got it, Steve.

Hey, Steve.

... why don't you go put a sock in it?

These midway folks were a lot cooler than Steve is. They actually looked out for each other. They cared for their elderly and injured people. And they probably didn't interrupt people while they were talking. At least I'm pretty sure they didn't. That's what my gut tells me. And that's because they were cool. They even helped people who were never again going to be productive members of the tribal group. And they knew it. But they still helped them anyway. That's pretty legit.

My friends and I weren't quite as ethical as this when we were kids. We never had a Homo erectus around that could teach us this stuff.

When we were about twelve we had this friend Mark who was kind of tall and goofy. He looked like Andy Dick a little. But he had one of those bowl-style haircuts, and it was cut way too low, so he just looked like some kind of medieval doofus.

His very modest height advantage also gave him a natural edge when it came to basketball, so he was always acting like God's gift to the world because of his basketball "skills."

He was really arrogant about it too, like being better at basketball just automatically made him better at everything and a higher quality human being. And he was always wearing his basketball clothes no matter what other thing we were doing - hiking in a basketball uniform; canoeing in a basketball uniform; riding horses in a basketball uniform. He was *slightly* above average at basketball, and he wanted the world to know it.

But one time we were playing rugby football and our little husky friend Martin just ran over and jumped up onto his shoulders. And he did it at the exact same moment Mark was running across some slippery wet grass, so Mark just started slipping and sliding like he was a deer trying to stagger across some ice. Husky little Martin's weight was bearing down on him and his feet started spreading apart like an A-frame, and he was just sliding across the wet grass like a water skier, twirling his arms around for balance, and his feet just kept going farther and farther apart. Then at some point his legs finally gave out and it just crushed him into the splits.

He looked like Jean-Claude Van Damme, but just sliding across the grass... and also if Jean-Claude Van Damme was screaming "GET OFF GET OFF GET OFF GET OOOOOOOFF!!!!!"

If you wanna get a visual of what it looked like, just do that thing with your fingers where you make it look like they're two little legs of a person walking around on the table, then start pulling one of the legs out like they're doing the splits, but just keep going until your fingers break and go completely flat against the table.

Like some kinda jacked up peace sign.

TV antenna peace sign.

That's what Mark looked like. Just two hairy legs all sickeningly at 90 degrees. ... It just looked unnatural.

It almost looked like some fake Halloween prank, like he was just standing in a hole with fake legs beside him, and some big billowy basketball shorts to cover the hole up.

It was funny too because everyone just freaked out and ran away. We just left him out there in the field, like some random Flintstones lawn mower.

Or like some meat helicopter crashed and rolled over onto its meat propellers.

You see, if Mark was a Homo erectus, there would've been other Homo erectuses out there trying to help him immediately.

As for us though, none of us wanted to get blamed for that, so we just left him out there doing the splits.

Don't worry. He was fine. We made sure there was an anonymous phone call placed to get some Homo sapiens out there to help him.

And he healed up mighty fine too.

After that, he always showed off his splits in front of the girls when we were at big tournaments. He'd conveniently end up doing his stretches right in front of the girls, and OF COURSE he would do the splits.

Funny how he was the only player that needed to get warmed up by using the splits.

And he would do it all serious too, like his flexibility had come from years of athletic discipline.

Yep, yep, yep. The human body is an amazing thing. Amazingly awkward. We've gotta be careful with these land-whale monkey bodies. There's about a million and one ways that we can break these things.

Some people think our contemporary frailness is just a side effect of our overly pampered lifestyles. They think we've gotten too domesticated. Well, there is definitely a lot of truth to that idea, but it's not as relevant as it might seem. There are plenty of modern day humans that are still living in a traditional tribal lifestyle, and we can tell from the nature of their bodies that they are basically just the same as all the rest of us.

Our bodies are suspiciously fragile, and they have been so since the midpoint of our evolution. This is obvious when you start to compare us to a normal land vertebrate. Even erectus were getting injured in the same ways that we are. The people that we think of as cavemen were getting just as messed up as the rest of us.

People often get "cavemen" confused with Neanderthals. They imagine how powerful a Neanderthal was, and they project that onto our lineage. But that was never what our predecessors actually looked like. We've been lighter in our structure and our musculature. Neanderthal was adapting back to a terrestrial lifestyle. Their bones were getting thicker to support stronger muscles. But this was never an actual feature of our ancestors. We've always been a lot more proportional.

One of the best illustrations of what sets us apart from the Neanderthals is our shoulders. Our shoulders have been getting injured since the days of Homo erectus. We've been vulnerable to the act of throwing stuff for literally millions of years now. We get injuries from common activities like playing sports a lot, just like erectus was getting injured from simply hunting.

So, throwing has been a problem for our ancestors. They were getting injured just like we are. But if we're supposed to be this big tough savannah caveman, then why would we be so fragile when compared to a Neanderthal?

Well, in order to answer that, we need to know some things about classic spear throwing. ... We know we need our shoulders just to throw them, but where did the actual activity originally come from?

Spear throwing initially got started by just mimicking certain cranes and tall wading birds. They slowly walk around in the water, and then they wait for an unsuspecting fish to come by. Then as soon as a fish gets too close to them, they explode out like a missile with their beaks in front. And even now we see jungle apes sometimes emulating this, by poking around in the water with a long skinny poking stick.

So, then spear use was just evolving from that simple stabbing motion. It all started by making these arm's length, impulsive jabs at the fish. This movement would then become a longer thrusting motion. And then they slowly began releasing it and letting it fly a little. Then the distance from which they would throw it was gradually getting longer. And this progression was incrementally becoming a legitimate spear throw. But it required a lot of development to make these things fly straight. So, that's why spear throwing with an atlatl only occurred much more recently in our evolution.

Homo erectus themselves would never accomplish long distance spear throwing. They were clever but not really there yet when it came to the aerodynamics – this means a dart shaft that's properly balanced with the right flexibility. And then you throw it over a large number of paces with an atlatl and something like a feather fletching.

This advancement only happened fairly recently. It was all within this last quarter of our evolution. That's about halfway between Homo erectus and modern humans. This was roughly the hominin species we call heidelbergensis.

So, three quarters is about the timeframe of heidelbergensis. And these guys were a little bit smarter than Homo erectus. They were using their newfound missiles for traditional spearfishing. And that's when throwing-spears were originally invented by our hominin genus.

This is the ultimate sophistication in primal development. This was the crescendo of our evolution in the natural world. When we invented these throwing projectiles with atlatls, we had summited the preeminent pinnacle of our indigenous environment.

If you wanna get a cool visual of what this looks like, go check out the Australian aboriginal spearfishers. They can nail a fish from over twenty paces away.

Yeah, there's your perfect paleo caveman right there. People always wanna know what a real caveman would look like. Well, there you go. We don't have to imagine it. We can just look at it.

Our forerunners were never cavemen. They were cOvemen.

If you watch those little aboriginal children, you can tell that they really enjoy their immaculate beach life. And you can see this in other populations too, like the Andaman islander children. Likewise, the little lion cubs are in love with the savannah, because everything deeply adores its intended habitat. This goes way beyond nostalgic imprinting - we're talking basic inclination and primal instincts.

But here's the point. Spearfishing requires grace and precision. It's the gentle hand to victory, as my great grandmother would say. It's very different from our sports and our land hunting. It's not an explosion or a violent eruption like you see in many land activities.

And spear throwing only happened in the latest quarter. Before that we were mainly just diving. And that's exactly what our shoulders are really built for - just some diving and some casual spearfishing. And so, this is why we're always getting injured. We're often doing things that our body was never designed to do. Our shoulders were never adapted for marathons and weightlifting. Nor were they well selected for ramming into rhinos.

If our ancestors had actually gone to the savannah and then they stayed there for literally several million years, then our species would be evolved to hunt big rhinos, and we'd be as strong as a full grown lion with far fewer frailties. But those people who left the beaches are not our ancestors. Quite frankly, they were an evolutionary dead end. The people who remained on the coastlines were the ones still evolving, and the diaspora that fled to the savannah were the ones slowly degenerating.

These tribes were thus devolving for two basic reasons. The savannah lacks key nutrients and the selection for intelligence. We know now from basic nutrition that a hominin needs marine food. And the savannah was selecting for simple hunters, not clever foragers.

Let's take these things one at a time. Let's just start with the problems surrounding nutrition.

As far as a hominin is concerned, the nutrition of a land life is impoverished. We don't have the same biochemistry as the animals that are exclusively land based. There are things that we simply cannot synthesize, like vitamin C and our own DHA. That's the legacy of not needing to make them, because of their abundance within the jungle-beach habitat.

The best things that the land tribes could've eaten were a little bit liver and some fish from the streams and lakes. But these supplements from the land web still don't cut it. They still don't fill the holes in our essential nutrient profile.

People will often call our brain a muscle, but this comparison is not even remotely accurate. Our brain is not a muscle of mostly protein. Rather it's mainly fats like cholesterols and fatty acids. Most of these are made inside our body, but some of them only come from a proper diet. We hominins must eat DHA and iodine. We just can't build our brains right on a rhino and squirrel diet.

In fact, this nutritional argument has become so ridiculously obvious at this point that many of the classic anthropologists have already abandoned the traditional model. Rather, they're saying that we must've come from the savannah when it was close to the seaside.

Well, they're getting there. At least they're paying closer attention to the nutrition now. Sometimes you just need a little patience. Anthropologists are a skittish bunch. Sometimes you have to be forceful with them, and sometimes you have to be tender. It's sort of a good cop, bad cop scenario. If they make a mistake you gotta get right in their face, but then just start randomly crying like a baby and go over and sit on their lap and pull your legs up in the fetal position, and bury your head in their chest and start feebly pounding on them.

And stay in character. Don't let 'm distract you by say things like "what the f***?" or "ow ow ow, you're crushing my legs."

Just stay focused and finish crying and then get really weirdly quiet, and start fiddling with something on their shirt with your free hand, and then start saying something in a really low creepy voice, like you're whispering something or you're possessed. Just say something like "you tricksed us... you stole the precious... filthy hominins."

Keep doing that until they call for security, then jump up really fast and act completely normal. Then, when the guard gets there and they're explaining what happened to them, just look at the guard with a bemused look on your face like they must be crazy for ever telling such a story.

I'm telling you, they're not like us. You have to use reverse psychology. And just have a little patience. They'll get it eventually. They've been force fed this savannah slop their whole life, since they were shabby little orphans begging for more porridge, and then they go and find some bones out in the desert and they think that that confirms their entire theory.

Well, there's always gonna be an inherent bias in where we find these fossils at, since a lot of the ancient coastlines are now below sea level. The erosive oceanside isn't exactly the best place for always laying down fossils.

Those bones out in the deserts are mostly the groups that went there to die. They weren't the heart of our civilization or some thriving pre-historic metropolis.

It's not entirely clear how long these inland groups would've lasted out there, but the longer that they continued to remain that way, the more divergent they would become as a phenotype. The seaside erectus would keep evolving, but the landlocked erectus would just keep backsliding.

After about seven hundred thousand years of prairie life, land erectus would've had a very strong back and a very weak mind. But not too weak though. You still would need to assemble a suitable bone spear. So, they would've just bottomed out there at that level. They would've frozen their basic intelligence for many thousands of generations. And that's precisely what Neanderthal was doing at a much later date in prehistory. They just kept repeating what they were doing, over and over and over, just running up to mammoths and shoving their spears at them.

But land erectus out on the scorching savannah would not have been more barrel shaped like a Neanderthal. Rather, they would be physically more slender. This is enormously helpful for discharging your body heat. It increases your surface area to volume ratio.

And the land erectus would no longer need our proboscis nose. They would end up with a forward facing nostril opening. And this would be enormously helpful for sprinting. It would force in lots of air like a ramjet. So, they would eventually get a snub nose like we had as a jungle ape. And that's why animals who zip around a lot have these forward facing nostril openings. They're facing forward so that they can inhale better while they're running.

Neanderthals themselves must've had noses like this. They were land-based for around a half a million years. And they were extremely good sprinters – far better than we are. And they also had a need to protect themselves from frostbite. So, the depictions of our fleet footed cousins need to be updated. Some Neanderthal would've definitely had snub noses.

There's an interesting side note here about noses. Modern humans have been considerably land based in our most recent prehistorical epochs. And yet, we haven't had the need to sprint like cheetahs all too often, so there just hasn't been a need to develop a sprinting nostril. Our technology and our creative intelligence have sort of buffered us from these pressures. We manipulate our environmental surroundings instead of just adapting to them. So, we basically just got *frozen* in our beach form, and we haven't had the need to change our nose a lot.

But erectus was still very much at the mercy of their environment, and they didn't really have the skills yet for generating fire. So, everything that they were consuming was entirely in its raw form, and the erectus who were living on their land meat had to consume it completely bloody. This meant their jaws would be extending much longer, and still their teeth would be adapting to get sharper. Then their face would be protruding like a dog's snout. And then this would help them slice through the rubbery land meat.

But this is one thing that the later Neanderthals would not really copy. Neanderthals already had fire, so these modifications for digestion and chewing were not really necessary for them. Fire, of course, makes things softer. It basically just predigests it. So, Neanderthal never had a necessity to develop a bigger jawline.

And that's about the last thing that you would notice about a land erectus, except of course that these landies would've been covered in a lighter colored of body fur. This would've kept them a lot cooler. Likewise, they would also be panting like a dog does. This is how a mammal normally cools itself. ... unless of course it's aquatic and it's always diving. ... which these erectus on the mainland were no longer doing.

Furthermore, there is also a very real chance that they would be walking around on their forefeet. They would essentially look like ladies in high heels, and this is exactly what you would desire as a serious land-based runner. This would make them run mASSIVEly faster. Their tendons would hold up their bodies as if their legs just had springs in them. And they could've developed this trait rather quickly if the rhinos were getting faster.

So, let's take a tally. Let's add it all up. What would a Homo erectus start to look like after seven hundred thousand years of rhino chasing?

Well, they would've seemed like they were getting a muzzle, like the snout that's on a cheetah, short and compact. Then they'd all be quickly panting in the daylight, and their mouths would all be open while they ventilated. Then they'd all be fully covered in a fur layer, and their coats would all be pungent with an odor. And, of course, their legs would all be zigzagged like a werewolf's, while their bodies would in general be skinny-buff, like those kangaroos who try to fight you. ... So, they would look just like a werewolf with a cheetah face. ... except of course they'd still have basic bone spears.

Furthermore, their intelligence would also be lower like a previous Homo habilis. But their strength would conversely be massive, like a much earlier jungle ape. They weren't strolling around the beaches looking for shellfish anymore - they were slamming around the herd lands getting pummeled.

And after seven hundred thousand years of prairie life, an erectus brought up this way would be a beast. Their tendons and their shoulders and their joints would all be tougher, and their knees and backs and ankles would not be so fragile. That's what all that brutal life will slowly do to you. And that's why Neanderthals were built like tanks when compared to us modern humans.

So, that's what seven hundred thousand years would look like. But what if they kept adapting even farther? Well, that's sort of a small diversion from what we're talking about, but we can still go there briefly, just for the fun of it.

In the arms race between predators and prey animals, these savannah guys would end up getting fast like a cheetah, nearly. They'd be able to catch a wildebeest if they kept getting faster. And they'd be able to bring it down by just getting it in a head lock.

This event would change everything. They would officially become a toolless animal once again. Thinking is a heck of a lot more difficult than just using your muscles, so they would eventually stop making their bone spears if they didn't really need to.

This would then select for claws and fangs again. Then they'd be back on all four legs before you knew it. This, of course, would make them much much faster. And so, convergent evolution would make them into a primate lion.

Then they wouldn't really be our chummy pals anymore. They'd wanna eat us just like every other predator. Then the spears that they once used for hunting rhinos, would be turned and used against them, in a strange twist of irony. ... The circle would be complete for the savannah erectus.

If our erectus ancestors had actually lingered on the savannah, then I wouldn't writing this book right now. I'd be howling at the moon and licking my crotch at the moment. ... Even if you took a sophisticated Homo erectus and you stranded them out there on the savannah, *they wouldn't end up looking like a modern human!!!* They would go the other way. They didn't have the circumstantial ingredients for selecting intelligence. And even if they did, they didn't have the food for it. Sure, they could've evolved like an elephant, to do it with a range of other foods, but the metabolic requirements would've been prohibitively expensive for them. ... But never mind all that. I'm getting ahead of myself.

So, that's what would happen to the savannah erectus after a couple million years of adaptation. We're not really worried about fast forwarding *that* far though. We just wanna know about a fraction of that - more like seven hundred thousand years of physiological changes.

We're mostly curious what would happen if these two groups had encountered one another. Would the land groups share their strengths with any beach tribe? Would a beach tribe even want what the land groups were offering? What if they had to fight and have a battle? Would there be a clear advantage for either party?

Well, the answers are - no, no, the nerds would win, and yes. So, let's talk about it. The land tribes would be conspicuously divergent. They'd have a snub nose, a lion's faces, and they'd be really good sprinters. They'd be very foreign looking to their ancient beach cousins. And they wouldn't be too bright, by coastal standards.

And then there'd be the issue of their body odor. It'd be pretty overwhelming for the coastal people. So, if you add up all that strangeness in these land folks, it would all be quite alarming for the beachies. Their faces and their odors and their general lack of intelligence would make them seem like *monsters* more so than cousins.

Coastal women wouldn't want a slow guy. It's not attractive when a man sounds like he's mentally challenged. And the savannah ladies would be intensely masculine. They'd all be buff and extremely hairy and profusely pungent. And they wouldn't actually have any breasts, unless they were actively breastfeeding. ... We'll talk more about the breast stuff later. ... You know we will.

So, it should be pretty obvious, in light of these details, that there wouldn't exactly be a lot of gene exchange here. And even if there was a bit of mingling, then it wouldn't really thrive in either gene pool. The genetics from these two different groups would actually be maladaptive to one another. And so, the strengths that time developed in these land groups would eventually go extinct with those *same* people. Those land groups never made it on the grasslands. They died out like the northern Neanderthals, millions of years later.

And frankly, it wouldn't take that many generations for a beach tribe to completely lose interest in a fledgling land group. Their cognition would be plummeting from the outset – within several generations the effect would be massive. They'd be barely scraping by on recycled nutrients, by just reusing what they could through the lines of mothers. But, of course, not every child is a future mother, so these stores would get depleted on these other babies. And so, they'd end up like some backwoods, malnourished, country folk. ... They'd have trouble with their health and their mental clarity, then these plagues would dull their wits and cloud their judgement. And then that would lead to *other* bad decisions. And then this of course would just *compound* their issues.

And not only would these third-generation land groups be extremely diminished, but their odor would be abrasive to every beach tribe. After several generations of rarely bathing, the land groups would not even notice that they smelled like heavy sex-panther musk. ... There are many different countries around the world, and a lot of 'm have some really pervasive aromas, and it just hits you like a wall as soon as you get there, but then the people that actually live there don't even notice. This is how it would be for these fledgling land groups.

So, that's what land erectus would've been like after a couple of generations away from the coastline. They'd be dull and uncouth and abrasive. And the beach tribes, with their intelligence and their bathing, would be repulsed, in no small measure, by their apparent degeneracy.

If we really *had* evolved as a land-based animal, then our species would not be obsessed with constantly bathing. Amphibious creatures LOVE to feel the refreshing dowse of water. But the species who evolved as purely land animals find this concept *appalling*. ... ... Go ask your dog. ... ... Go 'head... ... Do it! ... ... I dare you.

But that's no joke though. Land predators absolutely *love* to smell like body musk. When they find some other random creature's musk some-where, they just want it all over their body, like it some kind of wet t-shirt contest. So, then they act like it's some kind of catnip and start rolling all over it.

My dogs do this constantly. It's like heaven for them. Or some might say, it's like dog crack. They'll find some like random squirrel musk or something, and then they'll stop, and start to go into a dog trance, like they just hit the freakin' jackpot... *son*.

I used to try to stop 'm, but then I was like "eh, whatever... have your fun."

We've even been to people's houses that have things like goat skin rugs and beaver pelts. And my dogs go completely mental when they smell those. They just go into a frenzy and start rubbing all over it.

They also go bananas for the musk of turtles for some reason. When-ever someone who's patronizing the city park picks up one of the tur-tles from the turtle pond and then sets it on the grass beside the pond area, it leaves behind this smell that's apparently like heaven for a doggy critter.

... even though we're not supposed to be messing around with the tur-tles... whoever you are who keeps *doing that*.

You know what else dogs love? They love it when you follow the rules... and stop trying to grab the turtles out of the turtle pond... and stop trying to feed 'm your freakin' human food.

... whoever the fuck keeps trying to feed 'm your *cinnamon* rolls.

They're turtles!

They don't want your fuckin cinnamon rolls.

Animals don't want what we want! Turtles just want their natural pond food. And dogs just want their natural turtle stank. We all want different things!

And dogs just wanna keep those dank ass smells that they've worked so hard for. They went through ALL OF THAT trouble to go rolling around in some funky ass turtle stank, and they don't need some water-lovin', aquatic monkey, *jACKAss* trying to wash it off of'm. ... ... *Y'understan'*?!

And they sURE as heck don't want your stench of cleanliness all over'm. If you try to make 'm smell all nice and clean for you, then they're just gonna have to go roll it all off again. ... You *maniac.* Trying to wash an innocent dog like that. ... Are you sick in the head or somethin', mister? ... ... Hopefully he can find some sweaty foot odor on the carpet and get this nasty *cleanliness* off.

HOPEFULLY!

"*You might* understand that if you were actually a real *land* animal,

SIR!"

Well, we're sorry, dogs. Those ideas don't come natural to us. ... That's not even remotely how our species is. We don't lose our freakin' minds when we smell some musky goat rug. And we don't perceive that the aroma of cleanliness is utterly appalling. ... It's almost like we're not a real land animal.

Well, I'm sorry to have to break this to you, anthropologists, but if we really were a former rhino hunter, then we would *absolutely* love the smell of sweaty ruminant musk - no ifs, ands, or buts. ... This isn't even a debate - it's a dead giveaway. ... If we evolved on dry land, and dry land only, then the conception of washing our bodies off would be categorically atrocious.

Straight up. Flat out. Triple stampy, no erasies.

However, not only are we nOT a pungent, musky hydrophobe, but our reality is, *indubitably*, quite to the contrary. We are a furless, naked creature with an obsession for rinsing. We've been in love with the act of bathing since ancient prehistory. We even have these little water rooms in our houses where we twist a couple knobs and it starts some water flowing. Our obsession with this bathing and rinsing is SO pervasive that we've built these little water rooms directly into our houses.

Do you really think dogs would be building these little shower rooms into their houses if *they* were the ones in our position? ... Yeah right. ... They'd just have rooms with a bunch of goat rugs everywhere. ... And that's *exactly* what we would be doing if we were an actual land erectus.

But we're not. We are precisely the polar opposite of that. We are not a hydrophobic, water-hating, stink-lover. We are a hydrophilic, water-loving, shower monkey. ... Baths have always been a sign of great luxury... and so is living on the beach, in even a modest accommodation. When you hear expressions like "coastal" and "beachfront property," you immediately think of wealth and status and privilege. And rappers used to brag about eating crab legs and taking cruises... and sipping on Santana DVX. ... And surfing seems arouse our deepest yearnings, like some long-forgotten memory of an ancient love affair.

And our genus is not just obsessive about bathing and rinsing. We're also instinctively in love with being fresh and conditioned. The ocean is full of salt and little micro plankton. They clean it through the osmosis of salt and through the feeding of microbes. Every single drop is an entire universe. The ocean has been host to *many many* organisms for *truly* quite a while now, and so she's picked up a number of talents for staying radiant and presentable.

The water is also extremely clear in a healthy ecosystem. The coral and the various sponges are relentlessly cleaning it. In fact, if you find a natural waterway with poor visibility, it's usually because industrialized humans have disrupted it.

The ocean is full of salt, and that salt is good for us. It's great for eye drops and oral routines and spa treatments. We've been exploiting these wonderful properties for millions of years now. And this is precisely why our species is *so* obsessed with bathing.

We also insist that our partners be washed up and fresh smelling. We enjoy their musky odors and pheromones, but not when the effect is more than subtle. And when we formulate our perfumes and our fragrances, we unwittingly attempt to approximate the jungle beach essence. This means the rich and rousing middle notes of the ocean, with the earthy and floral accents of the forest, and then the light and sweet ambrosia of a freshly wetted lover. ... – not the year-without-a-shower Ode to Savannah Sweat.

This is what we tend to approximate without even knowing it. It's not just all that forest stuff and pheromones - it's also the freshness of a wind swept, breezy coastline. ... There's about an infinity of different formulations which you can concoct with this. Not to mention the terroir and the enigmatic top notes.

Our preference for a clean smelling partner isn't just cultural either. If we had *ever* been a mostly mainland animal, then we would still be more than happy to live downwind of a homeless encampment. We would *never* divest that inclination if we had *actually* had it. ... especially since regular bathing hasn't even been a thing in most land-locked civilizations. ... - not until recently.

We wash our dogs now, don't we? You don't see them losing it.

If we had ACTually come up on the savannah, then we would *still* enjoy the emanations of a homeless person. We would regard it like a wine that we were tasting. ... "Mmmmm... what is that? ... a Haven't Seen a Shower Since 64? ... Mmmmmm... that was a good year."

We would be just like our dogs. ... But instead, we're like, "NO!! ... GET AWAY FROM THAT!! ... NO, NO, NO, DON'T TOUCH THAT!!" ... That's the *beach* in us shrieking.

And the beach in us likes our colognes to be gentle and localized. Just a dab or an airy spritz is très bellissimo.

And, with all due respect, the key word here is "localized." Yeah, that means you, guy... who's apparently removing the cap and dumping an entire bottle of cologne down his pants every day.

You're not fooling anybody, mister. We know you don't have a hundred kilos of bergamot stuffed in your pants.

That shit's industrial, and it smells industrial.

We still luv yuh, doe, yuh prick yuh!

Just eassie does it.

Slow and eassie wins duh race.

And, no, I ain't implyin' nuttin' 'bout no Itayins. Yuh got dat?!

I'm just regressin' to my inner Italian to express belligerence. Italians are my spirit animal.

I wanna go live somewhere, where everybody is just yelling at each other all the time... but no one ever thinks there's anything wrong with it.

"HEY, WHO TAUGHT YOU HOW DUH DRIVE?! Yuh fuggin prick, yuh!!"

"Yoe!!! You talkin' UH me?! Yuh MOMz taught me how duh drive... right after I..."

I wonder if there were any Italian Homo erectuses. They couldn't have been the erectuses further inland - those people had to follow the herds of animals around. And that meant they had to be a lot more *quiet*. So, if those land guys ever got upset about something, or if they even just saw an attractive woman walking around, then they would have to keep their thoughts about it private, or they would have to just express themselves more calmly. Nope. There's no way those guys were Italian. If any erectuses ever *were* Italian, it would definitely be the tribes along the beaches. But then again, the beach tribes weren't very hairy. So, I don't know.

Alright, alright, alright. Eassie does it.

I'm just teasing Italians 'cuz I have a crush on'm. It's good to be loud and passionate. It helps us resolve things without it getting *craissy*.

Great apes basically have four different ways to deal with conflicts. They can make a lot of noise more like the gibbons; or they can have a giant orgy more like the bonobos; or they can kill each other off more like the chimpanzees; or they can thwart each other's efforts more like the humans. So, all things considered, I don't think makin' a bunch of noise is a terrible option. And it literally helps keep us healthy.

Suppressing our emotions is massively correlated with certain disease categories. It's called the mind-body connection. Thoughts affect hormones affect everything. That's why you can think of lemon and then your mouth starts watering. Thoughts are powerful.

Although, we *definitely* have that chimp gene for wanting to fight each other. Fights throughout evolution were unfortunately inevitable. And they wouldn't even have to be fights over anything rational either. It could just be like some random, stupid bullshit, or a basic misunderstanding about something. And then we just throw down and start going at it. ... *Lo and behold... the "highest" evolved lifeform on the planet.*

Our species and our genus are also very covetous. Whenever we want something, we tend to just take it. It's called conquest, and we've been doing it since forever. Conquests have a way of exchanging our genetics. So, if the land groups could beat the beach groups, or if the beach groups could beat the land groups, then whoever it was who kept on losing would eventually get bred out entirely. And then the side who was always winning would be our ancestors.

So, that being said, who would conquer whom? Which of these two groups is more likely to beat the other one? And how exactly would they manage to do that? Which side would have a clear advantage? Or would any of them?

Well, the groups throughout the mainland were stronger, and yet the tribes along the beaches were smarter. So, how would this affect the average battle? What would tend to happen when it's brains versus an extra dose of brawniness?

Well, it's all about the eagles and ravens. Eagles are way more powerful, but the ravens are way more intelligent. Ravens will just get right up in their faces. And they're not even worried about it. They know that they're about *one whole order of magnitude* more intelligent. So, they'll just walk right up to a massive bald eagle and take its food away. And they freakin' do it without even a shred of apprehension. So, if those ravens and those eagles were having a spear battle, then it's not even a question of who would win it.

You know, there's actually something funny about these corvid species. Humans have often imagined that we are somehow superior, but that's *also* how these corvids see the world. You can tell that they don't really take most animals very seriously. They're just like, "oh, it's just silly animal." ... Heck, they even think that about *us* sometimes. That's why the ancients used to call them the tricksters.

Anyway, this is a good example of how the erectus confrontations would play out. The beach groups would simply find a way to outmaneuver them. They would trounce them like the ravens trounce the eagles.

The landies might score some points in an initial skirmish, but the beach boys would learn their patterns and eventually outthink them. Revenge of the nerds is no joke.

And this is why we never quite adapted to the inland environments. We never really had the opportunity. Every time some group was finally adapting, some newer wave would come and just defeat them – some newer wave of exiles from the beach tribes.

If they adapted over many generations, then they would have the time to lose their beach intelligence - not just from a lack of coastal nutrients, but also from the selection of a land life. So, when some newer wave of coastal people got there, they likely would not want to intermix with them. Rather, they would find a way to conquer them. They would find a way to kick them off their hunting grounds, or they would find a way to wipe them out completely. ... And then *these* guys would start adapting until an even *newer* group came. And then the cycle would just repeat this basic pattern. Those who were adapting were never getting to stay there. There was always a smarter fish to come and eat them.

... It's almost like a scene in a movie where each member of the bank robbery crew just takes out the last guy, as soon as that guy's job is done, and then the only guy left standing is the most recent beach exile. And then he uses all his bank money to finance his war on batman... because his dad cut a smile into his face or something – I'm still not sure.

Well, there you have it. There's the answer. Everyone alive is from the beach groups. We barely have a trace of any land genetics. That is why our species is not more physical. But it's also why we're *stacked* above the eyebrows.

That's right. Ain't no one alive today nO country boy, son. Yuh hea' me, boy? Nature kuh care less one lick whach'you thank. You ain't no country... You uh *bEAch* slicker!

You can take the boy out'the beach, but you cain't take the beach out'the boy.

And our shoulders ain't made for no rhino huntin' neither. These shoulders are made for fishin'. And that's just what they'll do. One of these days these shoulders are gonna fish all over you.

And we ain't no Neanderthal neither. Yeah, it's fun to imagine that we came from some super strong uber caveman, but we just flat out *did not*. Are we strong? Sure. Are we Neanderthal level strong? Not quite. We came from nerds. Strong nerds, sure. But still nerds.

We actually *did* have a chance to get stronger though. There was a period of intermixing with Neanderthals. And we could've just borrowed some genes to get a lot more resilient. We could've just borrowed their genes for stronger shoulders. And then we could've bred out most of the other traits – just selecting for stuff we wanted and not the other stuff. And yet, somehow this just never happened. We only kept like stuff for better immunity.

But this practice didn't start with just Neanderthals. This trend has been a constant since the days of Homo erectus.

Erectus were on several different continents. They traveled by the shores and settled coastlines. But many were also spreading further inland, and they were adapting to those harsher inland conditions.

So, surely, there were some super strong erectus around. We even find their bones which display this robustness. They were limited to mostly hunting the really big megafauna. So, they had to get tough in these land areas ... And we know that they shared their genetics with us. We have traces of their species in our chromosomes.

So, clearly there was some mixing with these land groups, but we never did absorb their physical prowess. We simply bred for things like staying healthy. And then we bred out all the traits for hunting mammoths.

We have had many an opportunity to borrow some land strength, and yet time and time again we've just repelled it. And that's actually because it makes life much more difficult. This strength just works against you on the coastline. There's no good reason NOT to get some super strength... unless of course it's causing you some problems. But that tells you once again we're not a land tribe – we've mingled with all these species and yet *never* kept their land strengths.

Oh, and I should've already mentioned this, but I'm just gonna mention it here. And, no, I'm not gonna go back and just add it in retroactively to an earlier point. I'm trying to freestyle this beast. ... Needlessly so.

Anthropologist like to call those species "humans," and they say that these hominin varieties are all in the "human" family. But I don't really like that convention. Yes, we are all "hominins" within the "hominin" genus, but the word "human" is specifically *our* word. It's just for us - Homo sapiens - not squatty little Australopithecus. The word "human" has been around for a very long time, and we only ever used that term for us – not Neanderthals. As a matter of fact, the very first time we dug up a Neanderthal, we thought we found some primitive form of monster. We didn't say, "oh, look at this fellow 'human' we have found."

Nothing against Neanderthals. They're all in the same exact genus as us. We're all the fellow members of the stork legged whale chimp clan. But we already have specific names for ourselves. They are the Neanderthals, and we are the humans.

And besides, you can't just take away the only casual term for our people, and then leave us with this confusing latin "sapiens" term – this term that looks plural when it's actually singular.

"Look. There is a Homo sapiens over there. Not two or three of them - no - just one Homo sapiens."

"Hey! That Homo sapiens just threw something at me."

See? It's technically proper grammar. But it makes you sound retarded.

Figures that's what our scientific name would be.

No, you know what, screw this. We already have a name. We're the mothuh effen human beings, son!

Oh, and by the way, the term "sapiens" means "wise." So, it's like we're calling ourselves the "wise man."

Uuuuuuuuuuuuuh, WHAT?!

Did whoever named us that ever dabble in some human history, pray tell? Like even a little bit?

I'm not sure *wisdom* is the virtue I'de be going with.

"Yeah, we're the Homo badass-mofos-who-kick-a-lot-of-ass-and-get-all-the-chicks-and-take-no-prisoners-cuz-WE-ROCK!"

We need a new scientific term - one that actually makes sense and isn't confusing. We need something like Homo forehead or Homo chin-nychin. Or maybe something that expresses that we like to *get it on*. How 'bout Homo sexual?

And we'll just throw an "s" in there to make it more confusing. ... "Look, everyone. There is a Homo sexuals over there."

Makes about as much sense as the last term.

Oh well. At least we finally know where we came from. It's nice to finally know that we're not some hulked out, caveman, mammoth punching, marathon runner. Nope. That ain't us. Quit looking in the mirror and thinking you're just some shriveled up little chump compared to your ancestors. You're not. You're virtually indistinguishable from them. These anthropologists don't know SHIT!

Okay, maybe they do know some shit, but they don't know *ALL of the shit*!! And they still make miSTAKES sometimes! ... Next time you see someone from the anthropology department, or someone who even LOOKS like they could be an anthropologist, I want you to get right up in their face and say, "hey! I am *not* a pathetic shriveled mutant!"

"... you son of a bitch."

"I am ManApePig, child of the crow face clan, and the clans of my ancestors, raven brain, stork legs, pig body, duck feet, and whale chimp.

... I am all of those things, and I am proud. And I will have my vengeance... in this life... or the next."

Then when he's like "what the fuck?" just hurry up and call security before he does and then tell'm there's an anthropologist trying to harass you again.

That'll teach'm.

Everything we thought we knew about anthropology is wrong. And by "everything" I mean everything within a very small subset.

We were never built for marathons of heavy activity. Nor were we well equipped for any powerlifting. ... If you can get away with that stuff for a while, then great, but just know that you're pushing the envelope. You're doing a thing that is relatively new for our species. So, we just gotta be real about it and not think that we're failing our ancestors. We're only failing them by being stupid.

Sure, we can always try this stuff and experiment. And, of course, we enjoy our sports and olympics. But those athletes are actually quite often anomalous, as far as their genetics go. And they're also enormously revved up on the attention of the crowd and their adrenaline. So, they're doing things which are vastly beyond the capacity of our typical ancestor.

But I'll tell you what we ARE good at. Our species is *great* at walking around and sweating. Yep. You better believe it. We *sweat it up* and *walk around* like you wouldn't believe. We can walk around for *hours* with both our hands tied behind our back. And we can sit around *sweating* in the sunlight without even breaking a sweat.

And what's even cooler than that, is we can combine these two hominin super powers into one single ability - ... jogging. That's right. We can jog. And we can do it fairly decently also. We can shimmy and we can shamble and we can shuffle and we can ramble. All we need is some water and some electrolytes... that plants crave.".~

And so, the reason why we're sort of good at jogging is because we are *specifically* built for lots of subsistence foraging. In order to be constantly foraging, you have to be proficient at wandering. So, jogging is the logical next step. It's just a slightly more expedient form of meandering. And whereas walking around for hours is uncannily *good* for us, jogging that whole time instead is conversely quite *bad* for us.

Of course, some people have understandably been under the impression that our jogging is meant for chasing after antelope. But it turns out, in fact, we are *not* actually super-duper great at running marathons – not the vast majority of us. It's murder on our joints and our free radical damage. Technically what we're good at is not even jogging. The *thing* that we're actually good at, is instead, slow motion *shuffling*. This is where you jog and keep your feet really low. And even shuffling isn't really that great for us because of all of the free radical potential.

It would be infinitely better for us to simply walk for the majority of these marathons and then intersperse it with some intervals of full intensity sprinting. This is all incredibly good for us. And we can finish just as fast as if we were jogging.

But anyway, there's *another* major reason why we have to shuffle. This is how you have to do it if you're going barefoot. Our foot is not designed for pounding gravel. Our ancestors were only really jogging on sandy beaches. And if they had to quickly sprint for some emergency, then it was only very briefly on their forefeet.

And our sprinting was never an ability that was intended for being an ambush hunter, by the way. We are *way* too slow, by a lightyear, to have done this for many millennia. Our sprinting is almost entirely for emergency purposes only – like avoiding crocs and hippos for example.

We've been motivating our sprinters in the olympics using all the wrong logic. We keep telling them that they're running for the gold, and to imagine what they want in life. But that's completely wrong. That's how you motivate a greyhound. We're not a greyhound. We're a hominin. You have to release a lion onto the track as soon as the event starts.

Trust me.

We're not so much a "giving" chase type animal... as much as we are a "getting" chased type of animal.

It's sort of like how I get those fancy, official vests for my dogs that say "service animal." And then I can take them into places that discriminate against letting dogs in. But unbeknownst to the store owner, my pampered, entitled dogs are actually "*getting*" service animals, not "*giving*" service. ... The only exceptional training *they* have is to detect if you have any snacks on your person.

It's probably me that should be wearing the "service animal" vest. ... But that's not the point. The point is this - my hominin ability to sprint fast is not so I can *ambush* a prey animal - it's so I can *avoid* being a prey animal.

One of the ways I know this... is because I do not have the ability to "mark" my territory. That's not the way humans are built to urinate. We can't just go in little spurts and then stop it once we've started.

And yet, a predator is designed to do precisely that. They do it for a range of different reasons, one of which is for helping them to camouflage their body odor. When their natural musky odor is throughout the landscape already, then it is clearly much more difficult to know if they are close to you.

Another enormous clue that we are not a legitimate ambush predator is that our foreheads are freakin' massive for an animal. We've got stacks and stacks and stacks above our eyebrows. And our foreheads are almost *half* of our freakin' faces. ... When we're looking up to peek above an obstacle, you can see our giant foreheads rising over it. That's why our soldiers have to wear these big round tortoise helmets. Our brains should be *behind* our heads, not above them.

Legitimate stealthy predators do not have these giant foreheads. The last thing that they need is to have a massive, bulbous melon giving away their position. When they're crouching down and prowling with their nose up, their foreheads are strategically flattened, and their eyes are elevated. They are perfectly well designed to spring an ambush. They are quick and they are quiet and they are stealthy. But we are slow and we are awkward and we are smelly. We were never put together to be an ambush predator.

... We're more like a horse - ... clopping around and urinating super loudly. ... except without any of their advanced features for running, like their five-hearts effect.

And then we have our big, wide, obvious torsos. Our rib cage is really flat, because it's spread out laterally. This is precisely the polar opposite of what a stealth animal has. Your rib cage should be narrow like your dog's is... not all rectangular like a bunch of sponge Bob square humans.

And we should also have some elf ears that are nice and pointy. That would help us *hear* a whole lot better. What kind of predator doesn't have super hearing? ... What a joke. ... We have these rounded, little, mini, streamlined, human ears. And we have no ability to move them whatsoever. ... We USED to have the ability to move them. And we still retain those muscles to control their movement. But we have no control left, in our central nervous systems, to actually maneuver them. What kind of super stealthy, land hunter can't even move its ears around?

And what kind of super stealthy, land hunter doesn't even have night vision? That's nonsense. If we were really this hardcore survivalist, land predator, then we would absolutely have some decent night vision. It's an unequivocal advantage for hunting night quarry. It's not an ability that you would just forfeit as a serious stealth hunter.

But the point of all that is simply this, our sprinting is just a basic feature for emergency evasion flight. It was put there for averting grave danger. It's not in there to be charging around like that constantly. ... Just because we can do something *intermittently*, it doesn't mean that feature was our *specialty*.

And yet, sprinting, at least, is not as bad as jogging is. That's because we do it on our forefeet. This saves our legs and joints from all of that impact.

But whether we're jogging or we're sprinting or we're jumping, this action is slowly wearing out our human feet. Our feet are getting injured while we're running, and this is happening irrespective of our running style. ... It happens in big shoes and in little shoes and in minimalist shoes.

But this is nothing new for our big footed genus. ... The little hominins on the prairie were also having these issues. They had bunions and plantar fasciitis and all of our same exact problems. ... – not exactly the big tough cave people we were imaging. The land has been jacking our feet up for about seven million years now.

And so, we have to protect our big monkey feet. It's time to get serious about this problem. So, I'm filling up my socks everyday with beach sand. I just dump about several handfuls in each one. And I can't really fit my foot in a shoe after that, so I just wrap the whole thing up with a bunch of duct tape.

The only problem is that it makes your feet look homeless.

And it sounds like you're walking in diapers.

Kxxx... kxxx... kxxx... kxxx...

Okay, maybe I don't do that, but someone should. Someone should invent a shoe that's like walking on beach sand. I would do it myself, but then they would accuse me of writing books so I could sell my sand shoes. So, I can't. I just have to give it all away. Even my duct tape idea.

But if this sand shoe idea ever takes off, and you make a billion bucks, working from your small town factory in New England, just promise me one thing. Promise me you'll keep an eye out for your son, because Billy Jessup will try to beat him up with his bicycle gang.

And also, don't let him play any mystical magical board games about jungles and safaris. ... – ones with loud drums and rhythmic cyphers.

Just trust me on this one.

"In the jungle you must wait until the dice read five or eight."

Oh, and just thank your lucky stars that we were once a beach animal, because we wouldn't need your foot shoes if we were a land runner. If we were ever running so extensively on land like that, then we would never need protective footwear. We would simply have our own shoes in the form of foot pads. We'd be born with natural cushions automatically. And thus, we wouldn't need your shoes for our protection.

And this has got nothing to do with your physical conditioning either. Even the biggest, most religious enthusiasts, for the practice of running barefoot, are still wont to make use of minimalist running shoes. The resilience of their rugged, weathered calluses is *still* no match for pea gravel. ... even though my dogs are always running across it without even noticing it.

Normal land runners like dogs are born with nice big shock-absorbing foot cushions. And those cushions are well protected on the bottom. They have a layer of natural, edge-resistant callus pads. This is what you need to run on land a lot, and we have none of it. There is no way in a million years that we never got a foot pad from a million years of running around in the savannah. So, you can just forget about the idea of us ever having been a serious running animal. There is no amount of ambiguity to this point here. It's a dead giveaway. This is dead on balls accurate.

This is a really big point here to pay attention to, because it's so absurdly obvious when you simply *look at it*.

So, let's look at it. We had all of this time in evolution to shrink our face; to shrink our teeth; to get these big bulbous melon heads; to get these big hooded proboscis noses; to lose our coat of body fur; to get these big long stork legs; to get these big goofy clown feet; to get these big female chest tits; to get these big long whale dicks; to get these completely different muscle fibers; to develop the ability to sweat fluids out of our skin, across the entire outer surface of our body; to get this weird umbrella head hair; to get this subcutaneous fat layer like a whale's blubber; to get this massively altered shoulder structure, that's about forty five degrees lower than what it used to be; to get this huge diverse array aesthetic expressions around the world; to develop these big long paddle feet with a protruding toe appendage in front instead of a dexterous grasping thumb digit on the side; to get these big cushiony butt pads... aaaaaaaaaaaaaaaaaaand NOT get a shock absorbing foot cushion???

Uuuuuuuuuuh wwwwwhat?

So, we could do ALL OF THOSE massive morphological changes and then NOT just get a basic pad and foot cushion?!?!?!?!

Uuuuuuuuuuuuuuuuuuuuuuh, WHAT?!

No...

No, no... no, no... NO, NO... no, NO, no, NO, no, NO.

I don't care what *anyone* says. I don't care what kinda mental gymnastics, tortured logic, word salad, confidence act, self gas lighting, fancy jargon we wanna throw at this thing – the truth is absurdly obvious. We don't have these features, because we never needed them.

The science of evolution follows the principle of parsimony, AKA Occam's razor. It's a philosophical razor for separating off less likely arguments. It does this by simply stating that the most simple explanation is the most likely one.

Okay. Well. Then you tell me - which of these two things sounds the most simple? We spent millions and millions of years constantly stomping around barefooted across the rugged terrain of the savannah and developing an entire array of foot problems which still to this day are plaguing us, and then nature decided NOT to give us a basic foot pad, and it did so despite otherwise changing our *whole entire body structure*. Oooooooooooooooooooooooor... we never developed a foot pad because we simply never needed one. ... Which of those two things sounds most parsimonious???

Sadly, I know that the obviousness of this reality still won't phase a certain segment of people, but we all originally understood these things as a kid. We used to run around completely without shoes on, and then we'd step on some flaming hot embers, and then those flaming hot embers would not even phase us, but thEN we'd go step on a pile of lego toys, and it would ruin our life in an instant.

We all know the truth deep down in our kid's feet. Just quit suppressing the memory of that horrible lego incident.

And the next time some slick Rick in a van pulls up and he's like "hey, kid, you want some free evolution theories?" and he starts telling you about this savannah stuff, just be like, "okay, well, then what happened to our foot pads?"

BOOM!!!!!!!

End of conversation.

"I'm not supposed to talk to strangers that *don't make any SENSE*, mister."

Not to mention, if you're walking around on some beach sand with your human feet, it often feels like you're just spinning out your wheels on a mud slick. When you push off with your forefoot, as you would on a firmer surface, the sand just slides away, and you feel like you're going nowhere. But if you relax your calf and muscles and let your foot *roll* out, then it's almost as if you're walking again on normal ground.

In other words, our foot is like a big expansive snow shoe. It's better for sandy trekking than a normal land foot. It's almost as if we're built to walk on sand a lot. It's like our foot is meant for swimming and sandy beaches.

The coincidences just keep on rollin'.

Hey, do you remember that one time... when those aliens were sending us those radio wave messages, and we heard it at The Very Large Array of receiver dishes in New Mexico, and it was instructing us on how to build a huge gyroscope machine that looked super dangerous, but we still dropped that one blonde lady into it anyway, so she could make Contact, and she went through a wormhole through space-time, and then she eventually came out on a beach somewhere?! And then that one alien guy was disguising himself as her father, and he was like, "this is how we've been doing it for billions of years."

Do you remember that? ... That one time? ... In real life?

Well, why do you think those aliens chose to do it that way? Why do you think they brought her onto a tropical beach?

Coincidence?! ... Pfff. Yeah, right. Not after "billions of years." These aliens know what they're doing, alright?! ... And then their instructions even said that we're *not* supposed to have a chair inside of the passenger chamber, but we went ahead and built one in there anyway, even though we weren't supposed to do that, and then the chair was all shaking and rattling. Well, these aliens were trying to tell us something. They were trying to tell us... that when you're supposed to do something a certain way, that's the way you're supposed to do it.

Well, we haven't been using our bodies the way that we're supposed to. We're not even using our faces right. We've got these sinuses down below our cheek bones and they don't ever drain right. That's because they have to drain *upwards*, *against* gravity, in the totally wrong direction. So, then we get all these weird sinus problems and these super random nasal infections and all these sniveling, cowardly, head colds. ... We have all kinds of trouble with our faces that seems really really weird in the context of *other* animals.

But our sinuses are not an actual design flaw. We are simply not exploiting them as they are intended to be. We are supposed to be swimming down and diving vertically. This is how we drain our lower sinuses.

And it doesn't stop there. We are also supposed to be mechanically under pressure. This is what is happening when we are diving. And we're supposed to be submerged in salty water. This all affects our bodies, including our eyes. In other words, our eyes were specifically designed to be experiencing a little bit of pressure within a saline solution. So, who knows what this lack of uniform pressure has been doing to our eye health.

Our eyesight is also in need of the appropriate nutrients. We were designed to eat a jungle beach diet. This means DHA, vitamin C, crimson astaxanthin, and other things. These are immensely important, key nutrients for our body and our vision.

This diet is also important for our ongoing dental health. We've been consuming these power packed shellfish for nearly seven million years now. When you take them right off of their half shells, you get a shot of filtered saltwater full of minerals. Our teeth have thus evolved to expect this rinsing. Many cultures have been exploiting this for ages. And these humble little shellfish are also power houses of nutrition. They support our entire body, including our oral health, and our dental integrity.

Our teeth are an extremely big topic, of course, and most of us have done a little damage already, but we've also got some modern tricks to help us, and there are things we can all exploit in addition to nutrition. Sugar is best avoided in the daytime, but then a *small* amount after dinner will help your brain sleep. And, of course, we should be rinsing after eating it... or even skipping all sugar *entirely* if we're just not feeling very healthy for some reason.

Furthermore, processed food and carbs should all be moderated. If nothing else, then save them just for dinner... because these carbs will all break down and turn directly into blood sugar... and then these sugars will fuel the things that don't belong in us – ... everything from acne funguses to dandruff funguses to rogue cells to dental bacteria.

It's also good to brush with things like "tooth powders." They've got ingredients like clays and minerals and diatomaceous earth silica. And it's handy if you can procure for yourself some orthosilicic acid supplements. All of this stuff will assist you in remineralizing naturally.

Our teeth will thus last us forever, if we just help them stay mineralized. Our eyes will surely serve us much better, if we just feed them with maritime nutrition. Our sinuses will thus engender less drama, if we just invert them and provide them with immune support. And then our bodies will determine to flourish, if we just walk a lot and swim at regular intervals.

There is almost nothing inherently wrong with our modern human genomes. They are not distorted or perversely unnatural or severely maladapted. They are *extremely* well adapted for our native environment. ... And they just need a little love from the life aquatic again.

In other words, most of our problems are not from defects. ... They are usually just operator error.

We can't just go around not knowing how to use stuff. If you're gonna go messin' around with something, then you really gotta know what you're getting yourself into. Otherwise, all you're gonna do is get in trouble.

It's like that one time... when your one weird uncle was staying with your family for a while – that one who's always talking about how *he's* really the one who invented certain things, but he never got credit for it - and then we just cut out of school early so we could go back to your place and check out your new gaming system. But then we accidentally walked in on your uncle and he was using your computer to jack off to the internet.

And we were like "AAAAHHHHHHHH... what the !!FUCK!?!"

Then later he tried to lie about it.

So, we just told him... in the flurry of his panic, he didn't close out his windows properly, and we could see everything he was watching. Then we just started describing it to him and describing all the details, and he just instantly got enraged about it.

Then later we asked him how long this type of activity has been going on for and he just instantly got enraged again.

12

9 ¾ 3 -- DEFENSE AGAINST
THE SAFARI ARTS

Why was there always weird shit happening on every school bus?

I mean, weird stuff like that time that one super overweight girl a couple of seats in front of us just suddenly fell over sideways across the aisle, like a big fat bridge over the walkway, and then she was just shaking at a really high frequency and convulsing and having a seizure and you could see all the ripples going through her fat. And the kid that was sitting in the seat she fell into was like "OOOOHHHHH MMMMYYYYY GOOOOOOD!!"

Then later after the situation was resolved, we all felt really bad for her. And as soon as they finished bouncing her down the stairs on the stretcher, Tyler the jackass just turned around and said, "hey man... I'm sorry about your girl."

Not everything in life is always happy in life.

Sometimes things are sad.

I'm surprised you didn't know that.

And sometimes things weren't always happy for our monkey ancestors.

You know, our modern society has gotten this whole image in their heads of the noble caveman hominins all working together to take down a mammoth. We've attached so much prestige to this archetype. But the reality was, this lifestyle was a lot more disheartening than it was glamorous.

Most of these people that were living in these mainland areas were not living there and because they *wanted* to. Most of them were forced to. ... Having to hunt those massive behemoth megafauna, the size of walking mountains, was not a super fun fantastic festival. ... These Homo erectus were constantly getting themselves killed, and brutally injured.

Not a fun time *indeed.*

This isn't something that you would just *do* because you wanted to do it.

Young men throughout the ages have inherently perceived these sorts of adventures with a great deal of romanticism and revelry. At least, that's how they feel about it in the beginning. And then they finally get their chance to go and try it, and then they sadly get to find out what a shit show it is.

Young men have a perfectly natural instinct for adventure, but it's mainly so that we can go forage and discover new resources. And it's easy to confuse this inclination of ours with the desire to go rhino hunting and mammoth spearing. But then *that* confusion usually ends when your buddy gets his skull fractured.

It's also not exactly super helpful that our societies start adapting to these conditions. They start making it a source of manly honor that the young men show their bravery against adversity. Then they lavish recognition on the ones who have done this – the ones who have already fed their bodies into a meat grinder. Aaaaand, of course, they always shame and banish those who will not do it. ... So, as a social species, this is virtually irresistible to anyone. Society keeps this pressure on them constantly... so that their larger populations can benefit from their sacrifices.

And we also adapt to these insufferable conditions surprisingly easily. It's partly from those aforementioned cultural pressures, but we also have a quirk in our psychology. You see, we, as mammals, and as a fairly brainy species, have the ability to be trained in ways that are entirely arbitrary. If we witness our parents expressing fear of something, then we *ourselves* can begin to start baselessly fearing it. And if we see them ostensibly appearing to enjoy something, then we ourselves can also grow to just arbitrarily love it. Basically, anything we witness or get instructed on can quickly become engrained in our long-term psyches.

Indeed, we often end up with a bunch of random programming, some of which is highly sound and logical, but then much of which is largely vacuous artifice. This happens when an animal becomes encephalized. ... which is to say, its brain becomes much larger relative to its body. And this enables it to learn beyond its instincts. This enables it to learn all new behaviors. And in a world of constant threats and opportunities, this enables them to trounce their competition.

But these abilities are a double edged sword, in fact. On one hand, they can clearly make you smarter, but on the other, they can also make you crazy. You can learn things as a child that defy your instincts, buy you can also pick up behaviors that are self-defeating.

It's not enough to be big brained and encephalized - you also need the proper formative conditioning. It's one thing to be born with a little extra brain power, but if that power is not assembled and put together right, then it ultimately accomplishes nothing and just wastes that extra energy.

Erectus was pretty brainy for an animal. They had the same basic aptitude as a modern twelve year old. So, they were more than cerebral enough to finally get themselves into trouble, and they could easily establish a culture which romanticized self-sacrifice. All they had to do was teach their children this, and then just simply create a zeitgeist of rigid enforcement. And then once you've established that precedent of indoctrinating the children like that, the culture will become self-sustaining, and it will transfer from generation to generation.

Or, to put it another way, we're smart enough to be dumb. We're clever enough to brainwash ourselves. We can move into a world that just clearly isn't healthy for us, but we can delude ourselves that this suffering is instead somehow a good thing. ... And, yes, we will always have a sneaking suspicion of when our environment is not quite right for us, but we can easily override this with our powers of denial.

The simple truth is this - nature didn't make you to feel uncomfortable - it made you to be happy. But these people on the plains would not be happy. They would be as far from delighted and gleeful as humanly possible.

First of all, in order to leave the coasts and flee your homeland, you would have to rebel against your primal instincts. Not only was this the place that you were born in, but this is also where your soul was evolved to resonate. Every fiber of your being was built to love it here... - you are adapted to feel at home in this environment. So, leaving your fountain of genesis is a very real source of psychological torture. Even children who had never known the beaches would still miss it on a deep subconscious level.

Then everyone, without exception, would be woefully deficient in the various micro nutrients. They would, all of them, be suffering from things like thyroid diseases and cognitive impairment.

And yet, they wouldn't have the time to stop and think about it. They would all be busy with the task of dodging serial killers. Horrors from their deepest darkest nightmares would come and stalk them. ... These things would track them in the daytime... and then hunt them after nightfall.

Soooooooooo, that's always fun.

It would be like that time when you were writing that book and you were getting really tired because you work full time to support a family of dogs and your eyes were getting really really heavy. But then you opened them really fast and said "no! I can't fall asleep right now. I gotta stay awake and keep writing in a state of delirium!" But then you realized that you could actually just cheat the system a little by just closing your eyes for a second, and it's really no big deal because you'll just finish writing this book as soon you get done walking on this path in the mountains with a bunch of strangers that you've never met before. Then you guys walk into an enormous cave with a huge gigantic opening the size of an aircraft hanger, but you're pretty sure this isn't the right way to go, but then there's also a rumor going around amongst the hiking party that someone might partially know what they're doing. So, you just keep going deeper and deeper and deeper, and it's getting darker and darker and darker, until suddenly the path just falls away into darkness. So, you put your hand up really fast and say "wait! It's a cliff! This could be trouble!" And then some girl in the back who's moderately heavy set and has a really bad attitude says "bullshit, there's no cliff!" and she just shoves right past everyone and pushes you out of the way and then just walks right off the cliff, and everyone *gasps* in horror. Then you all just stand there waiting for her to hit the bottom, and you just keep waiting and waiting and waiting, but the sound never comes, and now everyone is going completely *insane*, wondering why she hasn't hit the bottom yet, and everyone is just absolutely mortified, wondering if she just immediately landed in some soft sand already and now she's just messing with us... or... ... is she STILL... FUCKING... FALLING?! ... But right when you absolutely can't stand the horror for even a single second longer you finally hear the sound of her impacting the bottom.

BUDOOSHHHH shuh shuh shuh shuh shuhhhhhhhhhh...

Then you cry out in horror "AAAAAAAAAAAAAAAAAAAAAAAAAAAA" until you realize you're home again in your house with no cliffs in it.

And also your dog is laying right next to you scratching his ear really loudly right next to *your* ear, shuh shuh shuh shuh shuh shuhhhh...

Oh man... I'm glad we're safe now. That was *messed* up...

Holy mackerel. That was like spending a night on the savannah.

But maybe I've got it all wrong though. Maybe if we just keep stacking on the horror, it'll finally add up and we'll just get Stockholm syndrome and start loving it weirdly.

Alrighty then. Well. Let's see. How else could we make this even more hellacious. Oh, I know. We can make it an absolute nightmare for every single hominin mother. That's right, this is just about the worst possible scenario for a pregnant or breastfeeding mother. Without the appropriate nutrition, the body of a hominin mother will just rob itself of resources to build her baby. If she isn't getting what she needs from her environment, then it'll *take* it from her brains and her tissues.

Most of a person's brain is formed when they're in utero and when they're breastfeeding. That's why babies have such big heads compared to their bodies. Nature wants to build that brain as soon as possible... so that the brain can be supplied by the mother's body. But if your mother isn't getting all her brainfood, then her body will just take it from *her own* brain. ... This will cause her brain to literally shrink a little. And this of course will give her some serious brain fog.

Thus, the mother will fall away, much like a rocket booster. She'll spend *seasons* just slowly recovering from this nutritional draining. That means *years* of feeling weaker and mentally disoriented. And that's why cultures have often spaced out their children a little. This helps their struggling mothers to have a chance to recuperate. And this is essential if you want a healthier child.

There have been a couple of populations that have somewhat adapted to this land-based DHA deprivation. They can now synthesize *a little* within their bodies. But even these groups still can't make it that efficiently. They have to synthesis it from ALA in plant foods... – *lots* of ALA in plant foods. ... And a few more populations can get it from raw dairy milk, because milk has some basic nutrition if it's still fresh and unadulterated. But even dairy isn't enough to support our brain growth. The milk of other animals is not designed for us. And neither milk nor any land-based vegetation will *ever* provide us with these minerals like iodine.

Not that either of these options was even available to the earlier hominins.

The first ones who could actually get around this though were the Homo erectus. They had axes made from stone which was knapped into a cutting edge. The edge was on a handle, about the length of their forearms. This handle then gives you a *massive* mechanical advantage. It enables you to build up momentum. And that momentum is what it requires to breach a ruminant skull.

So, erectus could get at the brains of certain species like rhinos. But even this is not a lot for an entire tribe of people. There would be approximately as many of them as it would take to eat an elephant. So, that's roughly around several dozen members including women and children. Furthermore, the brains of these rhinos and elephants are just a couple of little handfuls. So, you would have to be dividing up these couple of little handfuls amongst several dozen people. This would be barely a tiny spoonful for each member of the community. Realistically though, no past or present tribe would've even bothered with this absurdity. ... – this absurdity of scrupulously dividing it up like this.

Erectus would've been thoroughly oblivious to the utility of eating brain matter. Even WE were unaware of this utility before several centuries ago. And yet, erectus would've craved it instinctively despite not understanding this. They would've craved it because it had things they were deficient in. They would've ended up like an axe-wielding zombie with a taste for eating brain tissue.

And in a cruel twist of fate, not only would these deficiencies eventually turn you into a brain-craving zombie, but eating an infected animal brain will do this to you also. There are proteins in some brains which cause insanity. These proteins are little prions that'll eat your brain up. This condition is known as spongiform encephalopathy. And if your people keep making this gamble, then they'll eventually get a bad one and catch these prions.

Yep. It might take a while, but if you insist on playing gun games and spinning that cylinder, then you'll eventually get the chamber with a bullet in it.

This disease could easily decimate an entire population. It would wipe out all the people eating brain matter. And it would only spare the ones who were otherwise abstaining. But the ones who were thus abstaining would also be deficient. So, it would exclusively spare the individuals who had this cognitive impairment.

But then these mentally retarded survivors would need to make their own weapons. They would need to make spear heads in addition to pole shafts and artisanal hand axes. Because they might only get a few good uses out of the leftover weapons of their fallen brethren – especially the spear heads. And it would take more than just a few good uses for them to get up to brain speed. So, this is a pretty tall order for any would-be prion-plague survivors – those who were far less intelligent already. It takes an awful lot of skill to fully fabricate a hand axe.

These axes were sort of a big deal. If you wanna get a sense of how important they were, just try hammering something in without using a handle. It's not as effective as it might seem. It'll take you about a month to get that nail in. Or you could simply use a handle and get it done quickly.

Well, the same exact principle is in effect when you're trying to hack open a ruminant skull. It's one of the most reinforced structures in the natural world. If you tried to hack it open without any leverage, then you'd just be sitting there, whacking it constantly, burning up thousands of calories.

Even with our modern industrialized tools at our disposal, you even *still* will need a mallet with a chisel punch. ... You have to hammer the chisel down into it and pierce the skull plate. Then you go around the top until you've chiseled out a circle. Then you pop the circle off and remove the skull cap. Then you remove the brain inside in larger pieces. But this process will take you a while, even with modern tools, so you can forget about doing this efficiently with just a simple, broken, hand rock.

Like I said, this would not have been "a thing" before erectus. It would not have been a rigor that was worthy of the calories. Likewise, brain matter is extremely prone to rotting. So, it all would be quite spoiled by the time that you got to it.

If there actually *were* any simpler hominins ever doing this, back before the days of Homo erectus, then they would've had to do this very differently. For one, they wouldn't even have effective bone spears, so they would've had to simply find some rotting leftovers. And they would've had to get there fairly quickly, because the brains inside would likely have gone at least partially rancid already.

So, let's just assume, for a minute, that some early, primitive, jungle apes just happened across a carcass, and they also happened across it quite freshly such that the brain had not gone off already. Alright. That's a lot to assume there, but okay. Sooooooo, then how are they supposed to get this brain out?

Well... We're gonna have to give 'm the benefit of a doubt *again*. For this scenario, they would've had to use a stick that had a hook on the end. Then they could possibly just drag little pieces of brain out through that hole in the spinal opening. Or they could jab around repeatedly inside of it and shake the skull up. This would blend the whole thing up, and then they could drink it.

But this would likely be a trick that wasn't very common really. There might be one or two unorthodox entrepreneurs who were anomalously exploiting this. ... They would likely slurp it out right through the brain stem. Then the chunks of mush and goo would go sliding down the orifice through the spinal opening. ... Then they would gulp the whole thing down in several minutes.

That's fu****' disgusting.

Alright. So... as we've been discussing, if we reeeeeaaaaaally wanna torture the basic logic, then there might have been some times when these pre-erectus hominins were getting lucky. These individuals may have gotten some brains without a hand axe. But this truly would have been a rare occurrence. And it wouldn't even help but several people, because they likely wouldn't share these smaller features. And even if they did share, out of some deep-seated sense of necessity, then these portions were still too infrequent to significantly help them.

In other words, before erectus made their first real hand axes, there was just no pragmatic way to get their DHA out there. Our species can't so simply synthesize it like some others can. The savannah was a land of nutritional scarcity for us, and you just couldn't get those brains without a hand axe. So, when erectus finally showed up with their tool kits, they could finally breach those skulls with real efficiency. ... And it would *really* degrade those axe blades trying to do this with them, but at least they had a way to finally do it.

They finally had a way to axe you a question.

Doesn't this sound like a lovely little existence for an erectus mother? One minute you're on the beach, listening to the sounds of the ocean – the next you're all covered in flies, listening to the sounds of skulls being cracked open. ... and you haven't even bathed in a year yet, because the crocs'll freakin' snatch you by the river.

How lovely.

So, like we were talking about before, even with the occasional access to a brain or two, these mothers would still be deficient in a number of essentials. This would force them to give up their own supplies. They would sacrifice their own brains to build their children. And this of course would leave them feeling zombified. And it would take them several years to slowly recover. The welfare of these landlocked erectus women would be constantly diminished.

Those first few years of life are extremely important for us. If our mother does a really good job at it, then we can actually have an okay life no matter what happens after that. Even if we never get our brain-food thenceforth, we ironically can still have a surprisingly moderate existence. ... Granted, you will never be performing at your maximum, but you'll also never know that there's a problem. You'll just assume that this performance is how you should be.

Now, it's bad enough when these mothers lose their brain matter, but what's actually worse is when they also lose their minerals. Your brain'll grow back slowly if you start getting your omegas again, but there's simply no real way to just manifest your iodine. It's not in the plants; it's not in the animals; and it's not in the soil. It's not in the land in any way that we can assimilate. And this'll cost us about ten or more IQ points and give us health problems.

It's no joke. This is the single most critical weakness that we have as a genus. Right now, there are *billions* of modern-day people who are deficient in this halogen. They have been living way too far from the oceans – and they are just not getting enough iodine from anything. So, now *billions* of individuals around the world, at this very moment, are literally dumber and weaker and more erratic.

And yet, it isn't just our brains that get affected. It's also our bones and our bodily hormones and our women's monthly menstrual comfort. And when a mother hasn't gotten key nutrition, she starts craving random substances like clays and soils.

Women have it worse from this condition, and yet they keep it to themselves in many cases. They don't want their local men to think that they're having fertility problems. So, everything just appears normal, because the women are all suffering in silence.

This phenomenon would've been obscuring their nutritional deficiencies – at least for a single generation. But eventually their former momentum would've completely exhausted itself. And this is when they'd start to have some *real* problems. This is when their children would be born with health issues. And frankly... this would be an *awful* time to live through. They would've felt like they were stoned and sapped of energy. Not to mention the things like goiters and random hormonal issues. They would've had to work quite hard to find a remedy.

Maybe they'd get lucky and find a salt marsh. Or maybe they could fish in an inland river. Certain species will migrate inland from the oceans. So, if you've got a good technique, you can trap and spear them. But even then, you'd still be dodging all the crocodiles. And the nutrients that you did get would still be small. Sure, it would be enough to keep you moving forward, but you would still be diminished from where you should be.

So, there you have it. Malnutrition, lower IQs, brain fog, health problems, insatiable cravings, brain eating barbecues, pathogenic proteins, inter-tribal cannibalism, zombie apocalypse, and dodging river crocodiles. Did I miss anything?

Oh, wait. Yeah, I did miss something. There's one more little gem which derives from this lifestyle.

You see, when a deficiency is so catastrophic that it drives you to the point of cannibalism, these acts of desperation and madness can sometimes become a cultural habit.

This is different from when it's caused by straight up starvation. If a tribe had no more food and was then pushed into cannibalism, then at no point in this wretched ordeal would they ever be confused about its merits. They would acutely be aware without an ounce of ambiguity that this thing that they were doing was super twisted. And they would never in any way become deluded that this practice had somehow provided them some health advantage.

However, if something's really wrong with someone's body, and they're desperate to find a cure for what's been plaguing them, then they'll often start consuming lots of random stuff, especially if they're craving things that they're missing. And it's really quite simple to get a completely false positive - you might not be so sure which foods are actually helping you. So, if one thing that you found yourself consuming was the body of your local enemy, then you might be falsely led into assuming that this was actually helping you.

Humans tend to do this and eat lots of random stuff. It's sort of like a natural progression of our instincts for foraging. We've been designed to examine the landscape and eat a dizzying array of seafood. So, in our minds, we just assume we might find our salvation in some random exotic substance. And then we also have a throwback, psychological tendency to get a little culty about stuff. When we find some ostensible new savior, that we seem to believe is now helping us, then we often get a little bit obsessive over it.

People in the middle ages used to grind up narwhal tusks and drink them. They actually thought that these oversized spires were the horns of unicorns. But not to be outdone by this medieval ignorance, people in the current ages are still consuming copious amounts of random, animal body parts - things like shark fins and rhino horns and elephant tusks... which are basically just like eating a bunch of fingernails.

If this kind of insanity can still be going on *right now*, in an age when we can factcheck just about anything in a matter of minutes, just imagine how confusing it must've been for Homo erectus. You could easily get the totally wrong impression about things. You might imagine that your favorite lucky loin cloth could make the sky rain... or that something as completely arbitrary as inter-tribal cannibalism was some ameliorating your zombie cravings.

The logic of consuming other people actually seems to make sense from a primitive perspective. "Other animals are strong, so if I just put them inside of me, then I should get their strength too, right? ... Well, there's a tribe of other people over there by the river and they seem pretty healthy... and they're always talking a bunch of trash and calling us dumb-dumbs... maybe we should eat them and steal their life force."

"I want that tribe inside of me... "

Never mind. Okay, so... there have always been the witch doctors and the medicine people, and they have always had their prescriptions for various maladies. They likely had recipes for fixing goiter pains and zombie cravings, and they were surely filled with heaps of erroneous theatricality: three pinches of grey tinted mountain clay; two handfuls of bush tucker mushrooms; several large fistfuls of river salmon; aaaaaaaaaaaaaaaand a person's face.

In the early days of maritime scurvy, they would eat a bunch of useless tonic remedies. Most of these medicinal concoctions were doing absolutely nothing for them - except of course for the ingredients which had a little bit of vitamin C content.

Well, primitive tribes of people would do this also. And when you do these things over and over, they eventually become a custom. And then people start inventing made-up backstories. They start developing a mythical legend for why they're doing stuff.

"Then the spider god came down and told the sacred four-headed monkey lion that he must go and teach the people how to grow bañyañyas. But first, spider god said to monkey lion, you must also marry your brother, and conceive a child, with the body of a horse, and the chest of a man, and the head of a crocodile... because f*** you, that's why."

... according to legend.

Sooooo, this is when they lose the plot completely. This is when they lose the original pedigree. As soon as they start turning these practices into a bunch of mythological religious commandments, then the connection to where they came from is lost forever.

Or, if you wanna put it another way, when you put people into a very difficult environment that's making them sick all the time, they are inherently going to develop a bunch of random, erroneous coping mechanisms. And then these coping mechanisms will end up becoming "the will of God" at some point. And when you're dealing with a very primitive and oblivious population, there's gonna be a whole lot of extra stuff that ends up getting folded into that "will of God" of theirs. And when that happens, it'll only end up haunting them for generations. Cannibalism is just one example of the things that can go very very wrong in these situations.

And just for the record, no anthropologists can ever argue with me about this point here... beeeeeeeeeeeecause... God wills it.

GOD WILLS IT!!!

YOU DON'T QUESTION!!!

Now... I understand that you might end up waking up tomorrow morning and then rolling over to see my open book pages lying next to you in bed, and then think to yourself, "wait... did I just... agree with you last night... that there could be a link between savannah induced nutritional deficiencies and ritualized cannibalism?! Oh my god. I can't believe... you and I just... ... I mean... the two of us... just... agreed with each other all night long about that. What hell was I thinking!"

Well, first of all, I'm sorry to hear that, because your line of thinking still looks just as good to me today as it did last night, so I'm sorry... you... feel... ... that way... about... my thoughts... today... **D,x**

No, I'm so sorry. This is all my fault. Let me just gather my thoughts and I'll be on my... ... ...wait... hold on... have you seen my book cover? I was wearing a book cover last night when I came over. ... It was like dark, with like a jungle print, and like a really really sexy wood inlay on the edges.

Oh my god, you found it! EEEEEEEEEEEEE! Oh'my'god oh'my'god oh'my'god, you're my HERO again! Thank you sssssOOO much. Like... you don't even know. Oh my god.

Hey... we had fun last night, didn't we? Teehee. Okay, well, I guess I better get going... Hero! Oh, wait... there's just one more thing. Now... I know that you probably don't think my analysis looks as good to you right now as it did last night, but just in case you change your mind later, I'd still like you to knowwwwww thaaaaat... as far back as we can see in the archaeological record, there has always been ritualistic child sacrifice, since the very beginning.

Okay, just hear me out. It's been everywhere. And always. Humans have just constantly been sacrificing their children since at least the Neolithic times. Yep. They have been instinctively projecting all sorts of weird, abusive, parental psychology onto their deities... and then assuming that these imaginary deities are getting their jollies off, on some seriously sadistic sacrifice sh** - just like some sort of crazed parental figure with post traumatic stress disorder. So, then, naturally, every time there's a bad earthquake or something, they just freak out and start sacrificing their children.

I guess those ancient people never saw the movie Kingdom of Heaven with sexy deep voice Orlando Bloom - "If God (is a dick)... then he is not God."

You can't really blame 'm though. This dimension is a pretty jacked up and horrifying one. They had every right to be upset. ... However, can we at least agree on one thing, though, that sacrifice is uuuuuuuh not a legitimate uuuuuuuh... option? ... Just... like... as a rule? ... Is that... like... can we agree on that?

If these hypothetical deities were really dicks like that, then, by definition, they would *not* be praiseworthy. Wielding power doesn't make one worthy - wielding virtue does. Sometimes you gotta be like Perseus, played by sexy deep voice Sam Worthington, and do the right thing no matter what happens - no matter what the gods try to throw at you.

And why do you think his name is Sam WORTHI-ngton?

Do you really think that's just a coincidence?!

Is that what you think?!

Is that what you tell yourself?!

Is that what you tell yourself at night... to sleep better in your peaceful little bED AT NIGHT?!

IS THA...

Okay... The point of all that is just this – this is still a good mindset for us to strive for – refusing to abide these abusive parent "gods." If a lot more earthly mortals had this mindset, then our world would be like a trillion times better, without exaggeration. All we've ever really needed was each other... and a refusal to negotiate with these terrorist gods... who sacriFICE CHILDREN!

And besides, the whole entire point of a sacrifice is to give up something of value in order to prove your conviction. That's it. That's all it is. And there's about a million different ways that you can do that. It doesn't have to be like, something that's all barbaric or whatever. We can sacrifice in a way that's beautiful. We don't have to destroy things - we can preserve them - it's called *commitment*, and it's a *way* more intense form of sacrifice.

And if you think that you were told otherwise, then maybe G-d was just trying to test you. Not everything is supposed to be spelled out for us all the time - that's why we have a mind - we're supposed to be thinking about stuff... and using our judgement... - not just blindly following orders, like some good little automaton droids."~

Oh, and if you think that someone was already sacrificed in order to save you from your sins or whatever, then maybe their "sacrifice" was just the *natural* consequence of them trying to speak *truth to power*... in a deeply fallen world, in order to save you - ... not just some insane, sadistic sacrifice at the behest of some bloodlusting deity who's getting off on it. That wouldn't even make sense - what the hell kinda *love* would that be?

Oh, and slavery too. That's another one of those really big no-nos. Can we all just add sacrifice and slavery to the ol' Not Okay List? Is that... like... can we all agree on that?

You know, I actually had these three different guys at three different points in my life LITERALLY try to argue with me, unironically, that literal actual slavery was NOT a bad thing. Because, according to them, none of the world's major religions, including their own, had ever forbade it.

Uuuuuuuuh... WHAT?!

Seriously?! None of'm?

Wow... How 'bout the minor religions?

I didn't know they needed to write that one down for us.

Okay, soooooo, let me see if I got this. I'm not allowed to take your *stuff* from you... BUT... ... I AM allowed to take your *you* from you?

Hmmmmmm...

Sweet. I've been waiting my whole life for someone to tell me that.

I enslaved all three of those guys.

Yep. All three are my slaves now. Yeppers. All three. Total slavery. Yes indeed. It's pretty sweet. I just tell'm to do stuff. ... One time a cop even showed up at my house and he was like saying something about abdication or something, and I was like, "well, technically there's nothing wrong with what I'm doing here, your honor."

"... You see, none of the world's major religions ever..."

I couldn't remember the details, so I grabbed one of my slaves and I was like, "hey, tell this cop what you told me that one time about slavery."

Okay, so, obviously, cannibalism is a super weird activity for a social animal to be doing - naturally - but so is sACRIFICING your own children! ... We're a weird freakin' genus, man - even in the best of times. So, never mind what happens when you stick us in an unnatural, alien environment and then take away our brain food. Of course crazy sh** is gonna happen.

If you goen' *stick* us in some crazy savannah shit, then there gon' BE some crazy savannah shit. ... Yuh kn'a'm sayin'?!

So. That's the point I forgot to mention earlier. Unnatural environments... beget unnatural nutrition... begets unnatural behavior... begets ongoing cycles of calamity.

Isn't the savannah wonderful?

I know *I* sure love it. Okay okay, let's be serious. One of the main reasons I wanted to make this final point here is simply to address the cultural stigma that's been surrounding a lot of people's ancestors. There are plenty of documented examples of, let's call it, gratuitous superfluous homophagy.

Well, before we get too carried away with passing judgement on these ancient people, let us not fail to consider that these people often had very limited access to critical brain nutrition. ... That's all.

We hominins are more or less tethered to the coastal regions - at least as far as our nutrition is concerned. In a primitive setting, we can't be out to sea for very long it seems, or we'll end up getting things like scurvy from a vitamin C deficiency. But we *also* can't be too far out into the *mainland*, or we'll end up getting things like stupidity from an iodine deficiency.

Our human ancestors were slowly figuring this stuff out through a great deal of trial and error. They were finding different ways to cope with inland deficiencies. They could travel to the oceans somewhat regularly; or they could find some anadromous ocean fish that were migrating up through the rivers; or they could even just purchase some fish jerky from a long distance trader.

But it seems like every time we figured out these principles, the following generations would just eventually forget them again. Our bodies don't exactly come with an operator's manual. And our appearances don't really betray where we come from originally.

So, we've been doing this little kamikaze diaspora thing since forever. We keep on migrating away from the oceans, then we get dumb and deficient and irrational. Then we run around eating up everything, until we've found some useful substance that seems to help us. You can see this effect quite dramatically, starting roughly around ten thousand years ago.

Somewhere around ten thousand years ago, the ice age was slowly receding. The upper Paleolithic was ending. And the weather was getting warmer and wetter. Humanity had finally travelled just about everywhere. There were people from the Nile to the New World. They were living along the coasts and by the deltas. And some people had even traveled up the rivers.

But a few had sought to live within the salt marshes. This is where the deltas form a wetland. They lived in little cottage-style houses. And they fished on woven boats amongst the reed growth.

But one of these little fishing villages was about to change everything. They had figured out through trial and error that they could utilize certain grasses from their marshlands. If they collected them and then processed their seed kernels, it made them edible, and they scratch out a few more calories from their environment.

But then they realized that they could just plant them and this would make a lot more of 'm. And not only that, but they could just select them and this would alter their characteristics. ... And verily, *just* when you thought it couldn't get any better... they also realized that they could disperse them across a *much* wider area. All they needed was a sprawling wetland across the landscape. ... So, they just *made* one by digging out trenches and channeling off the river water. ... ...

These coastal foragers had made a wetland and a food supply. These humble fishers had made an ocean of starchy calories.

Aaaaaaaaaaaaaand... BOOM! ... Bob's your uncle and Sally's your misses. Their population just started exploding and boiling over. Their little cottages were at once supplanted by a massive city. And now their lives were a lot more complicated, and thus a lot to keep track of. So, they invented these little symbols that they drew with the river reeds.

We're still not really sure just where they came from though. The locals around this region had no idea either. They believed that these people got their skills from mermaids. This is how they thought they invented civilization.

You really gotta love that disparity though. These fisher-foragers were so incredibly industrious that the only way the hunter-gatherers in the surrounding region could make sense of it was to *literally* invoke the suspicion of magical creatures. ... Yeah, they're magical creatures alright. It's called crab legs and shrimp platters, my pedigree hunter-gatherer chums.

Or, if you're a British anthropologist - "moy pedigree huntuh-gavuhvuhvuhvuhvuh (mumbling off and trailing away because they can't pronounce it) chums."

Huntuh-gavuhvuh.',.'?vuh¿.'.v'v'v...

Where did all those Vs come from, Brittany?

¿

Alright, alright. Be that as it may, you might already recognize these dirty rotten mermaid cheats by another name. Yes indeed, these were the Sumerians... of ancient Mesopotamia. That's right. Long before wheat crops and ziggurats, these people were fishers and foragers throughout the wetlands. *ALL of the first civilizations were!!* The artisanal empires of neolithic China; the incredible hydraulic engineers of the Indus valley; the craftsmen and the cuneiform writers of ancient Sumer; and the ever-inspiring builders of Old Kingdom Egypt. ... But also the Minoans who would jump over bull horns; and the virtuosos of sustainable wildcraft along the North American upper Pacific coastal region; and the Olmecs who were masters of calendars; and at once, the Norte Chico who built a civilization at the end of the world.

You could go all day like this - Mekong, Mycenaeans, Mississippi, Mayans, Mochica - and maybe someday I will, but the point of all that is just this - all of these civilizations were originally marsh people - looking over the shoulders of mermaids to get the answers to civilization. These creators of cultivation were all people from the coastlines.

It's important to know where things come from. And it's important to know who invented them. Because it's a lot more cerebrally demanding to conceive of something that no one has ever thought of before. Inventors have always made this look easy, but it's not. They discover a brand new concept which has just never been contemplated. And then anyone can copy their model, and they can spread it throughout the world along trade networks.

Civilization was a completely new phenomenon on terra firma. Suddenly the stork-legged whale chimps were becoming innumerable. ... This could be trouble.

There was starting to become too many mouths to feed with respect to the amount of seafood in the area. Our numbers kept on growing precipitously, and yet the exploitation of fresh marine food was much less prodigious. So, the proportion of essential nutrition was beginning to plummet.

The more distorted this became within an empire, the more increasingly unstable their psychology became. Empires, of course, are a fairly complicated topic, but the role of this basic nutrition can only hardly be overstated.

Some people even think that these civilizations were getting sketchier mainly because they didn't have enough land meat to feed their peasant populations. But this isn't really the case. For one, this land meat is hardly anything more than just basic, bare protein. In the same way that cereal grains and starches are essentially just empty *calories*, land meat, in and of itself, is very nearly just an empty *protein*.

... It's just like the animal version of a wheat plant.

Secondly, very *rarely* were the people of these empires ever missing out on land meat to the extent they were missing out on seafood. ... especially for their rulers. Their rulers weren't ever missing out on anything from the lands around them.

Thirdly, there is also a long enduring suspicion about these cereal grains that these foods are essentially toxic because we were never adapted to eat them. However, these grains are not in any way more exotic to us than is land meat - our ancestors have been exploiting *both* of these since the upper Paleolithic. Eating land animals is just as alien to our species as are cereal grains. ... Neither one of these foods is any way more *paleo* than the other.

It's just that grain consumption within the Holocene has become relied upon to the point of excess. These massive quantities of simple carbohydrates will cause a very large array of serious pathologies. ... So... that is where the "toxic" effect is coming from - it's not so much the grains and any would-be poisonous proteins as much as it is the carbs in all of their excess.

And, frankly, I don't trust either one of'm – not the grains OR the land meats. Both of these "foods" are only *upper* Paleolithic, towards the latter extremity. If you wanna go true blue, evolutionary, ancestral, Paleolithic diet... then you gotta do it right and get the beach food. You gotta go full-on, blue zone, Mediterranean style. You gotta get the foods that we can acquire with just our ape hands – meaning, the foods that we can get without a projectile weapon. In other words, foods that we can forage in a natural, hospitable environment.

It's a good question though. Could there have been some sort of toxic effect on these burgeoning civilizations from just constantly consuming lots of early domesticated versions of cereal grains? Could there have been some sort of rogue protein, amino acid chains, just silently wreaking havoc throughout their bodies? ... Absolutely. It's entirely possible. In fact, I'd be *a lot* more be surprised if this somehow *didn't* happen. But it's also just as probable that any of these *livestock* were infused with these proteins - some sort of novel assembly of encoding which we just simply had not adapted to.

Animal cells produce all sorts of toxins. And as primates, we have been almost completely naive to these land meats until really quite recently in evolutionary terms. Our most recent primate ancestors have had *TENS* of millions of years to adapt to certain plant foods. And then our most immediate hominin ancestors have had upwards of *SEVEN* million years to adapt to these seafoods. And yet, our fully modern human ancestors have only had about three percent of that time to begin the process of adapting to these coarser land meats. There is just *noooo* comparison in which types of foods we were built to thrive on.

So, sure, were there ever some times when these empires weren't getting enough protein? ... Absolutely. ... And was it ever actually possible that their agriculture was clandestinely harming them in some way? ... Yep... more than likely. ... And does this explain why these civilizations were starting to crumble? ... Ehhhhhh, yeeeaaaah, a little bit. But things like this are not so decisively catastrophic. Chronic inflammation and bouts of protein deprivation are not even *remotely* the most pressing issue when you've got a brainfood shortage.

Another major clue which is well aligned with this overall pattern, is that there were cities, not too far from the oceans, that *well outlasted* their empires. For example, the blue lagoon metropolis of Lamanai, along with the aquaculture wizards of Byzantium, were each outlasting both the Mayans and the Romans by literally centuries respectively. The average Roman may have been getting a little fish garum in their diet, and the average Mayan may have been welcome to the occasional lake fish from time to time, but this was *nothing* when compared to the cities of Byzantium and Lamanai. And we can see from their bone structures and their stamina, that these people were a heck of a lot more healthy than their inland counterparts.

Oh, and just to be clear, when say "empire," I am only talking about people who actually built stuff like monumental architecture. I am not talking about barbarians and pirates who only attacked people and stole stuff. Sure, most of the empires of history were largely built on attacking people and stealing stuff, but at least some them were occasionally constructing things and not just breaking them.

Shoutout to the early years Carthaginians and the latter years Nabataeans for using mostly peaceful trade to build their empires.

Respect.

The other glaring correlation, in my experience, is the one in which empires who remained by the oceans were about a hundred times less sketchy than to the ones further inland.

The most crazy and ill-conceived populations were almost invariably the ones furthest from the oceans - specifically the herders and rangers. Historically speaking, these were the quintessential "barbarians." These guys barely got enough anadromous river fish to stave off goiters.

Then, way way waaaay over on the complete and polar opposite end of the spectrum, you had the empires who never even left the coastal seaside to begin with. In truth, you could say that their empires *were* the sea. They were real-life Atlanteans, in a manner of speaking. And I do dare say, quite indubitably, that these real-life Atlantean people were the absolute most level-headed, stable, calculated, fun-loving, and flourishing empires in the world.

For one, you had the sexy Minoans. They lived out in the Mediterranean on a sunny little island known as Crete.

These guys liked to party like no one's business. Everything about their culture was just brimming with life. Their art was full of swirling, fractal imagery, and they had festivals for things like swimming and playing chicken with an aurochs bull.

Their island population wasn't exactly massive, but their influence was widely felt throughout the ancient world. The Minoans were a popular culture - not because they brought the sword of war to people, but rather because they brought the world a shopping mall.

Yep. That's right. They were bout it, bout it. They brought all sorts of jewelry and clothing and cosmetics and incense and exotic foods and rare commodities and you name it. They delivered merchandise all over the Mediterranean. And then it flowed out through the trade networks of the ancient world.

So, you might've messed up somewhere along the line and done gotten yourself born into a little backwoods country settlement in the Bronze Age. And maybe all you had to play with in your little township was some copper. But when the Minoans would come a-sailin' up your harbor, it was exciting, because that surplus could then be traded. You could exchange your copper ore for an entire wonderland of different commodities.

And what's more is that their very own, ebony-locked women were also featured very prominently within their society. They were venerated priestesses and equals. And their images were adorned across everything. They appeared on vases and villas and statues. And they were depicted like Parisians on holiday... sauntering up and down the boulevards in their extravagant wardrobes.

Oooooookay... I think I see what's going on here. It's starting to make sense now why they built an *entire civilization* based on *shopping malls*.

Touché, Minoans... ... Touché.

And it might be easy for us to take this stuff for granted now, but they were living in a time of constant warfare. Civilizations were building walls as high as castles. And still, the scale of the Bronze Age atrocities was almost completely unimaginable. But the Minoans of this period seemed largely unaffected. They weren't really building massive defenses. And they weren't getting into fights and making enemies. They just kept trading and partying hardy for several millennia.

Sadly, however, their sunny little island was located on just about the worst possible location for seismic activity. After thousands of years of prosperity, they just got hammered and tossed about mercilessly. It was just one thing after another - volcanoes and earthquakes and tidal waves and ash clouds and famine and starvation and on and on and on and on and on.

Initially they were building back their cities, but as the troubles kept on coming, it eventually just broke them. So, their population had no other option, and they were slowly just absorbed by the early Greek predecessors.

But they never fell apart because of their own decisions. They played it cool and kept it real and only really wanted to battle people with dance-offs. And in a world of boring clothing and totes wack accessories, the Minoans were delivering hope to young ladies who couldn't even.

"Ug... As lugal of Assyria, I decide to attack Minoans."

"No, daddy! Don't do it! They bring me my bracelets!"

"Not to worry, my dove. I will bring bracelets."

"No, daddy. I've seen how you do stuff. *You won't!!*"

"Ughh. Fine. Who else to attack then?"

Alright. So, that's enough examples of these seaside civilizations. The next thing I wanna talk about is... ... ...oh shit... ... ...ohhhhhh shit... ... ...DAMNIT... ... This isn't gonna work. ... This isn't gonna work. The anthros are gonna talk *sh*** about it. They're gonna say it's all a pure coincidence because the Minoans were just an island nation.

DAMNIT!

Okay. Fine. Well. Then. First of all, being small doesn't mean anything. Professional traders have always commanded lucrative amounts of wealth, regardless of their size. And the gold-hungry, warmongering empires have always looked upon them and licked their lips in a weirdly excessive manner. Being smaller means nothing. Pirates of all sizes have been preying on merchants since the dawn of time. So, it is no less impressive that they were able to succeed the way they did.

Second of all, being on an island means precisely dick. Water doesn't stop a determined land empire. They just buy the boats and hire the

255 - BARTROLOMEW MC INNCEL

captains and borrow the technology. The only time that water in between you actually *does* protect you is if you have an active navy that is absolutely enormous. That's the only time a sovereign can repel a vast armada. And the Minoans were never equipped with this ability – not numerically anyway.

So, no, sorry. The Minoans were absolutely a big fat target. The only reason that they flourished and remained so successful was because they always played it *smart* and didn't get all sketchy and greedy and stuff. If the Minoans had really intended to become a military power, then they absolutely could have, with the capital and the human resources that they commanded. Their smaller size meant nothing in this equation. And, plus, I seem to recall a few other places that were sort of smaller like this, and I think they had some monikers that sort of sounded like "Portugal" and "Japan" and "Britain." And I seem to recall something happening at some point... It was like something about them taking over *half the planet.*

Was that it? ... Do you guys remember? ...

Oh. Wait. Yeah. That *was* it. Being smaller doesn't equal being weak; and being an island doesn't presuppose being overlooked by people. None of these counterarguments mean anything.

But now I'm pissed. Now that you future anthropologists have started viciously attacking me in my imagination... ... oh, now *I'm freakin' pyyi-iiiiiiiiiiiiiiiiiiiiiiiiiiiiiiissed.*

Alright. Forget about that freakin' island stuff... and that small population stuff. Let's talk about the Phoenician empire then... ... *shall we?!*

Alright, so, first of all!!! *Fffffirst of all!!!*

Alright, I can't sustain that level of aggression... alright... let's just talk about this normally. ... Okay. So. First of all, the Phoenician empire was all around the Mediterranean - not just on an island. Their capital was on the far east coastline - only like a two week's march from the Assyrian maniacs - definitely not a super-duper place to be located in those days.

And yet they persisted.

The Phoenicians were the absolute masters of early Iron Age sea trade. They were famous throughout the center the Old World, and they were admired for their products and their inventiveness. Their networks were servicing markets from the Levant to the Atlantic. And so, they connected these far distant regions and thereby flourished for many centuries.

Sadly, however, this wonderful empire of shopping malls and food courts was in no small measure a source of jealous envy. The Romans were never entirely comfortable with how much influence they wielded. And this was despite them not even being a warring culture. War was bad for business. They wanted to avoid that. They just wanted to make their products and accommodate their trade networks... even with the Romans.

And if some little fiefdom ever attempted to make a bunch of trouble for them, then they would simply pay some professional soldiers to go and facilitate a settlement agreement. Their citizens never even saw combat. ... But unfortunately, however, none of this was ever quite good enough for the Romans. As far as the Romans were concerned, the Mediterranean was just not going to be big enough for the both of them.

However, the key takeaway for our purposes is this. The Phoenicians never lost their empire because of any harebrained, erratic decisions on their own part. Quite the contrary. Before the Romans came along, the Phoenicians were the absolute picture of professionalism in the ancient world. And not only that, but they were SOOO incredibly good at their seafaring and networking, that even after the Romans had utterly decimated them, TWICE... they STILL bounced back in almost no time - just like the Minoans did after *their* disasters. In less than a single generation, their citizens were already living like kings again. They were living even better than the Romans were, and it was the Romans who were the ones who had supposedly conquered them.

Now, THAT is incredible.

That... is how a coastal civilization rolls.

Oh, and if you're reading these words right now in a language that uses an alphabet, and if you never had to memorize thousands of complicated picture-symbols, then you can thank our old, industrious friends the Phoenicians for that. That's why we call it the "phonetic" writing system.

High five, Phoenicians! ... Hwutish!

See? How would I even write that high five sound effect if all I had was a bunch of picture-symbols? Haha. See? How would I even... ... ... wait... ... ... wait... maybe there's a picture symbol for that high five sound already. And maybe it's even easier than what I did... ... ... DAMNIT!!

What if the Phoenicians just screwed us out of picture-words?! ... DAMNIT!!!

Now I'm pissed again! All my anger is back from earlier!

RrrrrrraaaaaaAAAAAAWkhhhh sorry I'm coughingkhhhhhh... khhh-hhh...

Okay KHHHHH... ... ...JEEEEEZ... enough!!

Alright. ... Alright. ... No more anger. Okay. SO... Before we go any further, there's just one more note which needs to be made here about these sea peoples. Just because a bunch of people start showing up on boats all the sudden it doesn't necessarily mean they represent a coastal civilization.

There have always been the pirates and the scallywags. They just sail around the seas always trying to steal stuff. And yet, these pirates were often just people who were desperate. They were people out searching for a homeland. Their countries were usually overpopulated. And then this in turn afforded them no opportunities. So, they were much more like a crew of fighting hobos, in actuality. They just melted down their grocery carts into stabbing weapons.

And that's why you can't tell the difference between a pirate and a homeless.

Take the vikings for example. They had a major population boom at one point. So, they were desperate to find some land to safely occupy. They had been farming right up to the edges of glacial ice sheets. That's how bad it finally got with overcrowding.

And that's how you know when someone's super hard up for some farm land - when you've got records of one guy getting all mad at his neighbor because he pushed over a huge wall of ice onto his land.

"HEY! ... This isn't *my* ice! ... ...This is *YOUR* ice, bitch!"

But the point of all that is just this, these roving bands of raiders were never representative of any maritime civilizations. They were like small little nomadic bands of refugees... - ones who were mainly just trying to get some resources in a sort of panicked desperation - vikings, pirates, Haida, Sea People, etc. Unfortunate? Yes. Indicative of maritime civilizations? Nnnnnyo...

Or, to put an even finer point on it, these were just little overflow groups - not entire civilizations, per se; not whole entire sketched out, zombified, land empires; and not whole entire regions of inland range-herder barbarians.

These range-herder barbarians were never attacking people because they needed to. They were attacking other people because they *wanted* to. They weren't just trapped on some barren little island; they weren't out there farming right up to the edge of some frozen polar ice sheet. They were all roaming across an endless expanse of prairie; and they had leagues and leagues and leagues of sprawling wilderness.

And yet, despite all that, they were *sTILL* extremely hostile and super sketchy. These guys are literally why the concept of *walls* was invented. Just about every famous wall you've ever heard of was specifically designed and constructed to keep these tweakers out. ... You know you're freakin' sketchy when an *already* super sketchy land-based empire is trying to keep *you* the hell out.

Yeah, these range-herders were getting protein alright, but that's about all they were getting. This is where the whole archetype of being a "meat head" comes from. ... or having "a strong back and a weak mind." And this is of course where the whole idea of being an "uncivilized barbarian" comes from. That all originally came from these nomadic landlocked people.

Even the classic stereotype of the oaf-like caveman comes from this traditional observation. This would've been the hunter-gatherers in the forests - sometimes quite literally living in cave openings.

Most of these people's basic protein was just coming from the land sources, but they would still go down and fish in the rivers. This was common with the land groups, like the steppe herder Yamnaya people. And then this would give them access to the anadromous migratory fish. And then this in turn would help them to stave off the very worst forms of iodine deficiencies.

To be clear though, they were basically just giving themselves a standard maintenance dose - far from the intended amounts that we are most ideally supposed to be getting. Therefore, this was not enough to keep them fully nourished. This was not enough to fulfill their true potential.

And yet, this dose was just enough to keep these land folks going, amazingly. The amount was just enough to keep them functioning. There's no rule that says you must be super clever. You simply need enough to keep surviving. And so, that's exactly what they did throughout these inland regions. They just followed the basic patterns laid down by their forebears, and they just didn't need to do a bunch of thinking.

It's like the old saying goes - coastal people create good technology; good technology creates land people; land people start getting all sketchy and attacking each other; hard times create coastal people.

You know that good ol' sayin'.

Anyway. The landies seemed to do a lot better, in general, when they were living in exceptional places with lots of river food. In Çatalhöyük, for example, about ten thousand years ago, there was a sort of inland marsh area that was crisscrossed with this network of little tributary channels. This whole entire wetland was replete with river clams... and they may have largely lacked the indispensable ocean minerals, but they still contained a ton of omega 3 fats.

Thus, the community of Çatalhöyük ended up lasting for well over a millennium. And they even did so while the majority of their contemporaries were busy flopping and flailing. That's because these marshes, like the one at Çatalhöyük, weren't exactly super common. These features are not afforded to many land areas. So, the average inland settlement didn't have this advantage. They only had their staples like crops and livestock.

That's a tough life though - for anyone - Homo erectus, Neolithic hunter-gatherers, Paleolithic nomads, anybody. You're never more than several years away from a major nutritional deficiency. It's not like in our times when they can just fortify our foods with certain nutrients.

There has always been this one consistent parallel, between the several major phases of our world. In each of these three epic epochs, we have forgotten about our basic dietary requirements.

In the hatchet-wielding age of Homo erectus, there were a few groups moving inland and hunting megafauna. These creatures were indeed a source of energy, and erectus had never seen so many calories.

When we hominins find a major source of calories, we often stop pursuing other small ones. And thus, we focus on the biggest baddest windfall, until we end developing deficiencies like pellagra from corn overreliance.

And that's when we start freaking out like zombies. That's when we start eating lots of random stuff. And this is how it felt to be a land erectus for a while. It was a desperate time of madness until they figured out the concept of anadromous river fish again.

These fish are not abundant in every season of course. So, they'd still be mostly lacking critical brainfood. But at least it'd be enough to keep them going. ... As long as they weren't dying, they wouldn't really be thinking about it.

Then several million years would pass from their time, whilst their cousins on the beach were getting smarter. And then the children of these beach people would master farming. And then soon they'd be making the same mistakes all over again.

The calories from these grains were quite abundant. So, once again, our people would start to focus on a land food. And then our numbers would get so large our fishers could not supply them. So, yet again, our wayward ancestors would be starved of critical brainfood. This is when our thinking starts to falter. This is when we make our worst decisions. This is what erodes our civilizational fault lines. And then THIS is when our empires start to crumble.

Moving forward. ... *Now* we find ourselves within the third age, where IQs have all been dropping for quite a while it turns out. Physical and mental impairments are unambiguously ubiquitous, alas. So, hereupon, something is portentously rotten within our current state of Denmark. ... They've been adding stuff to food to help these issues. But it's just a small amount to stave off goiters. We still need lots of brain food, especially as children. And you might feel pretty smart now, but you likely should be smarter.

It isn't just about our hominin brain power though. It's about energy; it's about exuberance; it's about our happiness. Rarely do I see these concurrently in any one person. Rarely do I see these when someone isn't on their fourth cup of coffee yet.

Rarely.

Well...

...there is...

...oooooone group.

But I don't know if...

I don't know if I should tell you.

I don't know if you're ready for that level of information.

I mean, you barely even paid one dollar over the printing costs for this book, so I don't even know if you really deserve, eh fuck it, I'll tell you.

But first... I want you to imagine something... I want you to imagine what it must have felt like to be a stone-age culture some *sixty thousand years* ago. I want you to picture in your mind what it must have, eh fuck it, let's just go look at it.

We actually *have* a stone-age tribe as a matter of fact. Yep, that's right. An *actual, genuine, no kiddin', real life, living, breathing, stone-age* culture... still preserved... after more than...
sssssssssssssssssSIXTY...
thhhhhhhhhhhhhhOUSAND...
yyyyyyyyyyyyyyyyyyyyyyyyEARS!!!

WhaaaaaaaaaaAAAAAAAAAAAAT?!!!!!!

SIXTY THOUSAND YEARS?!!!?!!!

Are you freakin' KIDDING ME?!?!

Nope. That ain't no clickbait, playuh. We actually have a REAL life, stone-age, cave tribe!

WhaaaaaaaaaaAAAAAAA... okay. Well, they don't actually live in "caves," per se, but they *are* completely stone-age. This tribe has been all but completely isolated since the upper Paleolithic. They were ancient and isolated before even pottery was invented.

Honestly, it's a freakin' miracle that they even exist out there.

Oh yeah, where is "out there?" Well, these guys are way out east of Africa on a tiny little island just south of India. They've been out there for so long that no one even knows their spoken language. It's not even remotely similar to anything else. Not even any other dialect around North Sentinel there.

What's really amazing too, quite frankly, is that we can actually observe these people without disrupting their lifestyle. It's like a portal
was somehow opened between our timelines, and we can look and see
the past and behold these cavemen.

I don't know, man... If that's not one of the most absolutely *mind-blow-
ing* things you've ever heard of...

... then ... I don't even know.

I don't even KNOW!!Well, actually, when I say "observe them," there's
sort of a caveat to that. Technically no one's supposed to go there. And
also, practically. Both technically and practically no one's allowed to
go there. But we can observe them from a distance and still take notes
from our previous encounters - previous encounters *before* they went
all This-is-SPARTA and kicked all of our anthropologists out.

These guys are *still* acutely aware of a previous incident, over a century
ago, when some explorer just decided he was gonna go kidnap a few of
'm for science.

It's okay though. He was only kidnapping them for science. It's not like
he was gonna keep 'm. It was just a brief, nonconsensual abduction for
a couple of little months of academic inquiry and involuntary detain-
ment. That's it! And then they got *all* sensitive about it.

And despite the fact that you or I may have had the sheer *unmitigated*
audacity to just completely forget about this enormously pivotal event
in world history... *they have not.* They will never forget. They will never
lose the record of this memory... of that weird, pasty alien who just
randomly showed up there one day... - the one who was speaking that

sort of whimsical, uncivilized jibber-jabber.

He clearly had not come from a very intelligent tribe. His language sounded like gobbledygook. And no matter how slowly they spoke for him, he still could not understand even the simplest of concepts.

So, needless to say, they were not *at all* impressed with our anthropologists when we showed up there again, over a century later, still communicating in some sort of primitive barbarian yak-yak. Apparently we had learned nothing in the last century. They still could not understand us. Obviously we were not that sophisticated. So, once again, they invited us to *kindly leave*.

Well, technically they shot one of our anthropologists in the leg. And also practically. Both technically and practically they shot 'im in the leg with a deadly ass arrow.

Soooooooooo, we don't go there anymore.

And their arrows are no joke either. They've got those really long ones that you use for classical bow fishing. That's not exactly the kinda thing you wanna have flying in and around your body area.

So, these days, we just leave'm alone out there... so they can keep on doin' what they're doin' and preserving their legacy.

It's not a shabby spot to have your Gilligan's island at either. The whole entire thing is surrounded by beaches. And the water is crystal clear with shades of indigo and turquoise. The forest is absurdly tall like some sort of giant green stadium. And it's about as close as you can get to a garden of Eden.

Their diet is mostly seafood and fruits and tubers. And sometimes they can snatch some eggs from seabirds. But they're mainly eating large amounts of reef food. ... So, it should come as no surprise that these people are on fire. They're super fit, and extremely alert, and hyper focused. You can see it all over their person when you get a good look at'm. They don't have that sort of downtrodden, dead eyed look that so many of the world's unfortunate land tribes have.

They're *quite* the opposite of that. These guys are practically doing cartwheels half the time. And they also have a wicked sense of humor, by the way. They're crackin' jokes, and havin' fun, and clearly trolling sometimes. Any simple-minded Neanderthal can just sit around and be *literal* all the time, but it takes a highly evolved and vibrant intellect to have fun with *hypotheticals* the way they're doing.

And, sure, yeah, I admit, maybe they're not always the most perfect little angels in the world all the time. ... But so what. ... They're humans!

Yeah, yeah, yeah, I know. They really like having giant sex parties on the beach sometimes. ... Who cares?! ... What, are you the judge of them?! Or sometheen?!?!

Yeah, yeah, yeah, I already know ALL about it. Maybe they might've shot ONE little missionary in the chest once... who wasn't supposed to be there. ... ...They're SORRY... ...OKAY?!

Probably.

I'de like to see how YOU behave the next time you get isolated on an island for the last sixty thousand years. Huh?! Let's see how good YOU do! Let's see how well YOU'RE able to control yourself the next time you wake up in the morning, feeling like you just drank an entire pot of coffee... even though you *didn't* just drink an entire pot of coffee! ... HUH?! Let's see how perfect YOU do!!

Alright. Yeah. But anyway. They're some pretty peppy folks alright. And that's usually how it happens. That's usually how it happens when you start living in the way that nature designed you to. You tend to start feeling pretty awesome. And enthusiastic. ... And trigger happy.

But it hasn't always been just rainbows and butterflies for them unfortunately. Not exactly.

Not that long ago, there was a major tsunami that swept across that region, wreaking havoc and destroying infrastructure all across the mainland.

Thankfully though, there were some very good-hearted people on the mainland and some really brave and thoughtful, real-life heroes who were good enough to send a relief helicopter out there. ... But sadly, however, the tidal wave which struck that entire coastal region was so incredibly devastating that no one really expected to find much.

And so it was. As soon as the helicopter flew over their village, it quickly became apparent that our worst fears had been realized. No, not that the islanders had been devastated - it didn't even phase 'm - but that the islanders could clearly see we were trying to attack them with a devil bird.

They quickly sprung into action. Within seconds, the devil helicopter was being penetrated and punctured by arrows. We were *immediately* forced to retreat in a hail of bow fire.

It was a stunning defeat for our forces.

Neither our wave attack, nor our demon bird, could do a *DAMN* thing to stop them.

They had won.

13

9 ¾ 4 -- HOMO ERECTUS MANIFEST'O,

... TURN THIS APE INTO A FELLOW

I used to have a friend whose last name was Shlonginhanz, so we used to say stuff to him like, "be careful not to finish your homework early, Josh, or you're just gonna be sittin' around with your shlong-in-hanz."

We thought that was pretty funny until he threatened our lives. So, we stopped saying it.

I was like, "jeez, Josh. Take it easy, man. You need to get laid, bro. By an actual woman, I mean. Not by your own hanz."

Man, this chapter has gone on for way too long. This chapter was only supposed to be one chapter long. Maybe I could finish this friggin' thing if the friggin' universe would quit putting ideas in my friggin' head.

Oh, and don't worry about Josh. We took good care of ol' Joshua. We wouldn't roast'm if we didn't love'm.

Besides, we had another friend whose last name was sounded like "Blow-machine." So, that worked out just fine.

Ohhhh myyyyy freaking god. I literally just looked up Josh's last name just now. You're not freakin' gonna believe this. His friggin' last name Shlong-in-hanz literally translates into "snake dance."

Ohhhhhhhhh myyyyyyyyyy fff********* gaaaaaud.

How...

could we...

POSSIBLY...

not KNOW that?!

The gods hath shown mercy unto Joshua.

Freakin' snake dance. I can't freakin' believe this. We would've come up with so much stuff! ... like, our own dance moves and all kinds of stuff.

Oh man! ... He must've lived in absolute fear that we would discover that. Well... looks like I DID discover it, Joshua. Time to get the ol' hazing crew back together. This is gonna be *lit*.

Oh, wait. He threatened all our lives. Never mind.

It's alright. Don't be mad at Josh. He can't help himself. He's a zee German. They're always a little high strung when they're not off on their annual month-long paid vacations to spread their sausage bodies across half the beach real estate of southern Europe. I get it. I would be cranky too. What kind of self-respecting German man would wanna go home again after an entire month of lying around in his favorite undersized micro-bikini?

Well, notwithstanding any aforementioned inappropriately small menswear, there is still one more little bitty reason for why you might not wanna go selling off your beach bungalow just yet. Just hear me out on this first, before you go running off to start your new life in a little mud hut on the prairie.

The last and final reason for why you might not wanna go moving out there just yet is because doing so would lead to an unraveling of your selection mechanism for intelligence.

As we discussed earlier, there is always a selection mechanism in nature which is constantly enforcing the genes for different types of intelligence. Getting the right types of nutrition for your species is undoubtedly of the utmost importance, but simply getting that nutrition alone is not good enough. Nutrition only loads the gun of progress – you still need some selection mechanism to pull the trigger.

This selection process starts in earnest whenever there is a brand new mother or father who has somehow gotten a novel mutation for an incremental increase in their intelligence. This feature will then be propagated whenever this parent can translate this advantage into making more than the average number of healthy children. Thus, with each new generation from that original advantageous mutation, many of the successive descendants will go on to repeat that pattern. They will carry that advantageous mutation, and they will go on to have more than the average number of healthy children.

This process will continue to propagate until their mutations have thoroughly spread throughout the population. Once this has come to fruition, the entire tribe will be carriers of that brand new mutation.

This fairly straightforward process is how the genes for intelligence become ubiquitous. The caveat, however, is that each of these mutations must indeed be legitimately useful. It doesn't really matter how magnificently beautiful or how wonderfully serene an original mutation may be, if it cannot get real-world results, well, then it probably isn't going to last. That's because if a new genetic feature doesn't actually help you to put a little extra food on the table, then it's just simply not gonna help you feed a bigger family.

The reason that you need a decent family size is because not all of your children will be inheritors of your new mutation. So, if each new generation just continues this pattern of not really having all that many children, then the mutation itself can gradually become less and less common until one day it simply vanishes altogether. The actual mutation, in and of itself, can ultimately disappear even though your progeny have not. The bottom line is this, passing on any new mutation within a primitive tribal context will always boil down to a simple numbers game.

This effect of spreading genes for various forms of productive intelligence is also quite dramatically amplified in males, because of course, men are just naturally more selected by females for their ability to be productive. This is why men, as a general rule, have tended to be more interested in working with *things*, while women in general have tended to be more interested in working with *people*.

Provider males have been going out into the world and extracting resources directly from nature for millions upon millions of years. While women, on the other hand, have been developing a very different set of primary talents. One of these inherent enhancements in our post-erectus women is a very powerful and endearing, pair-bonding instinct for protecting their provider males.

The jungle wasn't like this. It was mostly every man for himself. Only the biggest and baddest alpha males would even get a chance to breed. Individuals only really worried about themselves most of the time. And what's more is this, if one of those big, bad, alpha males ever got sick, or even injured to the point of infirmity, then the group would just move on and allow him to perish. The whole appeal of the big bad alpha was their strength. So, if an alpha ever showed a sign of weakness, then the appeal would just be lost, and the ladies would lose interest. Alphas get replaced at the drop of a hat.

This all changed however with the rise of the provider male. Suddenly your man actually meant something to you. He was now your own personal, hard-working provider and your own dedicated family defender. He was a committed and focused man who lived entirely for you. He wasn't just some expendable alpha. He was, in fact, your most valuable asset in the world, second only to your own biological children. You don't just leave a guy like that lying around to die. That's the kinda man you wanna get nursed back to health as soon as possible.

Way back in the early days, however, all the way back in the beginnings of this whole provider male revolution, there would, alas, be one more lingering hangover from the old jungle way of doing things. Even in this newly developing world, amidst its flurry of metamorphic and transformational energy, there would have still been a resounding lack of empathy for these provider males. If any of these original, cohabitating provider males would have found himself succumbing to injury, no female would have even thought to help him yet. They would have simply just expected another male to come along. And they would have only expended their energy on themselves and their children.

But then somewhere along the line something unexpected happened. At some point in our remote and distant prehistory, there was an entirely unprecedented revolution which was about to change everything... ... again. ... And this change was about to be ignited by a very special woman... ... again. ... Another Eve was on the verge of emerging.

This woman was being cared for by her provider male. And he could always be relied upon to bring her food 'n' stuff. But then one day he just didn't show up for some reason. He just never came around and brought 'er her clams and such. And this wasn't like him at all. It wasn't like him to just not show up like this. Something wasn't right. And indeed it wasn't. Earlier that day he had gotten injured.

And so, as was the way of things, he likely would have just wandered off to go die somewhere. Either that or he would *have* to save himself somehow.

But then, for some reason, something was just different about this day. Someway, somehow, and for some magically serendipitous reason that we may not ever fully appreciate, this man's woman just straight up refused to simply leave him like that. Maybe she missed him. Maybe she wanted to go and check up on him. Or maybe she wanted to go and make sure he wasn't just out there running around... and trying to become a provider for some other woman. Or maybe she just felt bad for him because she really liked him. Who knows? The truth was probably a little bit of everything.

Or maybe it was something altogether different. Maybe his illness or injuries were apparently moderate. And maybe her motherly instincts were just becoming a little more inclusive. Either way, the fact remains the same, this Eve of early altruistic reciprocity still took it upon herself to assist her partner. She accepted a *considerable* risk assisting her partner like that, instead of just expending her energies on herself.

Maybe she was successful. Maybe she wasn't. The exact details of her story are lost to antiquity. But one thing about this epic is for certain, this woman was about to develop an entirely new archetype. This was the beginning of the equal and opposite counterpart to the committed provider male. This was the emergence of the dedicated supporter female.

So, in the same basic way that the men were being selected to become ever more intelligent and resourceful, the women were also, themselves, now being selected to become ever more supportive and helpful. They were being selected by the overall degree to which their families were prospering, and indeed, the couples who were working together were exponentially more prosperous.

Teamwork, in and of itself, is already extremely powerful. It gives you a massive synergistic advantage. So, in the same basic way that a diet of land food is inherently inadequate, no other form of domestic organization can even hope to hold a candle to reciprocal pair-bonding.

Thus, generally speaking, masculine and feminine have literally evolved to be the complements of one another. They are designed to fit together like a jigsaw puzzle. Their abilities and their unique psychological perspectives have been meticulously tailored to be interdependent within the context of child rearing. Where one side tends to falter, the other tends to flourish. And when you put these two together, almost nothing is impossible.

Our genus has been astronomically successful, and these power-couples were the original jetpack of our meteoric blastoff.

This reproductive success was about much more than just turning out big numbers though. Certainly, having lots of children was an obvious and necessary component. This of course was needed in order to sustain many of the most useful mutations against the naturally attenuating effects of recombinant attrition. However, these parents were still in the process of discovering one very critical maxim. Sustainable genetic proliferation within an intelligent species, has been, and forever will be, about much more than just turning out big numbers.

Long term reproductive success is actually much more readily achieved by ensuring that a child's upbringing consists of an abundance healthy parental interactions. Every single child, individually, needs a great deal of personalized attention from each of their parents. If they do not receive this supportive attention throughout their formative years, then they will, without a doubt, be experiencing strategic disadvantages throughout their later years.

Children of more intelligent species have always needed to be surrounded by a full spectrum of healthy adult behaviors. They are in need of this variety of healthy adult influences so they can grow up with a well-adjusted temperament. This is what enables them to go into a peer environment with a total and absolute confidence and the ability to form mutually beneficial relationships. This is how they develop their quality friendships, and this is how they attract a compatible partner.

If, for example, a group of poachers goes and kills off a large portion of the adults within a group of otherwise natural and wild elephants, then this will absolutely cause problems for the young ones as they grow up. The unavoidable result of this destruction and having lost so many of their mature elders will be that these little ones will just simply not know how to act right. They will never be able to speak the language of healthy social interactions.

Mammal brains are far too big and complicated to just be automatically born, completely programmed, with everything they will ever need to function. We need to be coded by our environments and the world around us. Our mammalian children are just like little copy machines. They are built to faithfully and reliably mimic all of the behaviors and nuances surrounding them. If they are not impressed upon by the right types of influences, then they will inevitably become imprinted upon by something else. When this scenario unfortunately does play out, the actual imprinting that they do receive will be coming from all the wrong places - a few stressed out and disassociated adults from their personal life; a few sensationalized and unrealistic themes from their broader pop culture; and a few random and sketched out children in a very similar situation.

This blank slate aspect of our young mammal brain is really quite fascinating, but it can also become our enemy, because every tool can just as easily be used as a weapon. Yes, the programmable functionality of our consciousness is precisely the flexibility that we need to be capable of adapting to things, but it can also become our nemesis when our environment is full of insanity. We can literally become conditioned to think illogically. And we refer to this illogical conditioning with certain terms like "mental illness."

To be clear though, even if we do become entrained with a bunch of anti-logic and self-defeating behaviors, our subconscious will instinctively become aware of this – it will instinctively intuit the presence of these irrational inconsistencies. And then we'll feel this in our guts as a low-level anxiety. Our subconscious will attempt to call bullsh** if this insanity is causing us problems. Otherwise it'll just let it ride, because sometimes there's just so much of it around us that it would actually be *more* harmful to go against it. But that's another story. The point is that our deep and no-nonsense subconscious will instinctively try to fight this pathological programming.

However, with regard to these prepubescent elephants, as these little youngsters are slowly transitioning into their adolescence, they will need to begin socializing with their peers more. These will become their future friends and partners. So, it is indeed of the utmost importance that they develop a healthy social sphere. However, if these youngsters never had a decent role model, because their elders were all removed from their environment, then their behaviors will have these awkward little tells in them, and then their peers will sense this as a potential liability.

The language of healthy social interactions is a lot like any other language – it has subtleties and nuances and intonations. And if a youngster's inherent behavior doesn't have the right social accent, then their peers will automatically sense this as having come from an unhealthy background. ... This means a background without a healthy support network; this means a person without accountability; and this means an individual who is a lot more unpredictable. None of this is helpful for attracting new partnerships.

We are a finely tuned instrument for picking up on these little details. We can read facial expressions and vocal intonations and body language. We have needed these incredible senses for things like lie detection and outsider threat recognition. So, when we encounter another person who doesn't speak our same behavioral language, we can identify these people by their behavioral accents. This is how individuals from healthier backgrounds can always spot a person who is not... even if that person thinks they're incognito.

And so, this is why it is SO incredibly vital to just get it right the first time – to simply raise a child properly to begin with. ... Sure, they can always change their stripes later if they haven't been raised well, but this is like learning a whole new language for them. It's about a million times easier to just do this as a child. Children will pick stuff up... and they don't even have to think about it.

But luckily for mammals though, whatever can be initially programmed can also be reprogrammed. You just need to know the right tools to do the job.

The best way to help those younger orphan elephants, however, is to simply bring along an older bull to guide them. Sometimes there'll be an older elephant just meandering about without a herd for whatever reason. If you take this older bull and then you put him in with these younger elephants, his mere presence alone will condition them to grow up much more healthy again. They will follow his more mature lead and they will model his more well-rounded behavior. This even works if these younger laddies and laddettes are already in their adolescence.

The solutions for the various problems which are caused by a lack of something during childhood is to provide them with that very same energy that they were not given. This sort of latter supplementary attention is also generally more effectively provided by a more mature elder. The solution for a lack of attention is to *provide* them with attention. And the remedy for a lack of compassion is to *provide* them with compassion.

These principles apply to all mammalian and intelligent animals. If an animal has been consistently treated wrong, then it needs to be treated right. We aren't reptiles. We can't just be blithely disregarded into an egg clutch. You can't just leave us to survive on our own based on purely our own at-birth instincts. We don't work like that. We're not amoebas. We're not bacteria. We're not insects. We're not reptiles... We... are mammals... We *must* be cared for.

You don't have to be rocket scientists to be a good parent though. You just have to show up and have an honest intention to do well. ... And if you just do those simple things, then you automatically get a participation trophy.

Being a good parent is so easy a caveman can do it."~

And do it they did. Once these highly effective teams of mutually supportive Homo erectus parents had finally become a more frequent and common occurrence, they would have quickly come to realize just how many children they could start supporting. All they needed to do was consistently work together and support one another.

Once they had gotten that far, the very next revelation for them, ironically, was the fact that actually having such a large family like that was in many ways counterproductive. By turning out so many extra kids like that, you actually end up just defeating the whole purpose of what you're trying to do. Once the parent-to-child ratio ends up getting way too out of hand like this, the primary influence of the children starts to shift horizontally to their peers. It's just like the case with the elephants. The children then start to instinctively model their behaviors off of one another instead of modeling the adults.

When a young adult has grown up like this, their peers can nearly always sense this pedigree. This is not a good thing. Everyone can instinctively intuit that this immaturity corresponds with a lack of stability. People also just naturally understand that these sorts of characteristics always derive from an unhealthy background.

None of these things are very helpful, for either attracting, or retaining, a potential mate. This lack of fundamental stability is a major detractor when it comes to the process of pair bonding. People in general, but also women especially, are innately and inextricably drawn to the quality of stability. People wanna feel like their companion is someone whom they can rely on. They wanna know that their significant other will always be there for them – not just physically and resourcefully, but also psychologically.

Similarly, women in particular have a very well-justified instinct to be in a relationship with a partner who has a support network. They would prefer them to have a wholesome extraction, one with a family and a friends group, and a decent amount of popularity. All of these social interconnections amount to safety and accountability.

A paramount milestone in becoming an intelligent species is the point at which you discover the principle that just because something can be done, doesn't necessarily mean it should be. There is a definite balance to be struck for the families who were in the process of passing along intellectually progressive mutations. Just because you can have a million kids, doesn't mean you necessarily should be. With a more manageable total number of children, their biggest influences will in general remain their parents - not their siblings, or their, as yet, still immature peer group. This is how their character becomes more appealing. This is how your children become more attractive to potential life partners.

This isn't to say, though, that they should just never ever be experiencing any influence from their peer group. Certainly, they should be. And instinctively they will be – roughly about the time they start hitting puberty. This is only to say, rather, that they should never ever be looking up to someone who's essentially just the same exact size as them. They already have enough time to go out and play with their little neighborhood buddies - they don't need to be getting raised by them also.

So... there you have it. This is the absolute most comPETITIVE social structure for an intelligent hominin species... - a committed and extremely protective provider father... with a dedicated and predominantly synergistic supporter mother... and then a plentiful but still very manageable total number of children... aaaaaaaaaaaand... then a diet of essentially just beach food. ... ... When they finally got this general family recipe down, it was off to the everlovin' races for these earlier Homo erectus. ... No other form of Paleolithic lifestyle can even hope to approximate this effectiveness.

Of course, by the time of *early* Homo erectus, well before the midway point of our evolution, these familial instincts would all have been redundantly well encoded in our genetics already. They would all be busy with their Brady Bunch of children. They would all be teaching all their young ones how to do things. They would all be passing down their knowledge, with demonstrations and tutoring. And they would all be handing down their tips and tricks and secrets.

You can imagine an experienced father mentoring a couple of his more adventurous sons on his closely guarded secrets for offshore lobster diving, while another older child is off teaching his younger siblings about the finer points of crab foraging. Meanwhile, their mother is with them, and attending to a little one, while she demonstrates for an inquisitive toddler how to find her favorite mussels.

Children actually learn best from having to teach their younger siblings in small little incremental lessons. They just pass along the knowledge that their elders taught them. And then their parents answer questions and keep things moving.

Overall familial success can also be derived from the grandparents, or from the eccentric aunts and uncles, or from even the benefactors, but the original progenitor magic will always be the family unit. This has been the engine of our progress. *This has been the driver for our intelligence.* It's not only good for having lots of little smarty-pantses, it's also good for selecting for highly proficient mentors.

So, now, this finally brings us right to the heart of the matter. This finally sets the stage for our discussion. ... The reason that a life on the prairie would just straight up obliterate our intellectual progress, is because a life in this harrowing inferno would only serve to thwart our selection mechanisms. ... It wouldn't just destroy our precious brain-food fuel, it would also effectively destroy our selective reinforcement engine.

In a natural coastal setting, there are three primary and indispensable mechanisms for enhancing our mutations for intelligence. First, women will select whichever mate they ultimately decide is the most capable and compelling from amongst a pool of multifaceted competitors. Second, the more productively intelligent teams of mutually supportive parents will in general be able to provide for a few more children than the average family. And finally, these power pairs of exceptionally intelligent parents will also be able to pass along their knowledge and thereby accelerate their children's competitive advantage. Or, to put it really simply, women will in general be picking smarter guys; smarter guys will in general be able to feed more children; and then smarter parents will in general be able to super-boost their children with a higher quality education.

So... that's smarter dads, more food, and a better education. That's a threefold triple-decker for the beach life.

Now, this might seem like it's a little unfair. This might seem like it's an unfair advantage. But it's also a lot of work! More kids and more education equals *way* more work. So, everything balances out.

... And as an interesting side note, whenever these families started making things and selling their extra resources, they started accumulating early forms of wealth. But it eventually became a lot more wealth than they had ever needed, so they would just start taking care of their communities like a bunch of renaissance patrons. And then these families would also start competing with one another. They would try to build a better reputation... because they wanted their local communities to continue trading with them. It was a perfect symbiotic relationship. Everybody won. And it was all driven by these more productive families. So, yes, they certainly had a lot of advantages, but they also had a lot of responsibility. And this is actually how we evolved to arrange our societies, but that's another story.

Either way, this social order is absolutely essential... because this is precisely... how we became... the Brainy Bunch. ... – smarter providers, bigger families, and better mentorship.

But, alas, it is also these several mechanisms which get utterly decimated by the savannah.

Whenever a relatively primitive group of hominins such as Homo erectus ends up moving out and traveling inland to the savannah, the entire tradition of the nuclear family structure just ends up getting dissolved. The whole entire tribe just ends up having to huddle together and become one *giant* communal family.

Out on the plains, these Homo erectus would constantly have to stay in close contact with one another. They would no longer be able to just safely split off and go foraging on their own, lest one of them get picked off and eaten. They would essentially have to travel like a war party. Thus, the good ol' days of having independent foragers for individual families was gone. Now everyone would have to share everything. The days of privacy and self-determination were over.

And you probably wouldn't wanna be keeping a bunch of the extra food for yourself anyway. Your tribe of origin would be absolutely everything to you. You would be utterly and hopelessly dependent upon them. And your very existence would be like the field life of an embattled military. ... Soldiers in wartime become bonded. They develop a kinship which is almost universally unrivaled. So, one would be maximally incentivized to keep their entire tribe healthy. This means no more extra food for the children of smart families.

But it's not like you could even find any extra food anyway. ... – maybe like a few plant tubers, or some ambiguously edible insects. But even if you did, and you somehow got a few extra scorpions in your diet, this excess hoarding would not be seen as very tribesmanlike.

In this new world, it would definitely be the women who were doing most of the foraging. The men would now be hunting instead of foraging. So, even if you did somehow manage to get some extra food for yourself, the other women would just notice it immediately, and then this would soon result in a major scandal. Women and children are instinctively highly conscientious of what seems fair and proportional. This is due to the need to ensure that every single child is being cared for. But now this fairness, and this sense of equal outcomes, would become a yoke of uniformity that was foisted on *everyone*.

No longer would the more clever mothers be able to freely apply their talents towards bolstering their own children. This would just be seen as an attempt to be selfish.

And it wouldn't just be *mom* who was being thwarted by the tribal uniformity. Even if your dad was exceptionally clever, the spoils of all the hunts would still be equal. The fruits of your father's intelligence would simply be apportioned out evenly to everyone. So, just like with the women, the efforts of all the men would all be automatically homogenized.

Homo erectus could only really hunt the extremely large animals. They didn't have the tools yet to go after any of the medium size ruminants – at least not in any efficient way. Thus, the act of simply hunting was still a massive concerted effort for them. It required maximum possible participation. The men didn't have time for anything else. And once the hunt was over, the animal would just be divided up evenly amongst everyone.

Only the main alpha male and a few of his closest confidants would be able to take any extra for themselves. These reemerging alpha types would not be especially intelligent. Rather, this whole alpha male reemergence thing would effectively just be another retrograde relapse – just another degenerate regressive backslide... back to the old jungle way of doing things.

This is what happens when your environment is forcing everyone to become like a single, solitary, monolithic, military unit. Everyone just ends up becoming homogenized again. Everyone just turns into a jock animal. And in a world of mostly rhino-hunting jock animals, the strongest always end up becoming the leaders.

It's not like in a foraging environment where you can just follow your favorite patrons, because their ideas and their reputations are better. No. They didn't have time for that. They had to act in unison. They had to work together. They needed to respond in mere seconds. They needed a central leader like a rugby captain.

... And they weren't quite sophisticated enough yet to understand that community patrons can always just select specific people to serve as emergency response leaders and hunting captains. ... Yeah, they were definitely smart... but not quite *that* smart yet. ... They still needed a "strongman" leader like a lugal, warrior king.

Accordingly, from this point forward, these landlocked tribes would've start selecting again for the quintessential jock-like characteristics of the earlier jungle epochs. And as such, this would not only bring about an end to the selection for extra clever foragers, it would also begin the process of watering down the gene pool's reservoir of elevated aptitudes.

You can't just take a defenseless beach animal and then go shove it into a meat grinder full of lions and dinopithecus, unless of course it had a few extra tricks up its sleeve. It would have to be a little bit clever. It would have to have the ability to make a bone spear. But it also wouldn't have to be a genius. ... – just clever enough to make those primitive weapons.

Thus, their general intelligence would become sandwiched in between these two opposing forces. There would be the downward force being caused by the jock animal selection, and then there would be the upward force being caused by the bone spear requirement. So, they would just be pinned in between these two opposing forces – like a bowling ball that was stuck in a gutter lane.

And like we mentioned before, you can't just be a super simple, early, primitive, jungle hominid and then go waltzing right out there with a little pointy stick. You'll just get slaughtered immediately. ... - we know that this was happening to them. ... Trying to fight off something dangerous with a pike stick is like trying to use a toothpick as a hypodermic needle. It's just not practical. ... And trying to stab someone with a little toothpick would still be about a hundred times more feasible than trying to impale them with a pole stick.

Would a cheap wooden pole stick work if you were in medieval battle where a massive heavy cavalry was flying at you at an enormous velocity and you were trying to free the Scottish people from the evil king Longshanks??? ... Yeah. Yeah it would. It would work like a charm. ... And would a sharpened wooden pole stick be just the right ticket, in the off chance that a giant wooly cave bear was bearing down on top of you because "what one man can do, another can do???" Uhhhhhh, yeah. Yeah it would. It would work like a dandy. ... But is any of this Hollywood stuff actually practical in any real-life applications like lion defense or mammoth hunting? ... Mmmmmmmmmmmm no. Not even close. Again, think of the toothpick analogy, and now consider how ridiculously brutal it would be to try and force one of those things into somebody's arm. Now multiply that difficulty by about one hundred.

And as was stated earlier, these long-lost *land* erectus would have been pinned in between those two opposing forces. There would be the land-based jock selection always trying to push them ever lower, but then they could never get too low, in a realm of butchers and breakers and manapepig slayers – they would always have a need to make those bone spears.

The real adaptive pressures on them, *instead*, would be coming from the need to become ever more structurally robust. This is exactly what happened to the Neanderthal. They migrated into the continent and then they made their entire existence about ramming spears into mammoths. So, of course they became like a beast eventually.

Thus, the simple reality for these savannah tribes was that their physical bodies were becoming far more resilient at the same time that their minds were becoming far less resplendent. And yet, that waning would eventually abate when they had finally acquired an equilibrium. It was here that they would freeze in a state of suspended adaptation. ... – just like the Neanderthal, eternally arrested in their lands of frozen forests.

So, there you have it. If we were to remain within the confines of the interior, then the men would instead be selected for the qualities of jock strength. Furthermore, the talents of the more mutationally high aptitude parents would effectively become coopted. They would no longer be able to just provide a numerical advantage to their very own genetics. And there would no longer be any point in ever passing down your secret family recipes. They would no longer be able to just pass along their hard-earned proprietary knowledge. ... *So, the mutations for enhancing intelligence... simply... could not... reinforce themselves* - not through mating and not through feeding and not through mentoring.

We are not a savannah animal. We are not a deep jungle forest animal. We are not a landlocked, intracontinental, terrestrial animal of any sort. And if we would've ever attempted to do this, then our intelligence would have simply imploded. And now the weight of this absurdly obvious evidence is overwhelming. ... That is why the cold hard facts are increasingly drawing the official viewpoint ever nearer to this epiphany.

The very earliest beginnings of this revolution have already begun. Even now, there are anthropologists in the beginning stages of acknowledging this. They are beginning to acknowledge things like the inescapable nutritional arguments. And yet, they are still not quite getting it. They are still continuing to assume that the savannah was somehow important, and that the seaside interactions were just a side-note.

And there's actually one thing in particular they keep doing which is really pretty annoying. They keep trying to claim that if we ever really did adapt to swimming, then it must've only come from just one lit-tle "phase" of our evolution. ... Uuuh, no... Sorry. There was no little "phase." Okay?? And that word is really starting to get on my nerves.

Now, I really appreciate that these guys are at least beginning to look into this stuff. And I know that I really need to just calm down right now and just keep on being patient and understanding with them. I know that all they really need right now is just a little more time and support. And, of course, the very last thing that they need at this moment is to be unscrupulously berated for not having it all figured out quite yet. And that's fine. That's totally fine. I am completely okay with that. But they really need to quit using the word "phase" every time they start describing our aquatic evolution. I mean, I don't know if they're just doing it to mess with me or what.

But I'm not going to reprimand them harshly. I'm not going to im-peach their character. No. That would just be counterproductive. But, I mean, they seriously need to stop using the word "phase" right now. I mean, just seriously stop saying that word, okay?? You anthropologists are really starting to get on my last nerves with this. I mean, I've al-ready told you MULTIPLE times by now to stop. But you just keep on doing it anyway... again and again and again. Like, I'm just seriously trying to be a professional over here, but like how many times do I se-riously have to keep telling you, it wasn't a PHASE?! You understand?!

You UNDERSTAND ME?!?!

You FUCKIN UNDERSTAND ME?!?!

The NEXT one of you punk ass anthropologist CHUMPS who speaks that word from between your mouth... will do so... with MY FOOT up your... HEY!!! WHO JUST SAID that?! Which one of you sons of BITCHES just used that word again?!

You think I won't find you back there in the crowd?! I will find you... and when I do...

Oh...

HA!!

I see you back there... back there runnin' your mouth.

It's YOU again... Ssssssteve.

Alright, Ssssssssssssteve...

If you're so smart then, Ssssssssteve, then why don't YOU try to explain how we got so smart without being by the coastline. Go ahead, Steven. We're all listening. Why don't you amaze us with your great and powerful knowledge.

Is that it, Steve? Are you done? Is that what you believe? Is this what you honestly believe? Haha. ... Oh, Steven. ... Silly silly Steven.

What ol' Steve is trying to say back there, is, that sometimes there actually are inland tribes where the men function as dedicated providers. These tribes are more modern ones, however, where the individuals in question have considerably more intelligence than that of a typical Homo erectus.

Once you get well past the midway point of our evolution, you start to see our fully modern human brain emerging. From this point moving forward, you also begin to see a much more sophisticated level of hunting technology as well. The most fundamental of these technologies was the atlatl dart spear.

These newer atlatls and dart spears were a pretty big deal. They finally enabled our nearly-modern-human ancestors to begin hunting and going after the *medium* size fauna. They no longer needed those huge gigantic megafauna anymore. They could break off and start splintering out into smaller groups. They could just throw things at a smaller to medium size ruminant now. They no longer needed the whole entire tribe to go rush a massive rhino together.

And it was precisely this technology which enabled them to begin functioning more like individual families again. Sure, they would still need to be building their houses closer together for basic security, but the task of actually providing for their individual families would no longer be a communal one. It could largely become the responsibility of the individual families again.

And what that all means for us is this, once you've reached this level of extremely advanced hunting technology, you could once again have this selection process whereby the women were finally choosing their own provider males. You could finally now transplant this selection process *into* the continents.

So, what Steve is trying to argue here is this. He's claiming that once you've reached this level of extremely advanced hunting technology, you can finally start selecting for intelligence again, despite the fact you're living within the continent.

According to Steve.

Alright, well, where should we start dismantling Steve's hopes and dreams at? ... Oh, I know. Let's first talk about why this whole concept of hunting in small groups is even possible to begin with.

First, you had the beach tribes getting better and better and better at throwing projectile spears. This is that whole process we talked about earlier, where you originally start jabbing, and then that turns into thrusting, and then that turns into launching, and then that turns into actual, full-blown, projectile throwing.

Thus, the practice of throwing spears at stuff is born. And this only works with spearfishing, of course. ... It has to start somewhere simple, like trying to stab a fish which is just below the surface. There's simply nothing based on land which wouldn't be *better* resolved with a good ol' fashion *bashing* motion. And besides, it still needed our amphibious development so that we could get these swimming shoulders.

Now, at some point in your burgeoning discovery of spear throwing, you are going to be begin realizing that there's a problem. You see, you can't just throw any ol' type of stick through the air. Most sticks don't even have the right properties. However, through a great deal of trial and error, you are eventually going to figure out what really does work.

You basically need a lighter style dart spear. Its center of gravity needs to be about one third of the way from the front. And it also needs the perfect flexibility – too elastic, or too rigidly inflexible, and it won't even fly straight. And then you also need an atlatl. This is a completely separate, leverage handle that you use for throwing it. And if all *that* wasn't enough, you also need a massive amount of practice.

This level of super advanced hunting technology took until about half way through the second half of our evolution. So, if our entire evolution was on a twelve hour clock, then this development would have only just occurred at around nine o'clock on the dial. This would've been roughly about the time of late Homo erectus, or early Heidelbergensis, or ergaster, or whatever they end up changing the name to next, five minutes from now. ... Or we could even just denote it as "nearly-moder-human" – halfway between us and Homo erectus. This is when we finally got projectile spears.

Then, of course, as usual, there were the groups who were breaking away separately and then making their way inland. Except now they had their throwing spears and they knew how to use'm. They weren't just limited to the largest and smallest game anymore. They could actually split up into separate family units and go after anything. They could finally break away into smaller groups and just provide for their individual families again.

However, they generally would not've been going out solo though. Having multiple people gives you a huge advantage. It enables you to do all sorts of stuff, like having a few of your guys scare the prey out while the rest of your party just lies in waiting.

Additionally, if you are out there in the wilderness all alone like that, then you are definitely a much more attractive target for the predators. I mean, if you really truly had no other option, for whatever reason, then you could definitely break off and go solo. But this sort of thing was *not* advisable back in these earlier periods, back when every single ecosystem was fully loaded with mega-killers everywhere. Yeah, there were definitely some really big megafauna back then, but there were also some really massive super-predators that came along with that.

So, hunting back in those days was still a group affair. Sure, you didn't have to have your whole entire tribe with you, but you definitely needed a team of at least two or more, closely related, hunting buddies. And so, women still weren't choosing just one exclusive man yet. Technically they were choosing a whole entire, friends group. She might decide to be your mate, largely because *your friends* were diesel. So, this definitely blunts the edge of the whole, provider male selection mechanism.

Alright, but let's just go ahead and give this idea the most absolutely massive benefit of a doubt. Let's just give this whole theory some water skis and a jet pack, going downstream with an adrenaline needle in its chest. Let's just say we have a tribe of really hardcore, lone ranger hunters, where every single man goes off and hunts, completely solo, all by his lonesome. And let's just say that all the women of this tribe are completely free to choose between these men based on their own individual attributes. ... Alright, Steve. Does *that* sound a little more like what you were thinking?

Okay. Well, then the men of this tribe are actually going to have to be polygamous. They're gonna need about two or more wives apiece. They're gonna have to compensate for all of the men who are going missing – the ones getting ambushed by lions and tigers and bears aaaaaaaaand fracturing their ankles to the point where they can't even limp back to safety before nightfall and then getting eaten by a superior force of wolves or hyena.

Alright. So, now we've got our hypothetical tribe where the pretty young maidens are selecting from amongst the lone ranger hunters. Alrighty. So, what happens next?

Well, the pure and simple fact is this, the intelligence needed to be a good hunter is *nowhere near* the amount needed to be a competitive forager provider. *It's not even close.*

There is definitely a lot of nuance involved in hunting. It is generally not a pastime for dummies. But the barrier for basic entry *doesn't even compare* to competitive coastal foraging. It doesn't even approach the encyclopedic repertoire needed for basic foraging bushcraft.

And before anyone wearing camouflage right now goes and blows a gasket on me, let me just tell you a little story about a... wait. Hold on a minute. Why are you wearing camouflage pajamas right now? It's night time. You ain't huntin'! You don't need no camouflage pajamas in the comfort of your own home!! ...

No wonder your wife is always mad at you. She thinks you out there cheatin' on 'er, BOY!

Well, since you're obviously gettin' all warm and cozy for bed right now, in your little jam jams, why don't I tell you a little bedtime story, my little hunter. ... Always upon a time, there are crows and the eagles. Crows and eagles live in the magical kingdom of animals, and they actually have a lot in common. Many of their populations live along the coastlines and eat seafood. All of the princesses within those species select which handsome prince they want. They want a prince who can go flying off into the sunset on his trusty wings and then come flying back out of the sunset with some food for them. They will then form an exclusive monogamous relationship together and live happily for at least one year after, working together side-by-side and raising their little chicklets together.

And yet, despite having so much in common, there is still one *very* critical distinction which differentiates them. You see, the eagle is mainly a *hunter*, while the crow is mainly a *forager*. The eagle will spend his life swooping down and snatching little fish up, while the crow, by contrast, will spend his life mentally cataloging an almost endless variety of local food sources. He will learn how and where and when to find these local food items, and he will even learn how to secretly cache his findings in many disparate locations, and of course he will need to remember them.

The eagle needs to be clever, without a doubt, but the crow has to be everything that the eagle is and a million times more than that.

Indigenous peoples from all around the world and from every walk of life have long since known about the crows and the ravens. Even Neanderthal had a special sort of reverence for them. They have been known as the kings of tricksters. And not only will they trick humans and other animals, but they will also go so far as to play off completely different species against one another. They can literally understand how *two comPLETEly* different types of animals will respond to one another. They actually know what *each one* will be thinking, given a certain context. And then they *use* that understanding to their advantage! ... Hoooooooooooly ffffffffffreakin' mmmmmmmackerel!

They even get right up in the faces of these massive eagles. They are just completely unconcerned and unintimidated, despite the knowing that all of these eagles are about a hundred times stronger than them. The eagles could destroy them with one little flick of their talons if they could actually get ahold of'm. But the crows are unimpressed by this. They just trick them away from their food and then snatch it like it's child's play.

The bottom line is this, however. Women of the more advanced, Paleolithic, *land* tribes were generally *not* selecting for the trappings of intelligence. They were mainly just selecting for a team of hunters. And this just cannot push up against the boundaries like selecting for foragers does. ... It doesn't matter that these women had providers again – technology had finally made things far too simple. ... Steve!!

So, I hope that answered your verbal attacks and crude assertions, Steve. Thanks for actually letting me get to all that before you tried to interrupt me again. ... Yuh know, Steve, I think this could be the beginning of a beautiful frenemyship.

WOE! What did he just say?!?! ... Did Steve just shout something at me again? WHAT?! He called me a DICK?! That's it, you mother... I'm coming back there right now, you SON OF a BITCH!

Aaargh aaarrrhhh...

AAAAHH... AARRHH

AAAAAAAHHHHHH...

God! You tore my fuckin' shirt, STEVE! You son of a bitch...

Okay, I'm back. Alright... is this mic still on? ... Test, test.

What? Is he still talking shit back there?! PFFFFF Dude WHATEVER, STEVE! That was a LUCKY hit! You got a LUCKY hit, STEVE. ... PFFFFFFF uuuuuuhhh it's called the floor was slippery right there where I was standing right there!! PFFFFFFF look at Steve, thinkin' he's all Rambo now.

That's was a pure lucky hit, *Steve!!*

Pfffffff... I'm'a call you Lucky Luck Steve from now on. ... Your name is Lucky Luck Steve for all the luck you just received.

14

9 ¾ 5 -- OSHIYANDER'S FINE LOVE POTIONS & ELIXIRS...

... SINCE NEGATIVE 1892 THOUSAND

Every time I see some nice humble worker just doing their job and mopping a floor, I always wanna go and walk straight past their sign onto their wet floor and then deliberately fall down and pretend like I hurt myself and like I'm on the edge of coming to tears. Then when they freak out and look at me with concern, I wanna point my finger straight at'm, and then say, in the most high pitched little whiny dweeb voice, *"I'm* !!SUING!! you!!"

Not like I'm suing the company in general, but like I'm suing them personally.

"I'm suing !!!YOU!!!"

Still pointing at them.

Then you gotta smile and be like "ehhhhhhh, just kidding" and then get up with your wet ass and walk away.

I've had jobs before where I had to go through these like gated areas with lots of security, where you had to check in with these guards wearing bulletproof vests and stuff. I always wanted to go in there on April Fools and just be like "Hey, Larry! How are the kids?" and then whip out a gun and shoot'm in the chest.

Just shoot'm right off his chair.

BANG!!!

Or if it's one of those guard stations where you have to drive up in your car, just pull up immediately with a shotgun pointed at'm.

"What's up, Larry?" BLAM!!!

You know what though, I think I would much rather get shot in the chest by a rogue employee with a shotgun than have to live my whole entire life covered in flies on the savannah.

And we're not even gonna talk about a bunch of these other things either, like how you would just have these flies constantly landing on your mouth all day. And then there would be the ever-present plagues of biting and swarming insects, all over you throughout the day and night, driving you absolutely insane.

The real and true plains animals are each and every one of them automatically born with their very own insect defenses. Having natural protection against the eternal onslaught of these biting and swarming insects is not even an option out there - it's an absolute must.

There are many different schemes which have been worked out by mother nature to defend these poor animals against these pests. You've got things like protective body fur, armor plated skin, razzle dazzle zebra patterns, long whipping tails, automatically twitching muscles, and so on.

So, of course, guess who does NOT have any of these natural defenses.

Hmmmmmmmmm, I wonder who it could be.

Why didn't WE ever adapt any of these defenses? ... Hmmmmm-mmm...

It's a mystery.

Speaking of which. Welcome, all you ladies and gents, and all you wizards and woodland elves and wand wavers, to what has now become, the wholly fifth and most far-flung installment of Chapter Nine and Three Quarters. ... – a chapter which was only supposed to be one chapter long. ... in a book which was only supposed to be one book long. ... And yet, here we are.

This chapter is about love, and it's gonna be a thick one, so things are gonna get a little sweaty. It's gonna be twice as thick and twice as long as any other chapter you've ever had, and I'm not sure if you can handle all of this chapter yet. So, if you wanna go get limbered up first and then circle back around, then that's fine by me, because this chapter is also a bit of an intermission. We are going to be talking about... ... well, ... precisely what the chapter titles says it is.

Alright, let's have a quick review of what life would've been like for our dear ol' midway ancestors in exile. What would their lives really look like on the abysmal scorching plains of the early Pleistocene? Let's see, you'd be malnourished; mothers would suffer the most; you'd lose privacy and independence; you'd be subsumed into a collective, which itself would start to select for strength over brains; everything around you would want to kill you; biting and swarming insects would be eternally plaguing you; you'd be constantly brutalizing your feet on the terrain; you would be perpetually hot as hell, without any ubiquitous ocean water to dissipate your body heat; and finally, if you even so much as *thought* about the idea of ever keeping a little extra food for yourself, then Dolores would find out about it immediately and start telling everyone in the tribe that you were a vicious sneak thief and a total food-whore, which ironically would be an oddly weird compliment, in a way, because at least a food-whore would still be selecTING FOR INTELLIGENCE, DOLORES, YOU FFFFFFF...

But, I don't know if you would really wanna be calling out Dolores' like that, honestly. Your entire existence would be a never-ending mission of stealth. You would be existentially bound to a solemn vow of nearly perpetual silence by the very nature of having to always follow the migratory herds around. And the very *last* thing that you would ever wanna be doing out there is just blatantly making your position extremely obvious to the multitudinous mega predators.

The plains life would be a quite one for Homo erectus. And, unfortunately, you would have to keep your disgruntled thoughts about Dolores to yourself - just permanently stuffed down deeply inside of yourself. ... This would be an absolute hell for Italians.

But back on the beach, however, everyone would still be dancing and partying and having a good ol' time. You can drum the night away, as loud as you want. The clams and coconuts don't care.

And what's more is, their ability to dance and make music would not just be giving them a leg up, as far as blowing off some steam every once in a while is concerned, but it would also be giving them a myriad of other advantages as well.

Dancing is one of the best ways for a tribal woman to assess a man. She can make some very critical judgements about his temperament, his coordination, and his general fitness, just from observing the way he dances. Hence why men have adapted a generalized anxiety about dancing, in a way that women have not.

Dancing can also be a fairly acceptable way for a lot of young women to get up closer to a lot of eligible bachelors and subconsciously sample their pheromones.

Some mainstream schools of thought would also contend that the act of physically dancing also helps a woman see how well a man can fight. But this is a very odd speculation, however, because the movements involved in a street fight are actually quite ragged and frenzied and haphazard. Whereas the movements involved in a dance routine are in contrast quite flowing and rhythmic and mesmerizing.

Alright, now, hold that thought for a moment.

We've already talked before about the crows and the corvids and the ravens, and about how hominins have actually followed in their footsteps. Our males, just like their males, have surely developed these ritualistic habits of bringing fancy things to women. Our suitors offer gifts, like snacks and resources and pretty little valuable things, as a way of saying, "Look, I'm an excellent provider! I will definitely take care of you. No need to worry with me, toots."

Well, now, consider this. What was the *main* way in which these men were acquiring these food items for their favorite special ladies? What was the primary, day-to-day activity in which these men were irrevocably engaged to collect this sustenance? ... Well... they were diving. They were swimming around and snorkeling. They were moving about the ocean in a very fluid and rhythmic manner.

The very fastest way to swim through water is a method known as the underwater undulatory technique. This technique is also known as "the dolphin kick," or just the classic Aquaman swim. It works by keeping your arms pointed straight out in front of you, like a super hero, and then just waving your entire lower body like a dolphin. This method is, unhyperbolically, such a ridiculously fast way to swim that they literally had to restrict its allowable distances in the Olympics. Swimmers can only use it for a very limited, preset distance at the very beginnings of each length. And yet, this technique is actually one of the most ancient forms of bodily movement for us. It is simply no coincidence that this hypnotic undulation is reminiscent of dancing. Or, more precisely, that dancing is reminiscent of this movement.

Dancing was never about showing women how good of a fighter you were. You can already tell that about a guy just by looking at him most of the time. No. And there's no actual, major, causal correlation between dancing skills and fighting skills. No. That's not why this profoundly inextricable passion has forever remained a part of us. It was never about saying, "hey, baby, look how good I can *fight* for yuh" - it was always and forever about saying, "hey there, good lookin', look how well I can *provide* for you."

The species which are adapted for dancing do not do so in order to exhibit their fighting abilities. They essentially do it just to show off that they're healthy. If a species wants to demonstrate their toughness, then they simply do so by just, well, by demonstrating their toughness – they actually go and start a fight to display their prowess. They don't get into a dance off about it. They're not like, "Hey! Check out my moves, baby... ... But like, instead of my hips, just imagine it's like actually my fist going into that guy's face over there... ... ~~*Oolala*~~ ... ..."

Likewise, it has also been believed that the reason why our women like a good posture on a man is in general because this expansiveness is a sign of social dominance. And it sort of is, to a certain degree. But our attraction to this form, is in fact, far more universal than that. A good alignment is just as appealing on a woman. It was this tallness that helped us observe down into the water, originally. And then our swimming only served to reinforce this. It causes a strengthening of those upper body muscles that keep our shoulders back and our spines straight.

So, whether it's sensual tribal dancing or an elegant body posture, the original genesis of these mesmerizing turn-ons was a life of surf of sunshine – not a world of constant running, and getting into fist fights to show women how legit you were.

But dancing wasn't just for legit-ass men of course. The women have always loved to do it also. The women, however, have had far less trepidation about it and about dancing in public. It was never about being scrutinized for the ladies. A woman was never being judged as an athletic provider. She mainly just needed to be fertile. So, when a maiden would start to move her hips and body, it generally would just imply that she was ready.

And when that maiden in your tribe began to shake it, she instinctively knew exactly what she was doing. As soon as those sauntering circumferences started swinging into motion, the effect would just instantly alert the bachelors. And it would beguile your dumbstruck mind if you were watching.

So, naturally, you begin to make your way over to her, as if compelled by some imperceptible force. As you approach, she instinctively begins to giggle. This is done to further disarm you. She then follows it up with a series involuntarily exhibitions of naivety, entirely meant to provoke you into attempting to impress her.

So, you then fall for this unassuming giggle bait and thus proceed to bedazzle her. You start to wow her with your brilliance and your eminence. ... or so you believe. But really what she's doing, it turns out, is studying you. She's just observing your coherence and your quality. In other words, she tricked you. Not on purpose. It's just instinctive. She tricked you into showing off and flexing. She set you up to study you like an animal. ... And you just went for it.

But lucky for you, something you did back there actually worked, and she has now become authentically intrigued with at least one of the things she saw. She's never gonna tell you what it was though. But rest assured, it was at least ONE of those things you did. ...

So, now she starts to spend a lot more time with you. She wants to get to know you a little better. But before you know it, the two of you have become quite fond of one another. So, now that things are actually getting a little more serious, she promises, that during the next full moon, the two of you will go sneaking off together and then rendezvous on the bluffs overlooking the cove. And you can definitely tell by the way she said it that there's probably gonna be some second base action involved. ... – definitely probably!!

But then she instinctively breaks her promise and just randomly changes her mind all the sudden. She can't help it. Something very deep inside of her is just programmed to do this to you. It wants to see how you react to this. Her subconscious wants to test you. It wants to make sure that you're not a jellyfish that'll just say something submissive like "yes, dear." And it also wants to make sure that you're not a Tasmanian devil that'll just react by getting all stressed out and frustrated. Both of these two extremes are instinctively interpreted as a form of weakness by her subconscious. Rather, she just hopes you'll be assertive.

She wants you to be a rock. She wants a foundation to build her family on. She wants a man that's not just charming but who's also assertive – like a home that's nice and cozy but also stands up to hurricanes.

And lord help you though if you tell her "yes, dear." That's pretty much a one-way ticket to becoming one of her new girlfriends. She tests you like a hard place to see if you're a rock.

Don't get me wrong though about the importance of having fighting skills. Combat and battle have almost endlessly been a part of our world. A man has always needed to be a defender – not the biggest or the baddest or the toughest guy in town - just a man who's well equipped to defend his interests – just a man who's well resolved to not be a target. Bullies don't mess around with harder targets. It's not worth it. It's too much trouble. And apart from desperate times of dire scarcity, opportunists will only look for softer targets.

The very last thing that a tribal woman wants is to be standing next to is a soft target. Rather, she wants a rock. She wants an emotional rock to stand on; a good stamina rock to stand by her; a good intelligence rock to provide for her; and a good physical rock to protect her.

Now, this isn't to say that her man should necessarily have a personality like a rock. Women still like to play around a little. It's only to say, rather simply, that underneath it all, there needs to be a core of solid purchase.

Of course, things have changed a little since the days of our ancient kin. We are no longer living along a coastline, in a tribe of only several hundred people. And yet, our instincts still believe that that our world *still*. They still believe we're living in a village.

Thus, our subconscious and our instincts still think this is the Paleolithic. They think that just because we found a charming hearth-throb, that this is obviously the person we should start a family with. And up until a quarter million years ago, give or take, this impulse alone was almost always more than sufficient. If you met a local crush that made you dizzy, then you both could have your fun for many decades.

There really wasn't much to disagree on back then. But today we've got a billion things to choose from. So, we need to slow it down and have some patience. It's essential that we share a common vision. And yet, people often skip this step entirely. Their subconscious thinks we're somewhere in the Pleistocene still. But, frankly, if you skip it, then things won't end well. We cannot leave this all to chance like in the olden days, unfortunately.

Our bodies have no idea, really, that we're living in the current modern Holocene period. As far as our physical bodies and our psychology are concerned, this is still the ancient world where major life decisions were far more simple. And this is what our bodies are still expecting from us. They still want the accommodations of a beach resort, and they still want the social fabric of a tribal group. And most of these ancestral conditions are fairly easy to approximate, but the social interactions are not so simple.

Our instincts for forming relationships are still pretty viable now, but they just need a little help from our modern wisdom these days. ... Most things haven't changed though. Men still need to be the rock, generally speaking. And most women can instinctively sense this. Their feelings will seek it out like it's a beacon. Then they'll reward it with enormous dedication.

But there's something else that's really fascinating about women. They didn't used to have *any* loyalty to *any* man, but over several million years, they've been selected to develop it instinctively. And this loyalty is no small thing for her – its strength is only second to her bond with her child.

This instinct in a woman is a very pure and a very powerful one, and yet the story of its creation is not so simple. Unsurprisingly perhaps, its origins lie in a tangled web of oddities. One such peculiarity is our need to exchange genetics with the occasional neighboring tribe. Researchers designing the next generation of space colonies have determined that in order to create a completely autonomous gene pool, you have to start with a population of about forty times the numerical quantity of the largest tribal groups. But this *isn't* just an axiom for space travel. This is a general rule for just about any population. And it may not be an absolute requirement in the most critically strict sense of the word, but it is definitely a major necessity if you want the strongest gene pool.

Now, obviously, no tribe has anything like forty thousand individuals in it, so we inherently had a need to exchange our genetics. Technically we can do this in a number of different ways, but the most practical for our genus has been for a few of the more adventurous women to switch their tribal groups. ... Just as the males of many species are designed to transfer the maternal genetics between the various mitochondrial lineages, so too are the females of many pair-bonding species much better at transferring the population genetics between the various tribal groups.

The reason why this has been the case is because women are just automatically born valuable. They are a prized biological commodity for the mere act of existing. As soon as they reach their young adulthood, this extraordinary value becomes manifest. Furthermore, up until around a woman's middle years, she can still mystically and magically produce an entirely new person out of nowhere. ... ... How's that for a magic trick? :o oo'oo'oo'oo'oo... out of nooOooOoowheeeEeeEeereeEee... oo'oO'oo'Oo'oO'Oo...

... Well... when you consider how we hominins got our long stretchy legs, technically, it kind of *is* a stork that delivers the baby.

But this of course just makes a woman valuable. She doesn't need to prove herself to the bachelors; she doesn't need to build her reputation up; she doesn't need to acquire the most competitive provider skills; and she doesn't need to fight or become expendable. She just doesn't need to earn it like a man does. So, of course, a whole new tribe can very easily come to appreciate her.

This is the first of multiple reasons why a woman's ability to bond is so incredible now. It helped her fall in love with a hot new partner, even though this partner was from a whole new culture. This helped her realign her tribal instincts.

The rest of the story is a bit more complicated though. We *are* talking about women, after all. So, in order for all of that to come together, by complement, we also have to learn about the masculine.

Most of a man's traditional relationship value was derived from his provider skills. This of course is not something that he was just born with. He had to struggle for many years to achieve this notoriety. And there were *never* any guarantees that he would make it. ... In our more recent Paleolithic prehistory, only about half of the men ever passed on their genes, relative to the number of women who did so.

Indeed, every single generation of men have been descendent from a previous generation of hard-working provider males. But originally, this all started with just one single provider nerd... – nerd Adam if you will. But he wasn't just an ordinary nerd guy. There was something altogether very special about Adam.

Extra food doesn't just fall in your lap. You have to go ... *get*... it. You have to go ... *do*... something - something very difficult... something kind of scary... something sometimes dangerous. ... So... no... nerd Adam was not just some ordinary, boardgame nerd - he was also a very brave and extremely daring nerd.

It has never been enough for our bachelors to just be intelligent – we've also had to get a little crazy. We've had to search in all new areas for brand new food sources. We've had to dive into unknown reefs that no one's ever been to. We've had to try out new techniques and test new food items. And we've also had to fail and get rejected.

So, we've had to get a little bit rowdy. And we've had to take some really big chances. After all, you can't expect to win if you don't play. You have to play the game. You have to lose your shirt. You have to learn from your mistakes. ... And learn we have. Our men have been taking big risks since the days of Adam. And most of those dicey gambles were absolute failures, but some of these risky ventures were successes... and it's all about the ones where we succeeded. This is why we have a thing for gambling now. This is where we get our genes for risk taking. Some of them risked it all and actually made it, and then they made some little children with that same propensity.

So, every last man on the planet has a pedigree to be a risk taker. This is part of why we gamble at younger ages. We know when something's fairly ill-advisable, and yet we press ahead with vigor and give it the ol' college try anyway. Something in our head is like, "don't worry, bro, we got this. Bad things only happen to *chumps*, son. You and me, we ain't goin' NOwhere, yuh heard! ... Scientists are gonna be like '*whaaaaaaaaaat?!?!?! How do these guys live forever, bruh?!?!?! They don't even die, bruh!!! ... How do these guys kick so much anus, bruh?!?!?!'*"

But our women are well attuned to this reality though. And they have long since been selected to unironically desire this. When a man is not afraid to face a challenge, then he is also not afraid to always fight for her. And when a man is not afraid to always fight for something, then her instincts are well aware that he is going places. ... This man is a charging bullet train and she wants to be on it.

So, we were not just adaptively tailored to become calculated risk takers - we were also extensively selected to become partners who *expect* it.

That is also why our men are so determined. We are in need of this extra thrust to endure our trials. It takes *years* of risky gambling and experimental knowledge gathering, and *even then*, we even still might not master the art of providing.

The art of providing was actually pretty straight forward back in the early days of Adam the Australopithecus. All you really needed was some shellfish. But by the time of our midway ancestors, dear old Homo erectus, the art of impressing a lady was getting pretty darn serious. We had carvings and we had tool kits and we had beach houses, so doubtless we also had the simpler items. That means jewelry and bracelets and necklaces... and that means baskets and floor mats and sling pouches. The art of providing for women was getting serious.

Now, if you were sitting with a family of Homo erectus, just enjoying some mahi-mahi and a sunset, then you wouldn't really notice a lot of differences between you – maybe their chins and their foreheads because they were slightly sloped back a little. In fact, you couldn't even say that they looked like we do... rather, it is *us* who look like them. Plenty of modern humans have sloping foreheads, and you can't really hardly tell unless you're trying.

We've somehow just forgotten who we are now. We've been distracted by our clothes and our self-importance. But there indeed is an ancient erectus looking back in your mirror there. That face is yet from several million years ago. ... And it doesn't even matter where you came from. They had all the different ecotypes just like modern humans. Hopefully this paints a picture of what they looked like. They are *you*, and you are *them*, but for their edge slopes.

Instead what you would notice would be their personalities. They'd have the naivety of an innocent child, like one who frolics in the surf and builds their sandcastles. And although their totemic vocabulary was not so extravagant, their concentration and their ability to create things was recognizably human.

Now... they weren't exactly making high speed missiles yet - things like dart spears and atlatls and bows and arrows. The science of aerodynamics is not like the mechanics of axes. It requires a level of brilliance much nearer to a modern human's. But erectus still had tools unlike their predecessors. They could carve and make harpoons and weave things like basket traps. And they were well within their means to build cabanas. So, these men were now providing in ways that were formerly staggering.

The game had changed. There was *a lot* more food now. And so, their families could get a lot bigger. The art of competitive providing was now *way* more elaborate.

Don't worry though, we're still getting around to why women have such powerful bonding. Everything will eventually come together. It's just like trying to describe a massive spiderweb with all the little details. ... This IS women we're talking about.

Alright, so, women had already been pairing up with guys who were a little bit older at this point. This had been the norm for nearly two million years already. The sisters wanted misters who could provide for them better, so they naturally began to hook up with the men who were more experienced. Competitive beach foraging is pretty complicated. You don't just walk in off the streets and start providing. You've *gotta* know the tricks to find the good stuff. You've *gotta* learn a lot to become a sugar daddy.

But the learning curve was only getting longer, because erectus was indeed a tricksy hominin, and so the time it took to gain the skills that women wanted was only getting longer by the era. So, women kept on choosing men who were older, until the age gap was finally extended to about a half a generation. And this then slowly adapted the actual nature of our male-specific chromosomes... such that we men are now genetically programmed to spend our younger years competitively building up our resources. ...

It was a sugar daddy arms race. The ladies wanted more and more resources, so it adapted the male genetics to competitively accumulate it for them. And then they chose from these older bachelors which one they wanted.

As soon as a boy hits his puberty, nature starts dripping a chemical cocktail of rocket fuel into his veins. Then this *trickle* slowly turns into a *torrent*, and indeed this onslaught does not abate for several decades. This period is thus defined by its ferocity, as it continues from a male's adolescents throughout his early young adulthood. Then what began with an explosion of chaos will slowly crystallize into an order of woven tendrils.

Each new skill that he acquires will thread another strand throughout his Celtic knot of competence. And thus he gains the trident of several virtues - that of competence and competitiveness and confidence. These traits were indispensable for a tribal man. In fact, our women are specifically designed to seek them out in us - confidence being the most appealing of all of them.

As a matter of fact, confidence alone is irresistible. It's like spraying your entire body with a powerful pheromone. When you've known a person who suddenly became super confident, it's like they've cast an incantation of invincible attraction.

It's a little ironic, though, that these terms each sound alike - competence, confidence, and competitiveness. What a curiously convenient coincidence.

Oh well, let's not look a gift horse in the mouth, shall we.

This is good though - it'll make these terms a lot easier for us to remember. I'm gonna test you on all this stuff later, just so you know.

Don't worry, don't worry. I'll make it worth your time. ... For every question you get right, I'll give you a nickel.

Alright, where were we?

Oh yeah.

The reason why women are instinctively attracted to a man with these qualities is precisely because these qualities are the most potent ingredients for a success elixir.

For starters, we need a strong competitive nature that just won't quit. The *professional* side of a man can't just be worried about fairness all the time. ... - at least not when he's got a bunch of ravenous competition arrayed against him. And a woman's ancient primal genetics don't really care about it either, quite frankly - they just want a winner - whether she knows it or not. A winner will take care of her interests. And that's exactly why she finds competitiveness so appealing.

Next, we need competence. This one is obvious. There is indeed precious little upside to ever *not* being competent. It's kind of the most salient characteristic of what defines us as a species. I mean, we can't, in all fairness, expect total, all-knowing omniscience out of someone... like how a person in suspended prepubescence will still literally think their parents have all the answers to everything and have actual super powers... but it's definitely in our hominin genetics to be impressed by feats of competence. A man who can fix things and do stuff can always make a woman dizzy. ... even if it's just picking up the phone and calling the maintenance guy... like *a BOSS!!*

And then last but not least, we need confidence - real confidence - not that fake stuff. The more that you amass your store of *competence*, the more that you incite your natural *confidence*. And the more that you establish what you want in life, and the more that you are true to your authentic conscience, the more that you will find your self-assurance, and the more that you will trust your own decisions.

You don't really have to overthink it though. Just point yourself in the right direction and let it happen. For men, confidence is much more like an oak tree. It takes *years* to become truly substantial from a steady exposure to the elements. While for women it's inherently much more like a rose blossom. It blooms wider when it sees the sunlight of attention.

But, it doesn't necessarily have to work that way for a woman. She too can derive her confidence from well within her own identity. I used to work in this fast food joint, when I was a younger teen laddie whippersnapper. There was this one manager woman, Rosie, with a cleft lip, who was always making life harder for me, and barking orders like, "STOCK THE SAUCES!" But out of her mouth, with her speech impediment, it just sounded like "TALK DUH TOSSES." That was just her go-to phrase whenever she ran out of other crazy shit to say - "TALK DUH TOSSES!" So, I'd just be like "I can't talk them any farther! I already talked the tosses. If I keep talking them, they'll just start overflowing on the floor." And she knew what I was doing, so she'd just get madder and be like "TALK DUH TOOOOOOSSES!!!"

Anyway, Rosie was not what you would call... the most attractive woman... even before the cleft lip. But one day she just randomly showed up to work with all this confidence. I have no idea what the hell happened. And even she didn't realize it. She was just suddenly all upright with perfect posture, and was just like super calm and centered and not even screaming at anyone. She just had this like powerfully graceful gravitas radiating off of her. And I swear... we were like, "what the fff**k?! ... Did Rosie just get *hot* all the sudden?!" ... She was just suddenly all sexy without even changing her appearance. It was freakin' bizarre... like some spell had been cast or something.

I have no idea how Rosie got all confident all the sudden. I think it had something to do with her like graduating from some work program she was in. I have no idea. And it wasn't from getting laid. It was way beyond that. But whatever it was, it was freakin' stupefying. She just suddenly became this like bastion of humble intensity. ... and freakin' hot. ... like how a lot of older women will just slowly become super attractive because of their competence. ... except Rosie did it all in like a twenty four period somehow.

But whatever the case was, Rosie suddenly got hot. It was almost like that sexy school teacher effect. She could just walk right into the room and get everyone all hot under the collar. But it was still Rosie though, so it just made you feel all dirty inside.

But then the dirtiness made it even hotter.

So, you just had this mental image of the two of you goin' at it... and you were like "oh, this is filthy!"

Anyway. That's the way it goes. That's the bewitching power of authentic confidence.

Oh, and you can't really fake these things either - especially if you're a guy. Women have been finely tuned over thousands of millennia to systematically sort you out. If she ever misses something like a red flag, then it's only because she allowed it. She deliberately ignored her intuition because of some superficial infatuation or wishful thinking - it's not because you actually outwitted her with a mind game or something. Mind games are her realm. You're just a tourist there.

There are a couple of points in a male's development where our brains actually get washed over by a wave of testosterone, and it all but deadens us to a lot of little emotional cues. It's a lot like we lost a major portion of our sense of emotional smell or something. But she didn't. She's like the emotional version of a bloodhound - one that can smell that Gary was over in your house a few days ago, wearing a pair of pants he hasn't washed in a week... ... and he's also been infected with measles, mumps, AND rubella.

She will definitely know if you've got the trident of Cs - competitiveness, competence, and confidence. And the more you've got it naturally engrained in you, the more it'll be like catnip that drives her daffy.

The three Cs were actually pre-humanity's answer to the whole "bad boy" appeal. The whole bad boy thing is far older. It goes all the way back to the jungle - LONG before our ancestors ever got around to setting up shop on the beach zone. The whole bad boy thing is just that classic alpha archetype. It's just that age old thing of the aggressive guy who would beat people up. Females have been attracted to that forever.

That's also why the attraction to the bad boy doesn't really seem to be affected by the guy's financial situation. He's usually just some hunk with a really big attitude and an evident lack of prospects. So, it's always been baffling to a lot of people just precisely why a woman would be so attracted to that. ... Well, it's just that primitive old alpha thing. It makes her *feel* something. She *wants* to feel something. She doesn't want to be bored to death.

But an alpha never had to be a genius, of course - they just had to be an aggressive physical specimen. And yet, the fates would eventually decide to choose another path. They would raise up a standard against it, to now contend with the reigning alpha titans. And lo, the realm of Poseidon's blue ocean was thence appointed to bring forth its champion... - the c-man.

The c-man stepped onto the scene, ready to stare down the competition. He had the wit and he had the grit to exploit the calculated risks. And he was more than just confident and competitive, he was also now *competent*. That's *a hell of a* combination. "A science puke in a recon marine's body... ... 'nough to give me the goosewhillies."-

But this is precisely why so many people love the archetype of the James Bond character. He's not just dashing and daring, he's also brilliant. He was nature's final answer to the alpa male - the c-man... born of the sea.

That's a bit ironic too, since the term c-man also means "cool man," ... like a cool cat or a smooth operator.

Well, perfect! Then that settles it. That means all we need to do is become a c-man. And then that'll be enough to make women happy. Nice. Sounds easy enough.

I didn't realize it was gonna be this easy to make women happy. :D

Oh, wait.

Hang on.

Hang on a minute.

Wait, wait, wait, wait, wait, wait, wait. Hang on a minute. It's probably not that easy, is it? ... This IS women we're talking about here.

No no no no no no no. Why would they make it this easy? ... It's gotta be a trap. ... Why would they EVER wanna make it that simple?!

Yeah, it's never gonna be that simple with women. Not even close.

There's actually an old bit by Brian McFadden about all the different things that a woman wants in a man. It goes through this whole list of attributes that women instinctively look for in a man, and it's just this huge list of completely contradictory things, like how they want a man who's really tough but also sensitive... and very romantic but also realistic. ... Then, you finally get through the list and you boil it all down, and it arrives at the conclusion that *what* women really want is a gay football coach. ...

325 ~ BARTROLOMEW MC INNCEL

Brian gets it.

Well, there's actually a really good reason for why they do this. Women wouldn't just be experiencing this underlying impulse to go out and find an actual, real-life, nonfiction guy, who is, in all literal and unhyperbolic meanings of the term, *unironically ironic*, if they didn't really have a good excuse for it. They wouldn't just be expecting these contradictions for no good reason. ... ... *would they*, gals?

Well, the reason that they want an actual, anthropomorphized oxymoron as a partner, stems from the fact that our pair-bonding forager men actually have *two* puberties. Yep. That's right. Men actually have *two* puberties. The second one kicks in around our early to middle thirties, or a few years just before or just beyond that. It varies from individual to individual.

But when this roughly early-midlife, second puberty kicks in, it effectively turns on a much more fatherly and nurturing side of a man. So, you've got this wild man of youth who has now been tempered. This is nature saying, okay, we've been out here doing this whole workaholic, party life thing for a while, and we've been burning the midnight oil from both ends, so now it's time to start simmering down and creating a family.

And this is it. This is the magic moment. This is when a man becomes a living embodiment of antithetical characteristics. This is when a man becomes a living, breathing contradiction. And *this* is what the instincts of women want. ... Girls don't wanna have fun - they want everything. They want a wild man and a party guy and a responsible father. They want a listener and a compromiser and a decisive leader. They want a nurturer and a lover and a fighter. And they want a reliable and consistent provider who's not predictable or boring. ... "Fun" was not *merely* the beginning!

It's our own fault though. We've been showering them with resources for literally several million years at this point, so now they just expect it constantly.

Well, if you wanna go and place an order like that, then it's gonna take some doin' by mother nature. You don't just suddenly become a man who can make meaningful contributions to interior design decisions. You don't just walk in off the streets and become a guy who can simultaneously roll up both his shirt sleeves past his forearms, while also combing his fingers through his hair with his massive biceps, while also kneeling down to help a small child repair her broken doll house, while also riding a shimmering stallion without his shirt on. ... ... We're gonna need some time to get our stuff together.

This is why women mature much faster than men do. Men have this whole entire period where they're essentially in a preparation mode to become a family man. This is all because of that arms race to become a better sugar daddy. Nature was like, well, we don't need maturity yet – we just need resources. And so, younger men especially, are built to go go go... - go get resources; go get knowledge; go get a reputation.

And this is why younger men are still going through so many changes. Their personalities are changing; their looks are changing; their bodies are changing; their preferences are changing... a lot of stuff is changing. ... – nature hasn't finished with their identities yet.

So, this is just something to be aware of for all of the really young adults out there. If you're a guy, the things you think you want can dramatically change over this period. And if you lock yourself down with too many shortsighted life commitments, then you might be lookin' at a ye ol' midlife crisis at some point.

... We see these things in our modern era like skyrocketing divorce rates, and surely these are complicated issues, but disregarding the most fundamental aspects of our evolutionary psychology is indubitably not the least of our self-sabotage.

And it's the same thing for a really young woman. If you're looking for a guy to last a lifetime, then just know that a really young man is technically not at your level of stability yet. He's almost like a child that's psychologically a whole lot younger than you. Not that that's a bad thing or a good thing. It's just a naturally unavoidable feature of being such an intelligent species.

However, this isn't quite the end of all things puppy love, my fair lady, if you've already got your heart set on that really young whippersnapper. There are just *two* important things to simply know about. First of all, you wanna be very considerate, and just help him out with lots of little simple things. It's the simple things in life that make or break us. And secondly, you don't wanna be hassling him all the time. Just treat him like an actual human partner. Just be real with him. You're supposed to be partners that equally share a common vision.

It's honestly a pretty simple equation with a man. Men have simple needs. Doing these simple things will almost invariably lock in a more mature guy, and it'll definitely help your chances with a really really young man.

Usually, men tend to feel like they're in a battle for survival. They've gotta fight for the rest of their lives just to earn a steady living. And they've gotta do it a whole better than the next guy, or you won't even pick 'm as a worthy partner in the first place. And not only are they being judged on their physical attractiveness, but they are also being judged on their provider skills. And they can't just *switch* horses halfway through the race if things aren't going well - they *ARE the horse!* ... So, when this moment just amazingly occurs when a man can get some help with his daily toils, then it really could not feel much more delightful. These little favors are like an oasis in a desert.

One time I had this crazy Eastern European roommate girl who would just randomly do these little helpful chores and things. She didn't even have any ulterior motives - she just instinctively couldn't help herself. She would do things like go into my room without asking and then take all of my laundry and sheets and go throw it in the wash and then get all mad that I didn't leave my stuff in a more organized condition for her.

I was like "why are you getting mad? I didn't even ask you to take my stuff." But then she would just get even more mad. She was always barking orders at me, even though she had absolutely no authority whatsoever to be telling me to do anything. But then her favors were just *so* helpful that I eventually capitulated. I just gave in like a weak submissive prisoner.

That's how much guys like little favors. I swallowed a lot of pride back then. And I did it all for little favors.

I had to draw the line somewhere though. She would see me concentrating sometimes and doing this thing where I would twirl my beard hair with one of my fingers, and it would be really annoying to her, so she would just try to order me with her brutal accent, "you must stop thees. What you are doing? Quit fingering yourself."

I was like, "Uhhhhh... I'm a guy... That's im*POSSible!*"

So, that's the ticket. That's the key to a good man's heart. It's basically just being helpful with lots of small stuff... and then showing him the honor of dignity. That's not too bad. And if you simply choose a guy who shares your vision, and one whose natural traits are fairly compatible with yours, then these behaviors will inherently come natural to you.

You definitely want a partner who shares your vision. And what I mean by that is, you essentially want a partner who is *uncannily* simpatico with you... such that you feel like it is you against the world together. ... like you would rather take their side over literally anyone else's. ... like you could talk and talk for hours and still keep talking if you really wanted to. ... And if you don't quite feel that way about a person, or they don't quite feel that way about you either, then just know that any future with this person, might end up feeling much more like a business arrangement.

And don't worry if you're an odd duck or a strange one – the gene pool makes someone for everyone. Just keep on being true to your strangeness. There are definitely other people who match your crazy, as they say.

And like I said earlier, our instincts for forming relationships are still pretty viable now, but they just need a little help from our modern wisdom. They still believe that our populations are homogenous tribal units. They still believe that our core belief system is the same for everyone around us. And so, we don't have a very strong instinct for wanting to check up on a suitor's core belief system. This is where our instincts sort of fail us now. So, we gotta take the time to get to know people.

We gotta keep it casual until we know someone. Rushing through the courtship process is no longer an option. ... Just because it feels right, doesn't mean it is. And that's just what we have to do and teach our children now. ... unless we wanna start living in tribes again that are genetically and philosophically homogenous... where everyone's basically got a very compatible temperament... and where everyone shares the same exact belief system... and where everyone minds their manners because everyone knows everyone.

Alright. So, that's it. That's the basic enchantment for endearing a man to you. Just be sure to use it wisely. We don't want your fairytale turning into a fever dream.

A love potion, in and of itself, is just a magic trick. It is not true love. And anything untrue will not endure. That's why the word "spell" also means "for only a short period of time." For example, you can always arouse the interest of a love crush, if you can impress upon their mind that other parties are very interested in you, but alas, this is only a trick of the mind, and thus it will not last. There is simply no substitution for authentic romance. You must employ a much more powerful enchantment – one that actually affects the hearts of both you and your love interest.

And besides, if you ever try to fake authentic romance, you're essentially just holding your breath. And this is clearly not sustainable. You can only hold it in for so many seconds. You're just gonna end up turning blue and eventually blowing it. And then your partner's gonna be like, "I don't even know who you ARE anymore!" ... But, of course, they never knew... because you never showed them. You were too busy turning into a blueberry.

So, you're always gonna end up revealing everything. It's just that one way of doing it is perfectly sustainable, and the other way of doing it is perfectly disastrous.

The more open you are, the more relatable you are, and the more attractive and inviting your personality becomes.

If you want your special someone to grow much closer to you, then simply being your authentic self is just the ticket. People can sense it, and they like it, and they respond to it. It feels a lot safer when you know what you're dealing with. And if they decide that your authentic personality is just not their cup of jam, then you just saved yourself some trouble like you wouldn't believe it. ... So, it also doubles as a not-wasting-a-ton-of-your-time elixir.

This is also why we enjoy a lot of art and music. And this is ironically why we often end up bonding with certain people that we're fighting with. The great artists will instinctively pour their hearts out. And the conflicts will expose a person's character. In both situations you're getting more honesty.

And still another great effect of authenticity is that it enables you to stop taking yourself so seriously. And this makes things a lot more fun and easy going. There's a reason why people like happy endings, and why we love the types of music that either are more upbeat or liberating. We all yearn to be free and to just enjoy ourselves. ...

We all of us originated from a tiny little happy person. We just wanted to throw food at adults and play peekaboo. So, if you can offer a would-be partner that cool drink of water, then you can easily become an oasis in a desert of boring egos.

And there you have it. These are some of the most powerful, romantic enchantments for becoming a fount of enticement. If you wish to be seen as more alluring, then just unbridle the more honest and carefree side of yourself. ... And you need not become a paragon of brilliance. Just have a little fun and be less guarded. ... – but not too unguarded. Everyone still loves a good mystery. ... There's a balance. ... And seeking it... is what makes life... more interesting.

There are plenty of other habits that can make you more appealing, like the fact that just giving your partner a chance to be *heard* is a very important component of a good relationship. But nearly all of these things will come naturally when you simply follow the ancient rhythms of our prehistory.

Wait, what's that you said? ... You want *more*? ... You want something *stronger*?!

Are *you sure*?

I don't know... I don't if you're ready for the next level. ... Let's just leave this one alone, okay? ... I don't think you could even *handle* the next level of'fine'whatever'I'll'tell'you.

I hope you're ready for this. I'm about to introduce you to the mother of all love potions. Follow me. Right this way. We have to keep it in the back over here in case any kids try to... okay, I found it.

BEHOLD... I hold before you... the KEY... to love's deepest intrigue.

What thine eyes doth witness verily... is a spell so *powerful*... that it will test you... It will demand of you... and it will TAKE of you... But alas... if ye be so intrepid... ye may yet discover... the very quint*ESSENCE*... ... *of'love'itself*.

Hence, if ye intend to invoke the incantation of interminable intrigue, and ye endeavor to engender the enchantment of enduring entice-ment, then ye are going to need more than just parlor tricks. Ye are going to need the old way's high magic.

Yeee...

True love is but the inescapable adoration of inner beauty. We cannot help but fall in love with honest virtue. Compassion is attractive. Courage is compelling. Levity is inviting. And wisdom is enticing. But if you had to place it all within a touchstone, then one might simply say, benevolence is beautiful.

""""

Much ado has gone askew regarding the virtues, and the great debates have so long been waged on how best to attain them, but immersion in kindness will always serve to remind us what is supremely fulfilling, and immaculate courage will always seek out true wisdom when it is guided by levity.

"*"

Yea... ... ...

Lo... ... ...

VerILY... ... ...

AVAST!!! ... ...

~'*'~

When we see a thing's beauty, we thus adore it. This process is unintentional and automatic. And as we come to understand much more about it, we will see wherein within it resides its value.

~,*~

So, understanding unveils a deeper inner quality, and then this character, at once instills, a sense of wonder. Therefore, the key to ever truer states of fondness is yet uncovered by every richer shades of knowledge.

In other words...

Learning is the pathway to loving.

That's why light means both compassion and greater knowledge.

'''' ''*'' _'*'_ ~ _~,*_ ~

Fear is from a lack of understanding...

And love is from beholding inner beauty.

And this love - it does not dwindle or expire. To the contrary, it far exceeds its invocation. So, if your muse has thus beheld your inner beauty, then this effect, it will not wane, for many ages. ... ...

Abracadabra /~~*

And not only does this charm transcend the facile, but its delights are thus assured to quite verily everyone. Thus, not only is this not denied to the world of muggbloods, but they may use it as they see fit for spells and potions.

Yes, it is true. One does not exclusively need to hail from magic in order to use it. ... I mean, it helps. Certainly. But, strictly speaking, it isn't entirely necessary.

Yep... even those even those dirty dirty muggbloods can use it.

I know, I know... believe me, I know.

Hence.

A preponderance of muggbloods would disagree though. They would argue that this virtue is purely philosophical... and that nary was this a source of magic proper. ...

Ohhhhhhhhhhhhh, muggbloods...

My dearest deary muggbloods...

Verily... Like a sir... I must admit... you do amuse me so...

... with your silly silly words.

Muggbloods...

Indeed, when one doth gaze into the eyes of a goodly companion... and there beholds the purest fount of inner beauty... then not only is this not devoid of magic proper... but quite indubitably, there was nothing ever more so.

And so it is.

Alright. So, what happens next. What happens when you've finally found that lady or gentleman of your dreams? Or I guess, what happens when you've finally found that younger gentleman of your dreams? ... since we're still talking about a young lady who's going after a younger gentleman. ... – a pre-second-pubescent gentleman, if you will.

Well, the next thing you've gotta do for this young man is to start having lots and lots of... uhhhh... well, I'm not gonna say it, but you already know what it is... and he needs a lot of it...

Like you didn't know that.

But it's definitely worth pointing this out, though, because it's a really important concept if your goal is to snatch a younger guy. "Snatch" as in, like, to grab, with your hands. ... not like... ... *You'know'I'mean!*

This young man is specifically programmed to be capable of reconstituting an entire village by himself. ... – single handedly... in a manner of speaking. So, if you seriously intend to consume his entire attention, then you're gonna have to keep him pretty busy.

Like you didn't know that.

There's just one little caveat to this. And we have to preface it first with some biology.

The legacy of the animal that we are, is one of a lifeform in motion, but despite all these never-ending changes, there have been a few constants which have not faltered. One such undisrupted ancient feature is that our men have thus been treated as very expendable. So, there have often been these disproportionate imbalances, such that the women have greatly outnumbered all the menfolk. And then this has caused some awkward situations, whereby the women have had to share the same provider males every once in a while. And then this has kept alive that ancient feature, whereby the menfolk just don't bond from having intercourse.

They don't bond from having sex, on an emotional level, because this would never work with multiple partners. Such has been the legacy male expendability.

Now, as you might expect, and per the original question, the story of female bonding is a bit more complicated.

'Course it is.

Seven or so million years ago, our ancestors came across a beach, and as soon as they located the snack bar, they just settled in and started getting comfortable. Their bodies started changing and adapting, but their minds had not yet left the ancient jungles. They still had reigning alphas for a season, and the women just all mated with whichever one was dominant at the moment.

And this is where our women got this instinct, where nothing less than the best was ever fit to be their sire. And then even when our genus started pair-bonding, they still would choose their favorite man from the ones who were propositioning them.

And it's better for our genus that they've had these instincts. This is how we've stepped up our intelligence. But now this hypergamous instinct is getting hijacked, such that women are often wasting time on playboys. They're going after men of higher status, because of looks or because of wealth or because of power, and then they're sleeping with these men to create a bond with them, but that's simply not how these men are built to bond with you.

And men, of course, don't have these lofty standards. To a man, the average woman is entirely remarkable. A man isn't trying to bag some alpha female. He's just designed to enjoy them regardless, in whatever ways they're special and enjoyable.

So, this is why men of less scruples can just sleep around all over without a problem. And there are two types of men with less scruples – some who are higher in social status, and some who are yet to hit second puberty. Because when a man has finally hit his second puberty, he instinctively wants to stop, and start a family. But before he's actually hit that second milestone, he's still ready to jump at any opportunity.

So, this is basically the final caution for a really really young guy. Not only are we a lot less mature than our female counterparts at that age, and not only are we evolving our personalities still, but we also, much more readily, could become promiscuous.

This isn't to say, though, that we couldn't ever settle down and start a family yet. And this isn't to say, correspondingly, that a young lady couldn't ever lock down a super young guy. This is only to state, quite emphatically, that we might not feel the same about those decisions later. So, we just need to move ahead with all due caution.

A relationship that falls apart can be devastating. That is why we really want to make a good decision. We want to build our bonds based on our virtues. That is the only way to forge *authentic loyalty*.

A woman can also build her bonds from intercourse. ... generally speaking. And we're slowly building up to why that happens. But as we've said, this also doesn't affect a man the same way. For a man, it's just a high, like any drug. We generally don't fall in love with our neighborhood drug dealers. Nor do we ever establish any exclusive, brand loyalty with them. This is often a very difficult concept to really internalize for many women. It's never easy to understand it, when other people just don't feel the same way you do about something.

It's a very very different sort of experience for a man and a woman. As a matter of fact, I'm sure the term "love making" was originally mused from a woman - that's essentially how they feel about it, when it goes right. But a guy would never say something like that of his own accord. ... Simple cave words like "pork" and "****" will suffice quite nicely - thank you very much.

So, if a woman wants a man of any status, then the way to lock it up is not with intercourse, quite frankly. And no matter what you might presume about yourself right now, the day will eventually come when you insist on being treated like an actual thinking person. ... It *will* come. ... But if you've already established that precedent of simply *not* being respected like that, then he just plainly will never change the way he sees you - not for anything less than a miracle. He won't be able to. We have to establish our relationship boundaries from the very beginning.

And don't try to plan *LITERal* miracles into your life. Almost everyone does it. And it doesn't end well.

"Yeah, I'm gonna grow up, and then meet someone, and then miracles will happen, and I'll get everything I want." ...

And then we get all baffled when our life doesn't turn out the way we "planned" it. ...

"*What?!?!?!* ... ... *Why is this happening to me?!?!?!* ... ... I clearly wrote down that a *mIRACLE is supposed to happen at this point!!!!!*"

But that's the ol' paleo brain in our head still thinking this is the beach-tribe world. ... – "grow up; meet someone; miracles happen; everything works out."

We've gotta plan a little better than that nowadays. ... – just a little. ... You've gotta tell that dreamy steamy hot boy, "*you shall not pass*." Because here's a bit of wisdom that's been largely forgotten from the world. We are *suPPOSED* to be held to a standard. We are *supposed* to be expected to behave ourselves. That is how we established order before civilization came along. Everyone just expected everyone else to be chill. Because, if you *didn't*, then everyone else would know about it... and it would damage your reputation.

That's why weddings were also conducted in front of the entire community. It was the *community* that would ever after hold you to your wedding vows – not "the law" – there was no "law." The law was just your sacred reputation. The law was just your honor and your credit. And if you wanted to switch it up and change communities, then the people from your new community could still be asking around after you in your old community. Either that or you could be on a "suspicious newcomer" probation for the rest of your life. ... Reputation was key. ... ... But I digress.

The point is that we are *genetically* adapted to keep track of one another's reputations. We are *built* to be held to a standard. It makes us stronger. We need the pushback. Truly. We thrive under the pressure of a healthy expectation. And then we feel a sense of deep holistic fulfillment.

It's like trying to take a kid somewhere, when they're just complaining about everything, but then you finally get'm out there and they don't wanna leave the place. Well, that's what a healthy relationship is. It's the fun place. And you don't have to leave it... ever. You can just stay there in the waterpark overnight. But you have to make the road trip first to get there.

And the road trip is the healthy bonding based on virtue. ... – virtue; and quality time; and shared experiences; and inside jokes; and sharing a set of common immutable values; ... and especially, sharing a common vision for your future. That's the road trip. And it's fun. ... And then once you finally get there, you can enjoy the waterpark.

But if a partner doesn't wanna take the road trip, and they're just trying to jump straight into the wet stuff, but then they break it off completely because you're not willing to use the slip 'n' slide in front of their parent's house... well... then you just saved yourself a whole bunch of trouble later. ... – a WHOLE bunch. ...

We're supposed to be held to a standard. Did I mention that already? We are supposed to have a little bit of discipline. We are a highly intelligent, pair-bonding, forager animal – not some orgy of weirdo bonobos. We are a dolphinized stork ape with smarty pants. We are a raven legged, slithery skinned, puff body, with a pecker like a griffin and a wizard's brain! We are not some feral, uninhibited, *mOUNTAIn troll*!!

We're tough!! We're resilient! We're survivors. We survived our child-hood. We survived our school work. We survived the dark forest. We're tough as nails. We're tough as woodpecker lips. We can handle a few years of dignified courtship.

Medieval knights used to act like a bunch of barnyard animals back in the day - back before the high Middle Ages – even before the mid-dle Middle Ages. That is, they acted like a bunch of donkeys until women like Eleanor of Aquitaine showed up – women who would no longer abide it. They insisted that these knights behave with decency... at least in the presence of a lady.

And the results were so amazing it "spread like wildflowers." Now everyone in the Western World is all about it. We don't just go around acting all slovenly anymore. No. Now we have our standards of eti-quette. So, we can thank our lucky stars for these women like Eleanor.

These women had the ancient coastal potions – a pinch of holding hopefuls to a standard; with a sprig of making sure they share your values; and then a clover of using patience to prove their sincerity; and then your lover will ever after preserve your standards. ... ...

Alakazam /~~*

In fact, this is basically where our language originally came from – women just sort of gossiping about their suitors and sorting them out – sharing information that wasn't plainly visible at the moment. ... And who better to have invented talking than a teenage girl. ... I mean, really. ... How could it *possibly* not have come from a teenage girl. I mean, f'r serial, how could it *even?!*

But ladies weren't always so chit chatty though. Nope. Not in the ol' jungle days. ... back before we got all svelte and beach'ified. They used to be quiet. They used to be rugged. They used to be bro-chicks. ...

And these ladies didn't care about their intercourse back then. They were just as cool and casual as any male about it. They could copulate with any number of dominant alphas over the years. It didn't matter. There wasn't any special emotional bonding taking place. They were like bro chicks. They kuh care less.

So, what happened? What changed all this? What exactly was the biggest psychological catalyst? How in the world did these rough and tumble bro chicks, who couldn't even kuh care less about romance, end up turning into these classy lassies who think up fancy phrases like "love making?"

Well...

... nerds happened.

Yes, indeed. As we discussed before, the original Adam of our entire genus was a bit of a nerd. He was a clever little laddie, compared to most people. And he wasn't all that large compared to an alpha. So, getting in brutal fist fights just wasn't a winning strategy for him. Rather, he was free to do whatever.

And when you're free to do whatever, you've got some options. Sometimes we choose an option because it's simpler. And other times we choose an option because it's funner. But sometimes we just have no other option. Sometimes the only option is the hard one. And that's how we know that Adam wasn't the biggest guy around. Because if Adam had a hope of ever dominating, then he would've just been a fighter like all the other jocks.

But this would've wasted his talents. This would've misappropriated his intelligence. ... because primitive fighting is essentially just decided by raw ferocity – not tiny little jumps in a fighter's intelligence.

... because if this *wasn't* the case, then pretty much every other animal on the planet would be just as smart as we are. Their slightly smarter males would always dominate. Then they would always become the alphas and drive up their intelligence. And yet, this is *clearly* not what happens.

There's no debating this point. Adam was not a super jock. He was a bit of a nerd. ... Straight up... flat out... triple stampy... no erasies."~

Only a talented lad with nothing but time on his hands would have even attempted to actually feed another grown-up like that.

Maybe he wasn't the first though. Maybe his friend Drew tried it before him. And if that's true, then we owe Drew some praise too. But it doesn't matter though. Adam stuck with it. He got rejected, over and over and over. But he didn't stop. Maybe his best gal kept giggling every time he brought her some gifts. Maybe that's what kept him going. It doesn't matter. What matters is that he kept going.

And then one day... something... revolutionary happened... - she said... yes.

And then Adam got laid. Then he told his friends. But they didn't believe him. But then Drew actually saw it, and it freakin' blew his freakin' mind. Then he ran back to their other beta male friends and was like, "you guys! You guys! You guys! You guys! You guys! ... Adam just got FREAKIN' laid, dude! Like, he LITERALLY just got LIT-

ERALLY laid in front of my own two eyeballs! DUDE! I'm not even playing around right now. Adam just got luh'luh'luh'luh'luh'luh'luh'luh LAAAIIID, soooooooooooon!"

So, Adam used his gifts instead of his fists. He found a way. He discovered you could also be resourceful – you could win a woman over with a box of chocolates. Or, I guess in his case, it was more like a banana leaf full of fruits and nuts and oysters. And maybe he also brought her some tropical flowers. Who knows. The story is probably lost to this dimension. But whatever it was, there is *one* thing about it that we *do* know – *it...* was impressive.

Whatever Adam did was *pretty* seductive. He actually got her to go against her basic programming. She went with a nerd instead of a jock, which was completely against her instincts. So, this was nothing short of a total revolution.

And this is actually quite fascinating also, because Adam was the son of an alpha male. So, how did he end up with so much less body mass? He must've gotten it from his mama, or one of his grand-muh-mas. ... because there's basically *no way* you could be passing on these diminutive characteristics through the male lineages – only alphas got to breed. It had to coming from the female lineages.

And it makes sense that these diminutive foremothers would've been clever. They would've been relatively lower down in the female social hierarchy. So, they would've needed a special talent to help them compensate. They would've needed a predilection for cunning.

And then, at long last, they would've brought forth an exceptional boy child. This was Adam - the son of many talented women. The Kwissatz Hadderach.

Forthwith, through their line, the last would become the first; the meek would become the strong; and the genetics for enhancing their intelligence would come to dominate.

And so it began - the age of oceanic Romeos winning over their coastal Juliets with gift baskets.

And then the longer that these partners stayed together, the more ef-ficient, by orders of magnitude, their whole entire lives became. The women wouldn't have to keep on dating every year - they could fo-cus on their lives and on their children. They could keep their favorite providers who were already very accommodating to them. And then these partners could work together to raise their children. It was lo-gistically, exponentially more effective. ...

And so it was, dedication became adaptive; commitment became competitive; and fidelity became effective. If the men weren't being faithful, then their children would end up getting less resources, and then this would disadvantage them considerably. And if the women weren't being committed, then the males would end up raising an-other man's children. So, this is what weeded out the gene pool. This is what selected us for committed pair-bonding.

So, we had alphas that ruled with violence and intimidation, but then these fighters turned into fathers who were extremely protective. We had kingpins who had a monopoly on literally all of the women, but then these leaders evolved into partners who were enormously sup-portive. And we had women who would switch up their loyalties for a bigger badder chieftain, but then these ladies started getting their own men and started bonding quite exclusively.

And thus, what emerges as a dull and cloudy nebula of highly atom-ized individuals, slowly condenses into a star field of nuclear families.

And then these families will just loosely associate like a brilliant constellation. This is generally what is referred to as a "fission-fusion society," where the group is just a confederated union of many smaller little family units.

That is quite a grand progression from our original Adam. It's amazing to think that this entire transformation was already completed roughly four million years ago. In fact, we've been going at it for so long now, that this long-term pair bonding has long since become one of the most salient and defining characteristics of our genus.

That is a looooooooong looooooooong time to be pair bonding.

Thus, we are quite thoroughly bred at this point for pair-bonding. The engine of our meteoric success has been the efforts of these power couples. They have been charging right ahead, to the fore, for at least four million years now – fore for four. It's the rule of "fore for four." Not really. There's no rule. But our instincts for establishing these pair-bonds, now define us as much as anything.

Of course, there have always been the notable exceptions. Not everyone is necessarily meant for these classical pair-bonds. Nature doesn't always make everything one-size-fits-all. She likes to hedge her bets a little. So, some individuals are just generally better suited for assisting their family's genetics in a more tangential capacity. Upon being liberated to do so, they can fulfill their biological purpose as a benefactor, for example.

And, not only that, but, sadly to say, if a person has ever sustained enough shellshock, then they can easily start to slip into the old alpha paradigm. And they can also develop a sort of compensatory, consoling compulsion for that matter. So, for these and other reasons, not every single person will end up doing the whole pair-bonding thing.

Nor would it be productive for them to do so. Things would actually get pretty chaotic from a genetics standpoint if that were to ever happen. And there are *other ways* to project your immortal essence.

But nevertheless, these traditional pair-bonds have long since been our selective workhorses. They have been our primary engines of ascension – see the fore-for-four rule. So, it should come as no surprise that the actual "bonding" part of this equation is of paramount importance.

There has needed to be something which binds us - something which is *unique* to that person; something, for a man, beyond intercourse; and something, for a woman, beyond resources. ... We've had a need for something singular and intimate.

Lucky for us, we've got these diamonds in spades. The door to our hearts is through virtue. While the scepter which unites our purpose is indeed our values. Then the crown of our mutual interests is at once our children. And at last, the *swans* of a woman's *passions* are intertwined by *intimacy*.

Women experience bonding through intimacy. They develop deep personal feelings from the act of lovemaking. Then these feelings inspire a woman to be a lot more supportive. And then this supportiveness inspires these feelings in her pair-bonded partner. It's a bonding chain reaction.

Supportiveness, to a man, is what lovemaking is to a woman... just as resources, for a woman, are what intercourse is for a man. ... One is in essence more sensual, while the other is in general more emotional.

He gets his jollies from lady curves, while *she* gets her giddies from shiny objects. And then she gets her bonding from lovemaking, while he gets all emotional from sandwich making.

Generally.

Generally speaking.

Generally but not invariably.

You get the idea.

So, this is essentially the answer to why our women have such a powerful bonding mechanism. It kicks off a chain of occurrences which thereupon ingratiates a bonded couple to one another. This then intertwines them emotionally, and then that becomes the framework for a family. If a man is a steady foundation, then a woman is a powerful framework.

And, of course, these families are what select us for an elevated intelligence. So, our ascension to these atmospheric altitudes is all contingent upon her bonding from intimacy. It's all dependent upon her grasping to this soaring pegasus.

So, that is the ancestral concoction for a lifetime of romance – since negative 1892 thousand – balancing open lightheartedness with sexy mysteriousness; having confidence and competence and competitiveness, and a set of common values; fostering courage and kindness and levity as your core of higher virtue; establishing your well-defined standards from the very beginning; spending lots of time and having fun together; taking on a range of different challenges together; taking sober stock of your actual chemistry; committing when you finally know their nature; making sweet sweet love and being supportive; and then enjoying yourself for decades, ever mindful of what is truly fulfilling.

True, not every couple has wisely done this. Indeed, far fewer have than those who haven't. But it isn't quite the end of all the world yet. There are still some love elixirs for this predicament.

When your partner reaches out with some emotion - whether they're happy or they're laughing or upset – just make sure that you engage them with real attention, and never leave them feeling put out or neglected. Then look for things they do which are curious or special, or even try to find some labor that they take a little pride in. And then just give them a little praise as if in passing. That encouragement may echo inside of them for the rest of their lifetime.

Once you've done these things and spared some effort, just step back in quiet wonder and be amazed. Watch as they are fueled and driven forward. And witness how their spirits begin to elevate.

Finally, if you can manage, then there is yet one more thing. But you'll have to take some time to search your soul. Remember this fellow person is now your counterpart. They are your ally. They are your partner. They are your own. You may as well just feel some pride in having them. They may not be the moon, but at least they're yours. When they sense that you are standing tall beside them, it is sure to move them deeply and strike a chord.

There you have it - acknowledging emotional bids, giving little compliments, and just generally being proud of what you've got. That's the ticket. And if that's just not enough to stoke the flame again, then nothing will... nothing except cash, body modifications, and alcohol. But that's IT! ... ONLY those three things besides what I said.

You can always go either way with that one. I'm more of a natural guy, myself. And there ain't nothin' more exquisitely natural than some sweaty sweaty love makin'. Just fire up that kettle and go make yourself a nice warm cup of *sweet sweet* sweaty sweaty love makin'.

But don't make it too sweet though. Beneath it all we're still a primal animal, and people just want their partners to be obsessed with them sexually. We all just want our physical bodies to be the objects of passion. And we want them to want us voraciously, like we're some food and they've been starving.

... - like our body is some bread and they've been locked up in a dungeon... and now our body is the first real food they've seen in ages. ... so, they just start tearing in and gobbling up everything like a dirty ragged prisoner.

And that's not *all* people want. They also wanna hear about it. They wanna hear you say it. They wanna know *exactly* why their body is so compelling to you... and *exactly* what you plan to greedily do with it. ... Well... unless you're a fancy Victorian lady. ... Fancy Victorian ladies are different. They would never even dREEEEAAAAAM of such impropriety. ... ... except that they would. ... They would dream about it all night long.

But they'll just deny it though. They're never gonna admit that they're some freak'u'leaks. And they're not supposed to. It's not her job to broach the this topic. It's *your* job... boyuh! ... It's only her job to say, "no." ... And it's your job to say, "no mean yes." ... And it's her job to say, "eww, get away from me, you perve!" ... And it's your job to say, "mmm... that's right... I'm down for whatever." ... And it's her job to say, "hey, Chad!! Can you get over here? This *freak* won't leave me alone..." And it's your job to say, "okay, I think I misread that. ... Good evening, madam..."

And then you do a little curtsy twirl with your hand or something.

But it's still your job to go for it.

Boyuhhh...

And she just might come hit you up later when she's not embarrassed anymore.

But that being said, that previous sex scene is obviously a very primal one. ... – see prisoner-versus-bread. ... So, this is not some classical formal tea party. This is when the most ravenously rapacious part of you needs to come out. This is no time to be ladies and gentlemen, ladies and gentlemen.

You need to leave your courtly Victorian accoutrements at the door - if not for your own sake, then at least for your partner's. Many people are so environmentally conditioned to be formal that they carry those inhibitions right into the bedroom. Either that or they just end up needing to use alcohol to express their fun side. However, formal rigidity should not be bleeding its way into our innermost private lives. If our conditioning has pushed it to that point, then we've lost plot.

For example, let's say there's a couple on a date together, and now let's say that the guy is acting all timid and submissive. Well, in this scenario, the girl can tell already what his private side is going to be like. She knows on a subconscious level *just* what to expect from his bedroom performance. And, quite frankly, it's not exactly one of her all-time greatest hit fantasies.

On the other hand, however, if a guy is a total beast when he's in private, then this nature will be reflected when he's in public. His tells will betray this without him even knowing about it. And then her instincts will detect it automatically. And then she'll sense it as a *feeling* and she'll like it.

So, there it is. This is how certain individuals can always seem to exude an air of inexplicable sex appeal. They are embracing their innermost sex panther."~ And indeed, it is subconsciously apparent.

So, now, based on what we've been talking about, why do *you* think that the bad boys and French guys are so sexy?

Just take a moment to think about it. See if you can get it before I say it.

Hey, you're not supposed to be here yet. You're supposed to be thinking about it. Wait wait wait wait wait wait wait, go back go back go back go back go back, no no no *no no no no*, you're cheating you're cheating *you're cheating you're cheating* YOU'RE CHEATING!

Alright, fine, I'll just tell you.

... since you're actively cheating at this very moment.

Wait. Hold on. Before I tell you... I bid thee halt... and do answer, hence, these riddles five. ... Verily, doeth any of these elements pentamerous explainest why yon' bad boys and French guys are so hot right now?

Is it because bad boys and French guys are buff? Is it because rowdy chaps and Latin lads are wealthy? Is it because yobbish blokes and snobbish toads are rebels? Is it because nice guys always finish last? Is it because women like to be treated like garbage? ... ... Buff? Wealthy? Rebels? Nice guys always lose? Women want abuse?

Well, no, no... no, no... aaaaand no. It's not because of any of that. It's because they don't deny that they are an animal. They embrace their primal nature and their carnal appetites. They're not ashamed of what they are and what they're looking for. And a woman can detect this automatically. She perceives this as a feeling in her body. And a woman wants to *feel* like she's alive inside. ... She would rather *feel alive* than bland and boring – even it comes with other problems.

It's actually not a bad thing to be a nice guy. There's nothing inherently wrong with ever acting like a gentleman. It makes you seem a lot more slick and clever. But rather, niceness needs to come from a place of benevolence – not from acting like a nervous little servant.

Scurrying around and quivering are not erotic.

This is actually a source of frustration for many women. And they often have no idea why, but they can feel it deep down in their cockles. So, their instincts just tell them to start agitating you. They wanna see if you can man up like a bad boy.

It's not her fault though. It's not her fault you're not a real man.

Okay, maybe not. But if you lay it on 'er rowdy like a stallion, then she can simmer down and drift off like a damsel. She'll be floating on a cloud bank... when her instincts are finally satisfied.

And you don't have to be all over the top about embracing your inner French guy. You can express that promiscuous sentiment with just a compliment. All it takes is just a single provocative statement. You don't have to lay it on thick. In fact, it's even better if you don't. The light touch builds the tension. And so does eye contact. Tension is good. That's what makes it fun. People like fun. The world should be fun.

And it's important that you take the lead and own the naughtiness. This allows her to avoid the fear of being embarrassed. Sure, it involves the risk of be chastised, but it's also a sign of bravery, and that's erotic.

And this is what it means to be the rock. This is the essence of the c-man - the James Bond of the animal world - confident, competitive, and competent... and also carnal.

You don't really have to worry too much about this though, when you're a younger c-man. You're still out there experimenting and figuring out how the world works. And that's exactly how natured designed you. You're supposed to be out there learning and failing sometimes. This is all so we can build our skills and knowledge.

We're supposed to be out there gathering lots of data. We're supposed to be out there drinking from a firehose. And then it all will start to click when we approach our second puberty. You'll know that you're getting close when your brain starts coming up with dad jokes.

And you'll know that you're getting close *too* when you finally start to realize what the word "aesthetics" means.

Like what it *actually* means.

Ladies know.

We're not really built to grasp the concept of aesthetics when we're younger. Nor do we really care at that point. We're all about what works and what gets the job done. If we build a results-getting-machine, we're not really worried about what it looks like... or how that makes you feel inside. ... We might paint some sweet ass decals on the side, and give it some kick ass name like Laser Tits or something, but that's about it.

If the world was only dudes out there, then half of us would just be walking around all day in our bathrobes. And then the other half would just be wearing the same exact sets of shirts and pants for a month. There would be no such thing as a clothing store. You would just go down to wherever you get your power tools and pick up a box of shirts.

"Yeah, can I get fifty pack of regular, aaaaaaaaaaand le'me get uhhhhhh ten pack of front pockets."

There would just be one, *giant*, communal, washing machine. We wouldn't even care whose clothes we got back, as long as it was the right size.

"Okay, I got your forty two kilos of shirt. That's gonna be fifty two even... ... Alright, thanks Bob. See yuh next year."

And we would just have one soap for everything – doing our laundry; cleaning our driveways; rinsing our hair out. It would just be called

"Soap." ... The only difference would be what shape of a container it was in. ... "Hey Dale, I'm all outta the engine degreaser. You got any more uh dat mouth wash?"

Our refrigerators would just be a completely clear, plate glass door on the front... so we could just stand there for an hour deciding what to eat.

"Uuuuuuuuuuuuuuuuuuuh..."

Every movie would end in an epic fist fight for some reason. It wouldn't even matter what genre it was.

"Thanks for joining us on another documentary about cabin living... Please join us next week, when we talk about strategies for... Oh what the fu**?! This fu***n' bear wants to fist fight me!! Arrrrr ahhh agh ahhh..."

Hospitals would have to wrangle us down, just to treat our injuries properly, because most of us would just be walking around with duct tape bandages and improperly set wounds. ... They'd have to get all proactive about it and start rounding us up like planet of the apes.

"The horses are coming!!! ... RUUUUUUUUUN!!!"

Just snaring us in nets with weird sound-effects music in the background.

Bwomp... wyyaaaw... doo'toodoo...

Dave wouldn't be able to get away because of his injuries, but then you'd see him later walking around in a cast with some antibiotics.

"Doc said I better take these or they'll catch me again."

Welcome... to The Planet... of the Dudes...

Bwomp... wyyaaaw... *doo'toodoo...*

Yeah, guys would do well to be a little more mindful of their health status. Toughing it out has its limits. And guys should be getting their blood drawn every few years or so. We as humans should be losing a little bit of blood every once in a while. But a guy has to go out of his way to make this happen.

We actually store up certain things in our blood that can easily get up to toxic levels. Take iron for example. We have no internal mechanism for getting rid of it. So, it just builds up and builds up and build up, if we're never losing blood at all.

And this is partly because we developed such an extra long lifespan. Grandparents were becoming more helpful – they were helping out with children and continuing to forage. So, this selected us for living a whole lot longer.

We live longer than our cousins in the jungles. And this is really quite a feat since our metabolisms are so much higher than theirs. We should be keeling over *long* before they do. And yet, we don't, because of all of that former selection.

So, it's really quite important for us to be living how our bodies were intended to – eating and breathing like a beach tribe. ...

And if we're a dude, then we gotta be losing some blood every once in a while. There is simply no biological reason why we shouldn't be living for *at least* a century. ... and even getting there in amazing physical health, no less.

So, we gotta lose some blood. We gotta bring back the practice of putting leeches on people. Or you can just go and donate your blood every few years or so. Or you can schedule a therapeutic phlebotomy. They can even come and do house calls and make it simple.

Don't ask me what happens to all that blood though.

It may or may not somehow end up in an illuminati blood ritual.

I've said too much.

Anyway. Getting back... that's definitely where it ends up, shhhh. Alright, so as I was saying, we can always donate blood or have a therapeutic phlebotomy. Either way works. And there's so much other stuff that can increase our lifespans. We could go on forever about it. Sweating and fasting will detox us, and oxygen and astaxanthin will elevate our antioxidant game. And breathwork and diving will help the oxygenation. These are simple and extremely powerful modalities.

Okay. Well, how does that work? How did any of that make us live longer? And how is any of this related to the topic of romance? And why are we even talking about romance in an anthropology book? And what happens if I'm driving up to an intersection when I'm trying to get to work and then the light's starting to turn yellow in the intersection and then the car in front of me is starting to slow down, even though we have WAY more than enough of time to make it, but they just keep slowing down anyway, so I'm like *"fffuck this!!!"* and I just get up on their bumper and start shoving 'm through the intersection while they're trying to slow down?

What happens then?

... if I'm like *"fffuck this!!!! ...?"*

Uh, wow. Okay. Alright. Well, uhhhhhh that's a lot'uh question right there. Ohhhh man. Alright. I think we've been in this chapter for too long.

I think we're getting chapter fever.

That's alright.

Too much chapter is a good problem to have.

Alright, let's take it from the top.

Actually no. I'm not gonna answer any of that. Screw that. You'll just have to guess what the point of all that was. But what I will say is this. We should mention one more thing about men. Men have a very special adaptation - one which enables them to just be bachelors forever. We can party nonstop and be bachelors, or we can be shut-ins like total hermits with hyper focus.

The males of the animal kingdom are specifically designed to enjoy their bachelorhood. Some males love picking up new hobbies, and others have a million different pastimes. We got this trait about a half a billion years ago. So, now we don't obsess over dominance, and have endless deadly wars over alpha supremacy.

If we did, then the world would just be a nonstop blood bath. And this carnage would incessantly continue, until we literally couldn't fight another minute. ... - until every last survivor was just maimed and broken.

And if a tribe had ever done this and just destroyed themselves, then those last remaining males would all get conquered. And then those males would all get dispatched by those conquerors. So, clearly this obsession wasn't viable. Males would all need hobbies and a love of freedom. Males would all need pastimes and a love of bachelorhood.

And love it they did. Bachelorhood became an instant success. Even males that were successfully dominating would still look back and miss it every once in a while. ... – still look back and stare longingly, or want to join in the fun again.

This is actually one of those funny little attributes that a woman's subconscious mind is already scanning for. It's looking for those signs that a male is already perfectly happy with his life as a bachelor. Otherwise, she senses desperation. And nothing is more awkward for her than desperation. Her instincts want a male who's roaming freely. ... - one who's happily engaged in his purpose and has a mission.

And as we've stated earlier, women are much more built to be bonded emotionally. But if a *man* is just as eager, to be bonded sentimentally like that, then she finds this distinctly uncomfortable. When it's coming from the masculine, it seems needy. ... She's looking for a rock that

she can stand on – not an emotionally hyper sensitive she-male. ... ...
Generally speaking.

We are built to be more like a yin and yang symbol. Our attractions
to one another are complementary. A man is made to love a woman's
body, and a woman is made to crave a man's autonomy. ... Both our
men and women have a heart inside, and both our men and women
have strong libidos, but he should be the rock for her emotionally, and
she should be the garden that's welcoming sensually.

There's a famous old saying which sort of reflects this. It says that
men will use emotions to get to sex, and women will use their sex to
get to emotions. ... Well, that might not be the most scientifically in-
scrutable statement ever articulated, but it still sort of gets the basic
point across. ...

Surely, most people probably know by now, but in a romantic context,
the average adult male is *far* more driven by his libido than anything
else. And this is actually ironic, because it is precisely this physiolog-
ically which enables him to *NOT* need a steady-going woman. When
the compulsion for a mate is from libido, then this impulse can be "resolved" without them present. But when the attraction to a person is
from *emotions*, then this impulse demands the presence of that actual
person.

This is why so many women went totally bananas over Elvis. They had
an *emotional* attraction to The King. And that attraction wanted *that*
specific man in their lives - that hunk-a hunk-a burnin' burnin' man
in their lives. ... No, not man who simply looked like him, or moved
like him – nO, no, no, no, nO, no, no, no, nO – they wanted THAT
specific man for themselves.

And since their compulsions for The King were not just physical, then they could never just be relieved like a sexual desire. Or, to put it in a more crude and base vernacular, these women had some blue suede balls. It was an itch they couldn't scratch until they finally got literally Elvis in their lives.

They truly Beliebed, in their heart of hearts, that they were meant to be with *that* specific dream boat.

But of course, men don't really have this emotional fixation issue. They don't generally go completely bananas for just one specific girl like that. And if a girl has tripped their trigger and aroused their interest, then they can easily be refocused in a matter of minutes.

"Ooooooooooooh YEAH THAT'S good... ... Okay, what are we doin' about these expense reports? ... Tom, I want those estimates on desk *yesterday!* ... And WHO'S GOT MY COFFEE?"

So, the males of the animal kingdom have had a talent for staying focused. And that's because libido is a very easy thing to satiate. And most men are simply born to love their bachelorhoods.

And so, this is actually the *second* major reason for why the men of many species don't really bond so viscerally. It's not just simply because they needed to be polygamous every once in a while. It's also because they needed to stay *focused.* They needed to *not* be overwhelmed by romantic emotions constantly.

Otherwise, men would just never stop fighting. And there would be armies of obsessive stalkers who were even worse than the Elvis-girls. And the bachelors wouldn't be able to concentrate. ... – especially not the men in their twenties. ...

So, just think about that, the next time you find it annoying that men's libidos are so prominent. ... or that men's romantic sensibilities are much more "distant." ... I think you would greatly prefer that to the alternative. ... – namely, all of the intensity of men, and then the mania of a shrieking Elvis-girl.

Can you imagine that? ... - a bunch of dudes just going all nuts like that at a concert? ... but like with all of their strength and aggression? ... ... Yikes.

But this is also why our genus is so damn intelligent. Just like we talked about earlier, our men folk in our twenties need that window. We need that intervening interlude of inter-puberty incubation. We need to focus on our goals and build our skills up. We need an impulse for the opposite gender which can easily be alleviated. ... That's what we need to get our sugar daddy skills. That's what we need to reach the boss mode.

I think we need a new word for the men over second puberty. The only words we have are "boys" and "men." But what about the third stage? What about the guys over second puberty? I think it should be "boss-modes."

The two most important times... in a young male's life... are when a boy becomes a man... and when a man becomes a boss-mode.

"LLLLLLLLadies and geeeeeentlemen... and apparently I'm supposed to say gentle-boss-mooooooooooodes..."

Life is tough though, for us fellas... when we're still younger... when we're still growing up... when we're still not boss-modes yet... when we're still just little men. Nature sort of left us out of the whole, dating scene, back in the ol' evolution days. It sort of left us out of the whole, getting play from ladies, back in the old anthropology days. And it definitely wasn't easy to draw images on stuff back then, so we really needed to use our imaginations.

Yuh know, they still haven't figure out why our human women have tits. They're just these big, jutting, functionally unnecessary appendages most of the time. No other animal has them like that. But our species does. ... – just some big ol', obvious, easy to remember, easy to imagine tits. But it's *still* a total mystery why we have these.

Is it possible that women actually have these things for our sake? ... so we would have something to hold on to and cherish? ... like, when they're not around, I mean? ... ... Hmmmmmmmmm... ... probably not, actually. But I like where that idea is goin'. I like where your head's at.

So, why do our human women have these nonstop breasts all the time? Human females just get these fully developed breasts right out of the gate - years before they're actually able to have children properly. Then they just have these permanent, nonstop breasts, that just won't quit, regardless of whether they're actually using them or not. And if all that wasn't enough, they even keep these superfluous bad boys for the rest of their lives - decades beyond the point when they were last using them. ... for like babies and stuff.

This whole situation makes absolutely no sense though. No other animal does this.

We've been selected psychologically to simply take this stuff for granted now, but it's seriously not normal. It's about as weird as if our human men were just walking around all day with a bunch of permanent erections. Imagine that. Imagine if men were just going around all day with their crotches sticking out. We would need length support for running. And every dude would be wearing a junk bra.

It's about that level of weirdness, biologically speaking.

Oh. And, no, breasts do not exist because having bigger breasts helps you produce more milk. It doesn't work like that. Most of the actual volume in a human breast is just fat. It's not for milk production. There's no actual correlation between a woman's breasts size and its milk capacity. So, no, there were never any hominin males selecting bigger breasts to feed their babies better. There would be no reward for doing that, and if you keep wasting your limited calories on things that don't reward you, well, then nature will eventually deselect you. So, again, the theory of breasts getting bigger in order to feed babies more is just a big ol' negatory.

And this leads us to the second big misconception about breasts - no, human women were not selected to have permanent, mostly decorative breasts "because they're sexy." The logic goes the other way. We developed our instincts to like them because something else made them utilitarian.

Initially, believe it or not, we actually found them to be categorically *unsexy*. If a woman had her mammary glands inflated, then this was not a good time to be breeding with her. This was a time when this woman needed to focus on her child. And she needed to save her nutritional stores for breastfeeding – not for building a whole new baby in utero simultaneously. Likewise, males are inherently attracted to fertility, but breastfeeding gives off quite the opposite signal. So, breasts were initially a turn-off. And a fetish for swollen bosoms would've been maladaptive.

I know this one is really difficult to imagine. Most modern men can't even fathom it. But the only other way I can think to explain it, is to compare it to a very pregnant woman. Generally speaking, when you see a pregnant woman, you're not exactly inclined to be scoping her out. You're not checkin' her out like you would if she was *not* pregnant. You're just like, "oh, she's obviously busy." Well, that's what breasts used to be like. You'd see 'm and you'd be like, "oh, she's obviously in the middle of something."

You'd be like, "I wish this woman would freakin' move out of the way over here. I'm trying to check those babes out over there... but now I can't, because there's just this woman over here, blocking my freakin' view with her huge giant tits in my face."

Alright... so... if you are still unable to imagine what I'm saying, then I am rhetorically sorry for that, which is to say, that I am not sorry. But I have failed you nonetheless. Maybe someone else could have done a better job at it, but someone else isn't here. *I'm here*. It's just me. All the someone elses were too afraid to step up. They were too afraid to lose their academic funding.

And I spent YEARS beseeching them to investigate this topic. I was like "why don't you write a paper on this, Woodley?" and he was like "Myeh, I can't because I'm a freakin' dweeb and I'm too afraid and I'm too scared, and I don't wanna, because people might find out I have a tiny dick," and I was like "Jeez! Woe! Take it easy, dude. Alright. I'll do it myself then. And I'm sorry to hear about your condition. Your secret's safe with me. ... ... I promise I won't tell anybody."

It's Woodley. His name is Woodley. And he has a tiny dick. His name is Woodley, everybody.

But that's fine. I'll write the freakin' book myself. I don't care. I might as well, since I'm the one who figured it all out anyway. ... But don't come crying to me when Woodley has tiny dick.

What the hell were we talking about? Oh yeah. Titties. ... ... Titties, titties, titties, titties, titties. ... Why do our hominin women have these titties? It doesn't make a lick of sense. It's a total waste of metabolic resources. It's a completely nonsensical place to deposit your extra fat reserves. ... especially for a supposed runner animal. You would never see this in nature. You would never look at a wildebeest and be like, "wow, look at those huge tits." And yet, there they are on us... two human breasts... just stickin' right out there... in defiance of all logic and rationality. ...

Figures the most irrational body parts would be on a woman.

But that's not all. Our nipples are all wrong too. They're supposed to be on the bottom like a keg tap. They're not supposed to be way up in the center like a bull's-eye. ... at least not as far as natural conventions are concerned. Having them on the bottom makes them more like a baby's bottle. But then having them way up in the center makes them more like a beer can that you're shotgunning from the middle.

Downward facing nipples make sense. That's why nature does it. But not us though. We had to go and make things weird again.

And so, a number of mainline theories have attempted to explain all this stuff. But none of 'm has quite hit the spot yet. In fact, they're not even close. They're all like those super convoluted theories about our noses and our chinny chins. They're all like the anthropological equivalent of one of those old, nutty professor, Rube Goldberg contraptions. ... where you just kick the whole thing off, by like, pushing over a little domino piece or something. And then the domino hits a marble that rolls down and cooks your breakfast.

Except you don't get any breakfast.

It just catapults a plate into the back of your head.

Well, we shouldn't blame those other theorists. It's not their fault. In fact, I even *salute* them! It's not easy trying to make sense of the world, when your whole entire starting point is a bunch of old, grandfathered in, savannah banana gobbledygook.

It *ain't* easy. ... But persist we must. ... We must grasp the opportunity and seize the moment. So, somebody gimme a drum roll.

I'm sure you've already guessed where this is going by now. That's right. When the only tool you've got is just a big ol' beach theory, then everything around you starts to look like a pool toy. Yes, indeed. Lady breasts are pool toys. They are big, buoyant, floaty, rounded, pool toys.

Everything about them is a pool toy. Their inflated round volume is better for holding onto. And then the nipples up higher in the middle are much better for nursing above the water line. And of course, mom can sit or stand or simply float there. ... This design is more convenient no matter how she's nursing in the water.

And indeed, these same exact big bouncy breasticles... which are so preposterously ill-suited for a life of running... are at once, *the very same* big buoyant beach balls... which are so incredibly well adapted for a life of water.

So, I guess breasts aren't so irrational after all.

I guess breast were just *FEELING things!* ... Just because it's *a feeling* doesn't make it any less *rational.* ... ... than your *precious LOGIC!!* ...

... OKAY?!

Oh, and in case you missed it earlier, our babies are essentially just little porpoises. They are perfectly neutrally buoyant. And they have insulated thermal regulation. They can swim around instinctively and take little side breaths. And they are far more at home in the water... than they are just slowly crawling around on their bellies.

So, what better way to feed a little porpoise than with a big bouncy buoyant beautiful beach ball?

"Hooomoooo eeeeerectus,
swimming at the beach.
-k-i- -s-s- -i-n-g-
First comes love.
Then comes pairiage.
Then comes a baby...
which is essentially just a little porpoise."

And since our babies are indeed these little porpoises, we are going to need some hoisting cranes to help us get'm up out of the water. ... One of the most ancient and distinctly hominin inventions has been the nursing sling. Mothers have been using it since forever. We have been carrying our little porpoises around in these hoisting slings.

But what kind of an animal needs a sling to carry their own babies around?? ... ... No animals. That's who. None of them. ... None of them except us. Why would they? That's inconceivable. Their babies just latch on. So, why don't ours? ... Why is everything about us so unnatural? Why is everything about our species so random?

Well, the truth is, we're only random because we're a fish out of water. We're only unnatural because we're a creature which is out of its context. When you eliminate all of the alien foreign influences, a human actually does extremely well in our own native coastal environments.

Sure, we might have a few legacy flaws. Our eyeballs might be backwards with a couple of blind spots. But all in all, we're honestly a fairly well put together animal. And we're a *really* well put together animal for the environment we consider a vacation destination.

Alright, I'm pretty sure I covered most of the biggest things wrong with us throughout Chapter Nine and Three Quarters. If you noticed anything I missed, then please do let me know. I'm always interested to hear what people think.

One astute gentleman has already pointed something out that I just completely overlooked. According to him, we have a major design flaw with our finger placement. We should probably have all of our fingers on the left side so our right hand can just be a club for punching.

Oh, and that reminds me. We also have a lot of brain problems too.

I've seen it myself actually. I used to work in an insane asylum. ... Yes, a real-life insane asylum. I was one of the "men in the white coats." It would've been a cool job though, honestly, if the residents weren't always acting like a bunch of lunatics.

But I still met some really cool inmates while I was working there. Shout out to Scrambled Gregg.

Yep, yep, yep. I ain't no saint myself though. I just wanna be clear about that. I might've had some brain problems of my own in the past. I know nobody would suspect it, but I've also had my own fair share of near-miss head injuries.

When I was little, I wanted one of those cool Tony Hawk skateboards, but my parents just got me some cheap piece of fat bullshit instead. It had all these fat rounded dimensions like it was made for some mental handicap program or something. The wheels were such bullshit it wouldn't even roll down on an incline. It would just grind to a halt against gravity.

I fuckin' hated that skateboard. As soon as I unwrapped it I could tell right away that they bought me some total bullshit substitute. It was one of those stupid old asymmetrical ones too, that freakin' nobody has even used since like the seventies.

One time I got so mad at it for not rolling downhill I just lifted it up in the air and then smashed it into the ground. But it was so fat and retarded it just ricocheted back into my face.

The blow to my head didn't knock me out or anything, but for all know there might've been some kind of like Flowers for Algernon effect or something.

That fuckin' thing was indestructible. It was like some kind of safety board for idiots. And it was all these stupid bright colors... like red and yellow and blue.

Fuck yellow.

And it had like some stupid picture on it of a skateboarding dinosaur in sunglasses.

Fuck...

Dinosaurs...

15

9 ¾ 5 ¾ -- MY BEDROOM WAS UNDER THE STAIRS

Alright. I got one. So, a black, a jew, and a Mexican walk in a bar. The black turns to his friends and says, "AIN'T IT 'BOUT TIME WE START THIS CHAPTUH?!"

"The black"

He's a cat. He's a black cat. I should've mentioned that. He was hanging out with his Mexican and jew cat friends.

Sorry, I don't know any good ones from my corner of the world. But I do know a thing or two about cats. So, let's talk about cats. And I'll try my best to make fun of people's races.

You know what, screw it. Let's make fun of people - all the stuff they have no control over. Haha! This is gonna be fun. I've been bottling this up my whole life. It's time to get this hatred out in the open.

I HATE that people don't understand about cats. I HATE it!

Man, I really hope this isn't the first chapter you picked to read at random. You should know by now that I'm a bit... unique. My bedroom was actually a little tiny micro cupboard under the staircase when I was kid. So, who knows what kinda psychological impact that had on me.

True story. They used to tease me about it.

Alright, alright, alright. Easy does it. Everybody, just relax about the race stuff. Just calm down. You guys are all nervous and tensed up on me now. Don't worry, don't worry. It's just us in here. It's just our race.

Yep. It's only us. None of the other races are in here right now. This chapter is enchanted. It's ooooooooonly us. Yup, yup, yup. Just our race. You know, "our" race. You know the ones we are. Yup! Exactly.

"Those" ones. It's just us.

Yes, indeed. We're the ones that use those little rainbow flags to represent us. We stroll around all sassy and gay waving our rainbow flags and parading up and down the street at least once a year.

YEP! That's us! We use the rainbow flag to represent us because our big beautiful rainbows go streaming across the sky until they arc into a pot of gold. Then one of our little leprechauns goes and dances around it with his clovers and shamrocks. We just get out there one day a year and start wavin' the flag. We go parading down the street and havin' a gay ol' time. Yep. We might even pinch yuh if you're not wearin' green. Teehee.

YE'E'EP! That's us!! That's our race. It's just us. So, let's talk about our race.

You may already know by now... that those of us abroad had to flee the island once. It was uh'cuz of the famine. ... Well, I mean, we told'm it was uh'cuz of the famine, but really, it was uh'cuz of all the Irish people trying to get away from all of the Irish people.

You can only take so much, yuh know.

It's a lovely island though. It's been inhabited by leprechauns and fairies. And it was once the home of a magical race of elves. They lived there since the old times. But then they left when all of these selkie apes started showin' up.

They weren't the only fairies though. There's a little fairy in all of us - every mammal, every human, every race, every extraction - not just those rainbow people from their little green island.

Yes indeed. Every one of us has their own little fairy inside. Yours could be like Tinkerbell, or just a little tiny Peter Pann that Tink shrank down. Your little fairy is basically the real you. He or she is the real little core of your true personality. It's the way you used to be, when you were still a little munchkin just learning how to walk and stuff. ... You just wanted to play and splash around and climb trees and pet doggies. You liked animals and gardens and rainbows.

As one of the many mammal, bird, and intelligent empathic-life-forms on this planet, E.L.F.s, this is who we really are internally. This is our core personality. This is our executive consciousness. And this E.L.F. core is just an enthusiastic little fairy elf. ... It does not know fear; it does not know hate; and it does not know jealousy. ... It only knows happiness or sadness. And it only wants to play and have fun and make friends with people.

But, alas, survival within this world is not so simple. There is also a lot of work to do. That is why your elf-core needs a helper. Your elf-core needs a set of cat's pajamas.

Let's just go with the male example for now. And let's just say that your little fairy elf is Peter Pann. And little Peter, as we've already mentioned, is very pure. He is the heart of your authentic personality. So, he needs Truffles to help him. ... That's what he named his cat's pajamas - Truffles. And Truffles is tough. He has all of the basic instincts that are needed for survival. ... And so, these survival cat's pajamas are also equipped with one of those big, saber tooth tiger headdresses you can wear - like the shamanistic caveman thing - where your face is right where the mouth opening would be.

Alright. So, Truffles' basic instincts have three essential components. He has cravings and aversions and trainings - C.A.T. ... When everything is going well, so is he... and he's just a big ol' delicious truffle.

He has perfectly healthy cravings for things like tasty food, great company, and favorable climate conditions.

His only natural aversions are to loud noises, towering heights, and things that look like snakes. He is not born with fear - he only possesses these three specific aversions. ... Every other aversion he has to learn about.

And to that end, everything else he gets *trained* to do is a natural and adaptive skill or habit.

This is how Truffles is when he's healthy.

So, this pure little core of your personality is interacting with the world through these basic instincts - the cravings and aversions and the trainings. Little Peter is experiencing the world through Truffles. ... Or maybe you want your Peter to just be full size. That's fine. Choose your own adventure. But the point is this, this is how your mind is internally organized.

And once you have a blueprint for how our mind works, you can know yourself in ways that will truly amaze you. You can see things within people's behaviors which were previously invisible. And when someone has a trauma from their childhood... because their room was in the little cupboard under staircase, you can see the signs and symptoms quite conspicuously, and you can know just how to quell and slowly fix them.

But the primary reason for us broaching this topic in general is because it reveals the truth about our origins as a genus.

And to that end, we will start with this. As we mentioned before, when everything is normal, so is Truffles the C.A.T. We are naturally born to crave things like food and love and a purpose. And we are born automatically with cautions for heights and loud noises and serpents. We can also pick up new things along the way and discover that we have a thing for chocolate, and also a brand new aversion for grabbing sea urchins.

Alas, things can also go sideways just as well. As is often the case, if a mammal child is not sufficiently nurtured, then its instincts can be distorted from the very beginning.

The basic rule of thumb is this. If we are deprived of something that we need for our basic survival, then we will develop an unhealthy craving - one that persists into adulthood. And if somehow we are harmed as a child and we are not given the comforts that we are supposed to get, then we can develop an unhealthy aversion.

An unhealthy craving is what we call an addiction. And an unhealthy aversion is what we call a phobia.

Likewise, the other things which we are trained to do as a child, can obviously be unhealthy and counterproductive. And you can call these bad behaviors "eccentricities" if you want... so that way we can shoehorn this whole thing into a new acronym - addictions and phobias and eccentricities - A.P.E. So, if things aren't going right, then your C.A.T. can go A.P.E. shnizzle.

And we can talk more later about phobias and trainings. Those two are a lot more straight forward. But for now, our biggest concern is the concept of addictions.

There are certain things we need as little mammal pups. There are obvious ones like food and a comfortable climate. But then there are also the abstract ones - the needs of our soul. Every wee little mammal needs the following - love, attention, mentorship, and patience - L.A.M.P. We all need our lamp light. And we need it to shine nice and bright for us. Let's just call him Lamp.

So, you have your little Peter Pann, and he has his magical Truffles pajamas, and he's had Truffles since the time he was a kitten. And little Truffles has needs too. ... because he has to grow up strong and healthy also. And so, he needs his Lamp light, just like Peter.

When we are given all these things throughout our formative years, we will naturally feel fulfilled and very whole inside. We will go through life feeling normal and not even thinking about it. Even if we lose those amazing people who made us feel whole as a child, we will still continue to carry that wholeness for the rest of our lives, all thanks to them.

But alas, this isn't how it always happens. Sadly, many children and babies are simply not given what they need as little ones. They are just plainly not provided with what they require. Sometimes this happens quite subtly, and other times it manifests quite terribly.

When this does happen, and when a child is deprived of something, they never end up getting that sense of wholeness inside. Rather, it feels like a void. And whatever the thing was that they were deprived of, they now just become addicted to like a drug. If they were deprived of the sensation of an embracing parent, then they seek it in codependence or in booze or drugs or something. And if they didn't get approval from a parent, then they may crave all of the trappings of success with an irrational fervor.

We can get more into that stuff later. There's about a billion different examples of unhealthy addictions. And I know a lot about'm. ... and all the other mental illnesses... because I had'm all.

Yep... ... all of'm.

It was part of a graduate level psychology course. You had to take this injection that afflicted you with every mental illness in the textbook. ... - one by one of course.

The last one was possession. It was freaky. You just started foaming at the mouth and making all these weird hissing noises. And it was weird too, because you like, started twisting your body all painfully. ... You just looked like some sort of jacked up belly dancer with cerebral palsy.

And the other students had to make observations.

It's the only way to learn.

It's the only way to learn.

Oh, hey. That reminds me. There's also this thing called Munchausen syndrome. It's this somewhat amusing condition where people just like, fake their own medical health problems. And they do it so that they can get attention and stuff. So, like, you'll just see this guy who's totally fine one day, but then like, several days later he'll just have "cerebral palsy" all the sudden. ... which is not a condition you just suddenly get.

But I figured out a way that we could help this gentleman, and get'im to stop faking his condition. You throw hot coffee on'm.

Alright, alright. ... So, how do we know if a behavior is an actual addiction? How do we know it's not just some perfectly ordinary source of enjoyment?

Well, let's take dessert for example. I have it after dinner sometimes. It's nice. I enjoy it. It's a pleasant experience. But it doesn't put me through the roof with overwhelming ecstasy... compared to how I normally am – "I'm sweet enough, Turkish." But if you looked at my enjoyment on a graph, you would just see a nice little bump into the positive. And then I slowly ramp back down over a longer period.

And when I get back down to zero on the graph, I'm just sitting there at neutral. My dessert appreciation feelings are doing fine. I'm not even thinking about it. I can go for days without it even crossing my mind. And the reason that I can sit there at neutral like that... is because, whatever exact positive sensation I get from dessert enjoyment, it's actually a sensation that I already got as a child, or even more so in adulthood. It's not a void in my soul where I'm lacking. I am whole there - at least in that spot right there, where the sensations that you get from things like desserts go.

And, finally, I'm never willing to do harm to myself, or to the people that I care about, just so I can get my dessert fix. I don't have a compulsion for it that's over the top. If I ever found out that my favorite dessert had something really unhealthy in it, I would simply stop eating it - simply and sadly. But I would still stop eating it. The sensation that I get from sometimes eating it... isn't *nearly* enough for me to wanna harm myself. No freakin' way. Or, like, if I woke up in the middle of the night and I was craving it, and then I noticed I didn't have any, I wouldn't just go scampering down to the store in the middle of the night and harm my sleep over it.

I wouldn't do ANY type of harm to myself in order to get it. That's how you know it's not an addiction.

So, you could say that there are three basic components to addiction. Firstly, I still feel perfectly whole and neutral, even when I don't have it... which means I can easily go a very long time without it. Secondly, when I *do* get it eventually, I just experience a pleasant little high that sort of lasts for a while. And thirdly, I'm never ever willing to do harm to myself or to those close to me just so I can get some.

Now, of course, when it comes to the condition of addiction, all of these enjoyment parameters are distinctly distorted. ... because now there is a hole in your soul, right where you were deprived of something essential. So, you can't just sit there at neutral and feel normal without it. Your normal waking life feels downright crummy. Even if you've never even touched an addictive substance before, you still feel terrible, and constantly uncomfortable, when you're just living a normal existence. In other words, your graph is always sitting way down there in the negative.

And when you finally *do* get some sort of stimulus that spikes your graph up high, it basically just gets you up to slightly above where a healthy person sits at normally - so, just above neutral. And then it wears off really quickly and you come down hard.

And finally, you have such a pathologically intense craving for this stimulus... that you absolutely will cause harm to yourself to get it. You almost don't have a choice. It's like a constant discomfort or torture not to have it. So, normal things, like your long-term health, and your close personal relationships, don't really seem all that important... when you are actively in a state of pain or torment.

Our societies have all been acting as if these people with addictions have just been arbitrarily choosing to have them... ... but that is definitely not the case. We've been assuming that these people are just coasting along at neutral most of the time, but then they just stupidly go out and start doing a bunch of crazy stuff to get *extra* high. ... like they're just these pathetic little weaklings compared to the rest of us. But no. They are not just sitting around at neutral – they are sitting around at *negative*. And they've usually been feeling this way since their childhoods. Their normal waking life is almost excruciating. Or, at the very least, it is extremely uncomfortable.

And that is exactly what addiction is all about. It's not just some sad pathetic attempt to simply get high all the time. It is, quite conclusively to the contrary, a desperate attempt to not feel *low* for a moment. They're not out there questing for some lofty and epic nirvana. They're out there dying for just a moment of merciful pain relief. That's what addiction is. It's pain relief. That's why people pursue it with so much vigor. And it's all because of some sort of deprivation. Some part of the Lamp light was denied to them. And now it hurts, and they crave it desperately.

They crave someone who will actually notice that they even exist, every once in a while. ... or a mentor who will actually care about what they're feeling. They crave an environment where it's safe from chaos, and where they can finally control things. ... or they crave a bedroom where you can actually fit more than just a bed into it, and you can't hear people going up and down the stairs on top of you.

It's no coincidence though that a lot of these drugs just give people the same basic chemical sensation as getting a warm hug. And this isn't just some hypochondriac placebo where it's all up in their heads that they feel this way. Their actual brain has been structurally hardwired from childhood to feel like shit all the time.

And this seems like a really screwed up thing for nature to do to us.

After all, why not just make us psychologically bulletproof? ... But apparently this has worked a whole lot better to get us motivated to escape bad situations. But that's another story.

The main form of addiction that we're interested in... is the following. Whenever a little one is, very sadly, deprived of their own inherent bodily autonomy, or if they're ever made to fear for their own safety and security, then they end up craving this feeling with a vengeance. Which is to say, they crave what was taken from them - power.

The craving for power is one of the oldest and most nefarious mental plagues of our history. It comes out in about a million different ways with a million different intensities. It manifests absolutely everywhere, from the bedroom to the boardroom to the battlefield.

And a very close cousin to this addiction to power is the compulsion for attention. Usually, those who were made to feel powerless as a child were also neglected, and the only attention that they got was when abuse was occurring. So, all of that neglect just causes them to lust after basic attention. And then, because that neglect was also coupled with some form of active abuse, it trains them to associate this attention with sometimes degradating treatment. In other words, they get trained to get their fix of attention from submitting to a dominator.

When these two phenomena of both abuse and neglect are substantial enough, you end up getting really strong addictions to both domination and submission - however paradoxical that may seem. In other words, you end up getting high off of sadism and masochism.

In the nineteenth century, in and around the region that would eventually become known as Germany, they basically invented public school. ... in the form that we know it today. It wasn't like a little red school house with just one local teacher and then some older kids helping to teach the younger ones. Nope. It was hardcore. It was a very strict institution... designed to instill both discipline and submission. So, unsurprisingly, this region ended up becoming a major international hotspot for... "unusual behavior."

And a few generations later, they even came up with their very own word which specifically means to take pleasure in hurting others - schadenfreude.

It has even been lamented, "what a fearful thing it is that any language should have a word (which is) expressive of (such a) pleasure..."

So, anytime you have a general scenario where someone is getting off on the domination of others, whether it be in our species, or in any other genus of higher intelligence, this is a dead ringer that you are dealing with some sort of former disempowerment. ... - usually from their childhood.

And this is very unfortunate, naturally, but that's where it comes from. Maybe I can do a whole other book at some point on how to rectify those injuries... and how to replace those old psychological scar formations with some super strong and unintuitively more powerful soul muscle. ... - when the muscle has the tear, but not the recovery yet.

But a really gross oversimplification of this whole healing, and psychological rectification process, is this - just think about the details... of what's been bothering you.

Things only remain a problem because we run away from them. We just run away from terror and from the pain of injury. And then that running keeps us panicked and keeps us from thinking. And a state of panic is just about the *worst possible way* to do things. ... except for run from a tiger. But you're supposed to finish running from the tiger - not just living in a state of emotional panic.

This is why wild animals aren't out there getting post-traumatic stress disorders like we are. They get away. ... Or they don't. ... But they finish being panicked pretty quickly. And they also have a family who can help them - who can encourage them to feel a whole lot better about anything that's been bothering them. ... - something a lot of our own *human* children don't even *have*, unfortunately.

So, it can be helpful to do a set of exercise that are short-interval, high-intensity, power bursts. ... - any kind you want, really. ... It helps to complete that circuit of the fight-or-flight activation. ... And so does trembling, oddly enough. It helps alleviate that pent-up fight-or-flight energy.

And part of the reason that this happens, is because technically it's not just a fight or flight response - it's actually a fully blown, fight-or-flight-or-freeze-or-faint equation. If the leopard has already gotten you, then sometimes it's just better to faint. He might just assume that you're actually dead already, and then go stash you up high in a tree somewhere. Then later you can come roaring back to life again, and then go take off like a rocket and make your getaway. But we have to come roaring back to life at some point. We can't just stay there all frozen and nervous.

Or, if you're unfortunately grieving very deeply about the loss of someone, then... first of all... super freakin' sorry to here that... but, second of all, high endurance exercises, and immersion in freezing water, can at last be just the little death that you've been looking for.

But anyway, panic keeps us running from our thinking, and thinking is our bridge to solving everything. And you don't have to be a genius or a certified brain wizard. ... If you just allow yourself to contemplate a problem, then you will slowly start to realize where it's coming from. ... - like, what kind of primal fear is it avoiding; and what kind of ancient pain is it retaining. ... and what kind of simple logic will expose its irrationality; and what kind of consoling comfort will just make it less concerning. ... ... If you circle back and forth between these nodal points – between the problem and its equal-and-opposite comforting thought – then this winding back and forth will effect a sea change. Just have patience like the farmer with his crop fields. ... The implanted impurities of our embattled psychology are all transmuted by patience - ... by persistence, and perspective, and patience.

And there's a super sexy spell that you can cast for this. ... It's just two simple words -

~fff##K IT!!!~ /~~*

It helps if you use a wand when you say it. ...

~fff##K IT ALL!!!!~ /~~~*

... or it can be three simple words.

Those simple magic words... can make *all* the difference. ... ...

The more we make our peace with losing everything, the more we will get bored with fears and terrors. And the more we refuse to bow to a Bene Gesserit with a pain box, the more we will become the Kwissatz Haderach!!

So, that's the simple version. ... ... That's the basic blueprint for turning traumas into strengths.

But this process can also be amplified with a little help from an elder. It can often be incredibly insightful simply to have someone to talk to. This is part of why traditional therapy is so effective.

But not everyone can afford a good therapist. And not everyone can just as easily find one locally. So, in that case, I have a different recommendation.

Just go down and find like the blackest possible barber shop in your local area. Then go in there and look around and find the oldest, old, grey dude holding scissors. Then just walk right up to him and start telling him your life's story. ...

Don't even pay for a haircut. Just go in there and start spilling your guts all over.

Or just keep going there, day after day, and getting like, one little millimeter of hair trimmed.

... but like evenly around the whole surface so it takes forever.

... - at like the same rate your hair's growing.

Otis got that old time wisdom. He can learn you 'bout all kINDs of stuff you ain't never knowin' nuttin' about.

... like the priorities in old banjo songs.

"In The Big Rock Caaandy Moouuntains...
all the dogs have ruuuubber teeth...
and then the people whooo beat... you up... for being homeless... are
all crippled and stupid and blind..."

Otis knows. ... ...

He gets it. ... ...

He may not have been there...

... but he gets it.

"... iiiiiiiiiiiiiiiiiiiiiiiiiiiiiiiiiiin...
The Big Rock Can Dee Moun Tains..."

Someone should start a therapy business like that. You just go in there
and it's like a regular psychiatrist's office, with like a couch and stuff,
but then this little old black dude just comes in and drapes a salon
apron over you.

And he's not even licensed or anything. ... he's just got that old time
wisdom.

"Well, I ain't sure what you should tell yuh mothuh 'bout that... but
how you goen' want this back part cut?"

Aaaaaaaaalrighty then... ... ... So, what was the point of all that? What does any of that have to do with this chapter?

Finally. I thought you'd never ask. I thought you were just gonna keep on letting me ramble there. None of that had anything to do with this chapter. I was just waiting for you to cut me off. That was *all* part of the intro.

Oh, wait. Hold on a minute. There might actually be something...

Okay. How 'bout this? If we as a society have to hide something from ourselves - something that, by all rights, should just be a normal, everyday, above-board aspect of living - then that tells you right away that something unnatural is going on.

Okay, well, what are some examples? Well, two things that we always hide from ourselves are child abuse and animal slaughter. We don't make movies just showing child abuse like it ain't no thang. If a movie ever does show it, it's always something sad or disturbing.

Likewise, we never show scenes in a movie where a guy just walks up and shoots a horse in the face... like it's normal. It's always a very sad moment, and they just leave it out of frame or simply imply it. Or, like, if they *do* show hunting in a movie, they never actually show you the gory details. And that's because nobody wants to see that. It's always just been considered an unfortunate aspect of land survival.

Child abuse and animal slaughter are inherently abhorrent to us.

Alright, well, what are some of the other things that we block from our own view? Well, we block nudity and cursing, along with just about everything that a human being has been specifically selected to wanna keep away from their mouths. But the nudity and cursing are only being blocked from the children - adults actually seek it out.

So, again, that just leaves the acts of child abuse and animal slaughter as these conspicuous oddball outliers. If those two acts of mayhem were perfectly natural for us, then we wouldn't have to block them from our vision. They wouldn't even phase us, in fact. We could just show these images of animals writhing around in pain after getting their throats slit and it wouldn't even phase us. It would just be like plucking an apple. But that's *not* how we see it. We see it as f***en disturbing. ... inherently.

And the only way that we as human beings have ever been able to overcome this... is simply by using schadenfreude. In order for us to exploit these land-based animal sources, we've literally had to abuse our little children. And that's really something to think about.

We've had to expose our little children to things which are naturally horrifying to them. We've had to teach them how to kill and how to dismember. And we all did it historically, because as far as we could tell, we just straight up had to. Even I had to do it. I had to work in a slaughter house. It's fu**en disturbing. And now I get off on hurting anthropologists.

And so, all this sort of unnatural living for us is what gave us this idea that harming children "made them stronger." ... and the idea that if you mollycoddled a child it would make them weaker.

Even up to this very day, they still recommend just straight up leaving crying little infants all by themselves to "get tough" or to "self-soothe" or some bullsh**. And this is *absolutely mind boggling*.

We... ... are... ... mammals. ... We don't do that **** to our little ones.

That's ******* insane. We protect them. We love them and attend to them and mentor them and give them patience. We don't punch a pregnant woman in her stomach to toughen the baby up. That's insanity. No mammal does stuff like that. ... just leaving their crying little baby all alone all night. ... ***kin' 'ell.

Yeah, sure, we're not supposed to be babying the hell out of'm their entire life, but there's a limit. There is a balance. Our children are supposed to be thoroughly protected until they're about two thirds of the way to puberty. ... and then, and only then, do we slowly start allowing them to start engaging in some less forgiving challenges. But even then, we still stand by their side and fully support them. This is just biology.

And *that...* is how you create strength - not abusing them.

Sure, you can really make a child super callous, if you expose them off and on to some unnatural conditioning, and surely this can help them to survive in an unnatural environment, but wherever you see a scar there's also a fracture. Something is broken. And it just stays that way. And then they turn into adults who try to soothe this pain. They turn to comfort foods and booze and other substances. And then they try to act like it's all macho to drink booze... even though booze, by its very nature, is a painkiller. So, technically, it's actually a lot more super badass to simply not drink booze. But I get it. No judgement.

"RAAAAA!!! Who can drink the most shots of this psychological painkiller that's actually making it feel like you're getting a warm hug from your mommy inside?!?!?! RAAAAAAAA!!!!"

Alright, alright...

Alright... ... so... ... Where were we? ... Oh yeah. Alright, so... ... yeah... really what's going on with all of this "toughening up" as a child... is that we are making these little children profoundly weaker. We are unmetaphorically hardwiring their brains to be constantly in pain. ... Yeesh. ... That sucks.

And, yeah, I get it, it was necessary to awaken this ancient, primordial demon so that we could survive inside of the continents away from our home-world, but it's time to start waking up now. It's time to start discovering our own biology. We can't just keep on repeating these patterns without ever understanding them. We can't just keep on in-fecting new generations like we're some vampires of trauma. Or, at the very least, we surely need to be consciously aware of what we're doing.

So, yeah, protecting children is not a sissy thing - it's a mammal thing. It is precisely the polar opposite of weakness. Think of it like the elves. Yes, they are beautiful; yes, they are elegant; yes, they are intelligent, but it would be folly to presume that their armies are anything less than capable.

If you wanna toughen a kid up, just spend time with them. Talk with them. Reason with them. Problem solve with them. Work with them. Work out with them. Exercise with them. Take on major challenges together. That's what makes a mammal child stronger. And long long long... loooong after you're gone, that strength will still be with them and your descendants.

So, child abuse and slaughtering higher-sentience creatures are not just alien and unnatural to us, they are also connected to one another. That's right. They have been feeding off of one another since well be-fore the Holocene. Overpopulation and warfare forced us into the continents, and then these shellshocked early humans just started slaughtering animals. And of course, they put this on their children and created these cycles.

Warfare, slaughtering, and child abuse are all connected to one another.

But we've doing this stuff for so long now, that we don't really think about it very much. We just force it away from our consciousness... or try to make it look all heroic or something.

There are some old sayings that've been around for a pretty long time now, and they basically say that... you don't wanna know what goes on behind the walls of a slaughterhouse. ... And that is exactly correct. You do not. It is not pleasant. It is like actual, literal hell. So, people just block it out of their minds. They pretend like it doesn't exist. They do what they think they have to.

And there are only a few types of people who actually condition themselves to be able to deal with stuff more directly. And that would be certain types of scientists and medical workers and slaughterhouse employees. And then there are a few classes of stalkers that like to get into this kind of stuff.

However, regarding the scientists and the medical professionals, they are usually just people who are driven by an insatiable and overwhelming urge to want to learn things and help people. And it is precisely these emotional incentives that help them to recontextualize the disagreeable stuff. And then we happily pay them generously for their effort. ... because most of us would rather not have to think about it.

And then you have all of these poor sons 'u' guns who have to work in the slaughterhouses. These are basically just people who need a job somewhere. And a lot of'm will just straight up tell you how disturbing it is. ... and how they deliberately invoke a state of schadenfreude...

just so they can make it through the workday. And they're not exactly at liberty to go on record about it either.

It's sort of like these people who have to sift through flagged and censored content all day long. And then they often just end up with post traumatic stress disorder. ... Similar concept.

And then you have the third category of people - the stalkers. ... For reasons I will never understand, there have always been these men out there... who just loooove sneaking around in the bushes. And they do this, apparently, so that they can try to get a good look at an exposed three year old. And once they're finally in position, they just pull out their shaft and hold it in one hand, while they reach down and grab their balls with the other hand.

That's if they're using a musket.

All of the stalkers that I've interviewed are usually pretty quick to tell me that they prefer the term "hunters."

And then they say I'm describing what they do really weird.

But I'm like, "is that not correct? Do you not do those things?"

And then they say, "if you keep sayin' this shit, then we ain't goen' be friends no moe."

But be that as it may, these gentlemen have obviously chosen a very strange lifestyle. So, it leaves the rest of us to wonder, what exactly is going on here? What on earth would possess these men to go sneaking around in the bushes like this?

Well, it turns out... that these stalkers are basically just being driven by a very deep craving to be survivors. They get off on the idea of supporting themselves. They get aroused by the thought of self-reliance. And so, the bottom line is this, these "hunter" friends of mine... are very much inspired by a passion for survival.

... ... "survival."

And not only that, but there's almost always a nostalgia component as well. Usually something happened to these men as young boys. Usually there was a history there... where their father used to do things to them when no one was watching. He used to take them out stalking with him and then touch them on their public areas when they did a good job for him.

And so, then, you end up with this highly inspired passion for survival... along with a deeply imprinted childhood nostalgia. And then the two of these emotions combine together. And then this of course is a very powerful psychological incentive. And what this does... is it effectively changes the way we see the situation. It alters our basic perspective and somewhat recontextualizes it. And so, this is a major component of how we sometimes are able to overlook certain things.

This happens a lot in war. People become highly emotionally motivated. And it's not only about protecting their way of life - it's also about making their country proud. So, it's just like the passion for survivalism and the childhood bonding nostalgia. ... - same basic concept. But when this happens in war, though, it just straight up enables some people to literally murder other people. It severely distorts our view of their natural humanity. That's why even in war, a lot of ordinary soldiers would just never even fire their weapons. ... and even on the medieval battle fields, a lot of ordinary foot soldiers may have only been *pretending* to fight one another. Or even modern police officers.

If they end up, for whatever reason, having to shoot another person in the line of duty, then they usually have to get some sort of counseling for it.

It just goes against our nature to straight up murder someone. ... especially if we just got drafted into the military involuntarily. But when there are powerful enough emotional incentives involved, this can finally be enough to push us over. ... especially if these incentives are to protect ourselves, or even more so, to gain another person's approval.

Of course, we might end up regretting it later, but it can still drive us in the moment to push the envelope. That's why a lot of former soldiers can't get the faces of their victims out of their minds. It just goes against our basic human nature.

But not every type of soldier is just a weak little sissy like those guys. Some of them can just start slaughtering people. And they can do it over and over and over and over. But these guys didn't just get like this naturally. It took *a lot* of traumatic conditioning. Otherwise, they would simply be like every other soldier.

So, whether you're talking about hunters or professional soldiers, or pretty much any other type of dismembering profession, all of these different categories of people need at least one of two major factors - either a powerful emotional incentive, and/or a good amount of latent schadenfreude.

Humans were never born to delight in butchering one another and our fellow mammals and land animals. And a lot of the ancient traditions could inherently sense this. So, they would always say a prayer and thank the animal for its sacrifice.

Any child can tell us this. If you give a kid a bunny or a little hamster, then they always try to pet it and share their food with it. That's what kind of animal we are.

The only butchering I know about is when I have to give my dogs a haircut. They've got some long fluffy coats like Alaskan sled dogs. So, sometimes they gotta come on down to dad's butcher shop and get a trim done.

I should start a barber shop like that. We just give you these really fast and cheap haircuts, because no one there has any idea what they're doing. You just come out with this like, really warped version of what you wanted.

But, yeah, sure, I get it, our human kids really like to eat a bunch of random stuff. I'm well aware of that. But it's always, like, little snack size things that are barely moving. ... ...

... You know, ... like shellfish.

That's who we are. That's what kind of animal we evolved to be. We are a snacking animal.

"Out of the mouths of babes (shall ye know) the truth." ... ... and sometimes into the mouths.

So, is it any wonder, then, really, that the most healthy and the most long-lived people on the planet... are also eating these diets? ... - these blue zone, Mediterranean style pescatarian diets? ... and then straight up living well beyond a century? Should we really be so surprised by that?

You know, that reminds me. If you're living in the western world and if you just so happen to be religious, then there's a very good chance that you're already a fan of ol' J-dog... of ol' J-town... of ol' J-dizzle... of Nazarizzle. Seriously though, if Jesus is your boy, then with all due respect, your boy J-town... was ALSO... eating... this... diet... ... ... WHAT?!?!?!

Yup!!! ... That's right! ... Ol' J-school was a Mediterranean style, blue zone pescatarian, son.

Yes sir. He ate nearly the exact same exact diet... as the longest living... and most healthy people... on the everlovin' planet!!! WhaaaaaAAAAAAAAAT?!?!?!?!?!?!

Are you front-door kidding me?!?!?!

Nope... I ain't playin', my son. That's your boy... He was eating fish; he was feeding crowds of people with fish; and he was even helping them catch fish. And yet, he never did any such thing with any land animals. ...

... Skadoosh...

How's that for an insane coincidence? ... Or was it a coincidence? Maybe it was a little... wink, wink... a little... hint, hint... a little... nudge, nudge... in the ol' rib cage.

Okay, well, then why wouldn't he just tell you about this then?

Well... let's see... maybe he knew that you still had twenty one centuries of brutality ahead of you... and instead of just giving you a bunch of critical directives that he knew would just be a really difficult task for you, he simply left it to the later generations to put all the clues together. I'm mean, his friggin' symbol was a friggin' fish, for prayin' out loud. The pope's friggin' hat is a friggin' fish... for the love 'u' cryin' out loud.

Just follow the clues. It's all hidden in plain sight. And just because he didn't tell you about something doesn't mean it's not relevant. He never told you about science. He never told you about quantum physics. He never told you about how a rogue group of time travelers in the future was gonna go around kidnaping famous people like Herman Melville in the prime of their youth so they could put'm in a ring together and force'm to fight each other, then erase their memories of the event and put'm right back where they found'm as if nothing ever happened, except their face is all messed up.

... with a bunch of black eyes and shit.

"U'u'u'u'gh... ... What the f*** just happened?! ... ... "

"... What the devil has bedeviled me?!"

"ELIZABEEEEEETH! ... ... "

I bet he never told you about that one... ... ... now dID he?!?!

... DID 'E?!?!

I rest my case.

Yuh know, though, I actually had like three more full length pages where I was just gonna keep on playing around with this topic, and it just keeps getting better and better and better. But then I realized... nah, some people might get the wrong idea. They might think I'm just taking their beliefs too lightly or something. ... ... too... in vain... if you will. ... So, I thought, nah, forget it.

So, instead... let us just bow our heads now... in silence... ... ... But let us also break that silence... for a second... to ask ourselves this question... ... what would Jesus do for lunch?

- WWJD-FL -

Oh, and I'm pretty sure he likes me better.

Alright, alright, alright, easy does it.

Eeeeaaaasy does it... ... I meant no harm intended... I was just makin' an observation.

... that like... if you and I were both stuck on a train track... and, like, there was a train coming... and like... there was a switch that controlled which way the train would go... and like... he had to make a decision at the very last second... ... on pure instinct... ... I mean... I feel like I know... like... which way he would be leaning with that... because like... he and I both have the same exact diet basically... and like... you don't... ... ... uhiihiihiihiiii... ...

... soooooooooo... ... ...

... and we both like holding little sheeps and stuff.

16

INTERMISSION

ALRIGHT, Let's get it ONE MOE TIME!!! ... ...

YEEEEEEEEEEEEEEEEHAAAAAAAAW...

In The Big Rock Caaandy Moouuntains... all the tree'ees are maaade of cigarettes... and there's a big ol' lake... full 'u' whiskey and beer... and you can row around and drink it for free...

In The Big Rock Caaandy Moouuntains... all the weather is always nice. And there ain't never no need... to haftu wrap up your feet... with trash for iiiiiinsulation.

In The Big Rock Caaandy Moouuntains... all the cops have wooden legs... And all the jails are just maaade outta syrup and cakes... and you can eat your way out, if you take several vomit breaks.

In The Big Rock Caaandy Moouuntains... all the bullets are maaade of candy... So, if a farmer gets maaad, cuz you're sleepin' in his barn, it's okay if he puts you on your knees and puts a gun in your mooouuuth...

So, I'm'a gonna go...
where there ain't no snow...
and where the wind don't blow...
and all the trains run slow...
and you can sleep out in the night...
cuz ain't no dogs'll never bite...
and ain't nobody tries tuh beat yuh...
cuz they're blind and they can't see yuh...

In The Big Rock Can Dee Moun Tains!!!!

And I'm'a gonna smirk...
where there ain't no work...
cuz there's money in the dirt...
and where they hand out free desserts...
but then you start to feel sick...
because there ain't no regular food...
and then you try to get out...
and then you realize it's a trap...

iN tHe BiG ROck caN Dee moUn TainS!!!!

YEEEEEEEEEEEEEEEEEHAAAAAAAAW!!!

17

9 ¾ 6 -- LITTLE MAGIC GIRL...

... WHO KNOWS ALL THE ANSWERS

Alright. So, in the last chapter we just got done talking about how Jesus probably likes me better.

It's okay though. Just because someone may or may not be the favorite, it doesn't mean that...

Alright. 'Nuff said. I think we get the picture. I think we both know...
... I think we both know...

Alright, alright, alright... Eeeeaaaasy does it. Eeeeeeeeeeaaaaaaaaaasyyyyyyy does it. I don't want anyone out there thinking I'm making fun of 'm... or their beliefs... or degrading them...

or profaning them... or dehumanizing them in any way except for anthropologists.

After all, Christmas is one of my favorite holidays. And there is one veeeeery special guy... who just neeeever gets enough credit... for being the reason... for the season. ... ... Yep... That's right... You know who I'm talkin' about... ... I'm talkin' about Punxsutawney Phil. ... the world famous groundhog. ... the reason for the season tO CONTINUE!!! ... - Groundhog Day - February 2nd. ... Put those damn Christmas lights back up, you sONS 'u' BI****ES!!! ... ... This sh** ain't over!!! It's still cold outside!!! Put them damn lights back!!! ... Those are hog lights now!! ...

And buy more presents!! ... for Pete's sake!!! Santa's comin' back with Phil and the Festivuss fairy!!! ... No, don't spread out the original presents. ... BUY MORE!!

... It's freakin' Groundhog's Day!!!

Jeez!!!!

... I shouldn't have to tell you this.

Alright. I forgive you. ... But get them lights back up!

And maybe we can change the name of this holiday, to like, "Punxsutawney" ... so it's just, like, one big long word like "Christmas."

We wISH you a Punxsutawney!
We wISH you a Punxsutawney!

We wISH you a Punxsutawney!
Aaaaaaaaand aaaaa haaaaappyyyyyy Feeeeestiiiiiiivuuuuuuuuuuuuuu-uss!

Oh yeah, and since it's technically Groundhog's Day, that means we get to keep celebrating it over and over and over and over and over...
... ...

... As long as it's still cold out there, campers, we get to keep cele-brating. ... So, don't forget your booties, woodchuck chuckers, 'cuz it's party time!

I freakin' love the colder holidays. ... even when it's summer... I just crank my AC up all insanely high. And I just keep blasting it until it starts snowing in my house. ... then I light a fire in my fireplace for survival.

... - when it's hot as fuck outside.

Oh... hey... you know what? ... speaking of all this prophetic stuff... there's this old saying out there that basically just says that the humble will inherit the Earth. ... ...

So... I guess that means, like, people who had to grow up under the stairs and stuff.

And now, just to be clear, it doesn't mean like, weak or obsequious or something - it just means, essentially, humble.

And, of course, just because something's an old folk saying doesn't just automatically make it correct or anything. But let me riddle you thusly.

Let me escort you now... ... ...

... on a journey... ... ...

... of life!!

Long long ago... long before there were any multicellular lifeforms mucking about, the only things around yet were some little tiny single celled organisms. And these guys were just floating around at the beach resorts, gulping each other up.

Now, as far as the fossil records indicate, we can tell that one of these individuals was named Otis. And one day, Otis was just floating around, as usual, when he came across a little tic-tac shaped bacteria. And as far as prehistorical data suggests, his name was pronounced something like Milo.

But Otis payed little mind to such formalities and simply ate Milo... as was the custom. He just gobbled him down in one swallow.

But then Otis began to feel a bit unusual. Something he had recently eaten just didn't agree with him.

So, Otis thought to himself, "uh oh. ... You've done it now, old boy. You're gonna be sick again."

And just when Otis was about to lean over lose his breakfast, something down deep in his gut just told him, "hang on, Otis... not this time, old boy."

So, then Otis just decided that he would listen to this gut feeling. ... which of course was Milo. ... because Milo was still down there. ... -

alive as ever. ... and happy to be so, I might add.

And not only that, but something about this new arrangement was fortuitous for the both of them. ... because, as you see, not only was this new predicament a lot safer for Milo, but it was also enabling little Milo to make some extra energy for Otis. Yes, indeed. Milo was making extra energy and Otis was absorbing it.

And Otis felt gREAT! He was always jogging in place really fast. He was always talking to his friends all the time about all these excellent ideas he had for a business. Like, for example, like basically, like, what he basically wanted to do, right, was like, make a morning beverage, right, that you could drink, right, and it would, like, give you all this extra energy, you know? And like, we could like market it to like...

He was on fire. ... He never came down... He never had to sleep. ... He never stopped shaking... And he never stopped grinding his teeth... He felt amazing!!!

And he was like, "Milo,ol'buddy,we'got'a'good'thing'here,little'buddy. It's'just'you'and'me,little'buddy.   We'got'a'good'thing'goin'here,buddy. This'is'great,man!   It's'just'you'and'me,man.   We're'doin'it'man. We'got'this,bro. We'don't'need'Candace.Screw'Candace..."

And so, from that day forward, Milo and Otis were inseparable. They were the best of concentric friends. Otis had the guts to stick up for Milo, and Milo had a glut of extra energy.

You may have heard of these two already. But you may know them more precisely by their formal titles - Sir Otis Eukaryote and Lord Milo Mitochondria. These two are the first original ancestors of every single animal on the planet. Every cell within your body is one of their many-greats grandchildren. Yes indeed. You are trillions of little Milos and Otises.

The adventures of Milo and Otis have truly been legendary.

Alright. So, let us fast forward. Let us go, now, to a few of their single celled descendants. It had been a while since that very first party where the original Milo and Otis met, and now their descendants were a lot more adapted to this high energy partnership. They weren't grinding their teeth and always talking about business plans anymore. They were just zipping around like little tadpoles, because they developed some little tails with all of that extra energy.

But it was still a glob-eat-glob world back then. And it was just every little tadpole for himself on those early beaches. ... that is... until a couple of little siblings became quite famous. The original records of their births were unfortunately lost to time, but what we do know from secondary sources is that the two of them went by the names Siam and Misa.

Siam and Misa were born stuck together. Their progenitor cell had attempted split, as was the custom, but then somehow, something unusual happened, and the two of them just became stuck together. And yet, they made no apologies for their condition, such as it was. One of their famous sayings was - "we are Siam and Misa, if you please... ... r'r'r'rooowwww... ... we are Siam and Misa, if you don't please... ... bu'dun dun dunnn..."

They developed a lot of this confidence at a very young age. On their very first day of peewee swimming classes, it soon became obvious that they were fast. They could swim circles around the other kids. And this was because their surface area was effectively lower. And it gave'm a bit of a god complex. They used to swim around and sing this one song. "Now we sniffing 'round at our littoral zone... ... ... If we like, we stay for maybe quite a while... ... brr'neert nart neert..."

But, of course, all of this swimming makes you hungry - so, a tadpole gotta eat. And eat they did, because these cats could run down a fish like nobody's business. They could just gobble up everything around 'm and nothing could get away fast enough. And this led to them putting on a bit of "holiday" weight. So, some of the other little tadpoles started calling them the "fat twins."

But this little snark attack would soon prove fatal... ... because Siam and Misa also had another song.

"Do you smell that tadpole swimming round and round? ... ... Maybe we could reaching in and make it drown'... ... ... If we sneaking up upon it carefully... ... ... There will be a head for you... a tail for me... ... ... buh'doink donk doink..."

So, Siam and Misa started dominating the playing field. ... using teamwork - just like their
great great great great
great'great'great'great
greatgreatgreatgreat
gr'r'r'r'r'r'r'r'r'r'r'r'r'r'r'r'r'r'r'r'r'r'r'r
""""""""""""""""""""""""""""""r'r'r'r'reat great great great great great great
great grandparents... ... ... Otis and Milo.

And so, they passed on their sticky mutations to a whole new generation. And before long, all sorts of new stickiness was afoot. More and more people started sticking together. They were forming up in all these different cluster shapes. Some were shaped like rods and others like saucers. While some were shaped like tear drops and still others like just random globules.

But the shape, now, that we are most interested in... is the rope structure. This is just a simple twisted set of interlocking helices, like a braided rope. So, it could be like a double helix, or a triple helix, or higher. And these structures are a stable form that occurs in nature.

So, these helices were like a rope with a bunch of little strands sticking off of 'm. And these strands were all the tails of the little tadpole cells. They would wriggle these little tails back and forth so they could push their whole rope structure forward. And they were basically like a team of little rowers on one of those super fast, tandem row boats.

So, these ropes were fast - really fast - faster than any other structure out there. But when you start to go really really fast like that, you start to get some turbulence around the back end. So, when these ropes would get too long and their speed was eventually excessive, their back ends would start to unravel like a ribbon in a windstorm.

So, then a few of these tail-end cells would get ejected. Then they would go off and start a whole new rope structure and repeat the process.

And this was basically the birth of multicellular procreation - a lot of shaking and wild movement around the back end.

And this was also the birth of specialization. ... because it was only these little, tail section cells who were responsible for multiplying. Everyone else was just mainly a rower.

And then these ropes were also learning a whole new party trick. They were learning how to steer their big long bodies. If a cell up near the nose was beginning to smell something, then that cell would just contract and his neighbors would copy him. And this contraction would bend the nose cone and it would arc towards the food smell.

So, that's the way that they became the food seeking missile bunch.

This was basically the beginning of what we would call muscle cell activity - cells beginning to flex their little structures to produce a larger movement throughout the tissue.

But they also learned another helpful party trick - they learned how to share their resources by exchanging them with their neighbors. Because, not every single cell could always catch their own food particles.

But this sharing also helped them to become more synchronized, because everyone could rest when the cells were all satiated. And so, this harmony made their efforts more efficient.

And then they also began to catch food particles much more efficiently when their nose area began to spread open and they developed a mouth hole. The helical coils would just loosen a little... so they could open up slightly and form an opening. Or they could also just add another helix which would naturally make the structure a little bit wider.

And one way or another, this mouth hole would continue getting deeper. It got deeper and deeper and deeper until it emerged out the back end. So, now their rope structure was no longer just a rope structure, but instead an actual tube pipe. And now the food 'n' stuff that was caught by the mouth part could make it all the way to the back end.

These first multicellular lifeforms were essentially just an intestinal tract.

So... ... listen to your gut... because that's where it all started.

But that's not where it all stopped though. Roughly around this same general time period, there was a younger generation trying a new dance craze. These younger little tubules started shaking their backsides. ... because there was a whole bunch of turbulence around this back section, and all of that pounding and smacking was blowing their backs out. So, they needed a simple solution to address this problem.

And so it was, the younger little tubules started moving to the rhythm. They started shaking their little backsides in a waving motion. And then this simple little movement would harmonize the turbulence.

So, they finally found a way to break the speed limit. They could finally grow much larger without being damaged. But this simple adaptation had yet another advantage. It could help them to push themselves forward without using their little tadpole tails.

So, they finally had, what we would call, muscle powered locomotion. They could finally move in a way that was more ef'fish'cient.

So, they were off to the races.

And these little tiny swimming intestines were just barely getting started.

And what's also pretty cool about this swimming motion... is that *our* intestines never stopped doing this – they are still in there doing it to this very day. They are still in there undulating in that same exact, primordial, swimming motion. ... - pretty amazing honestly.

But it was precisely this adaptation for swimming that was about to free these little lifeforms from their microscopic confines. After all, if the only way you have to move your body around is with your fuzzy little fur of a million tadpole tails, then you can only get so wide before you're just too heavy, and you can only get so long before you're just running into your own tail again.

So, it was *swimming* that had finally released them... to expand and become a lot more complicated.

And, of course, as we all know, there were about a million bazillion things that started happening at this point. But the one that we'll just mention here briefly was the one that caused the next big revolution.

As these lifeforms were getting bigger and more sophisticated, they developed an internal network for delivering electrical signals. So, if their nose was ever smelling something tasty, it would quickly inform their muscles to go and swim that way.

But the world was only becoming increasingly complicated, and so the fragrances that they had to keep track of were only growing in abundance. There were ever more complex aromas which were a signal for opportunity, and there were also more chemical signatures which were a spelling of danger. So, they just needed some sort of biological storage system to keep a record of these smells and their impressions.

So, their nose just began to develop an organic method for storing information. Thus, if they encountered some sort of distinctive aroma, one they hadn't really seen for quite a while, then this system would just recover those previous impressions.

This was essentially the beginnings of what we would call consciousness.

So, if they found themselves surrounded by some ephemeral fragrance, and it was just like those fresh baked lemon cakes that mama fish used to make... then woo'OOSH!! ... - right back to their childhood.

... - those fresh baked lemon cakes she used to make underwater.

So, even though we've just become this predominately visual primate, and our sense of smell is extremely weak within the animal kingdom, we've still retained this ancient feature from some of our earliest and most archaic primordial ancestors.

So, when those aromas from your far distant memories transport you right back to your childhood, just remember, this is also your ancestors, reaching out to say hello to you. ... ...

They're proud of you, kid. ... You've got legs now. ... ... So, go get'm, tiger.

Just follow your nose, and listen to your gut, and keep on moving to the rhythm. ... ... is what your fifty quadrillion ancestors would say. ... ...

But we can also get that flashback effect from our other senses. Like... for example... if you hear that old song playing from long long ago, it can take you right back to that weird old relationship that was kind of fun but also doomed before it even started. ... ... And it's a similar sort of thing with our sense of vision. But this really cool flashback effect just sort of dampens out and gets weaker as you start moving up into these more recently acquired senses.

So, that's what happened. Our original nose-brain just kept on stacking up these new features. And the rest is history. ... or prehistory.

And then these little minnows just kept on turning into all sorts of different creatures. Some were trading their fins for little leg appendages, and others were trading their dermis for a protective, outer, armor layer. But some just kept on swimming and exploring new places.

And then some of these professional swimmers started going up to the ocean's surface. They started gulping in the air up above it and slowly learning how to breathe it and hold it in their air sacks. And then some of these developing breathers started flopping around on the beach areas.

The first two fish ever to do this were a couple named Harold and Maude. They really really liked it on the beach area. It was totally non-conformist. There was no one even up there. ... except like a few crustaceans and vegetations. ... And there was like a weird, upside down car for some reason, but that might've just fallen through a time slip."~

But then people started blowing up their spot later. And before you knew it, there were all sorts of people up there, just flopping around on the beaches. But then time went on, and a lot of these little floppers started getting good at it. And then, before you knew it, once again, there were floppers who were turning into walkers. And now we call these little walkers the first amphibians.

Three of these little fellas were a set of salamander brothers. Their names were Shadrach, Meshach, and Abednego. And these brothers made a pact. They *swore* that one day... each one of 'm... would evolve into an entire new branch of the animal kingdom.

Shadrach just decided to stay a salamander. However... he made this solemn promise. He *swore*, that if he, Shadrach the salamander, or truly any of his salamander descendants, ever saw a child being bullied, especially if that child was a little magic girl... who could move stuff with her mind... and who always knew all the answers, then on their sacred honor, they, as the keepers of the oath, would launch into a vicious attack against the bully. ... ... especially if that bully had some crazy name like The Trunchbull.

It's called The Salamander's Oath.

Alright. Abednego was next. He swore that he, Abednego the sala-mander, and his entire family line, would one day... evolve into the dinosaurs and the reptiles. ... He just made that solemn promise on behalf of his wife. ... without consulting her.

... It was gonna be dinosaurs.

And finally it was Meshach's turn. But Meshach just decided to mix it up a little.

He decided that he, and his entire family line, would one day... be-come some really weird animals... that basically just looked like they were mammals wearing dinosaur pajamas.

So, like, one of his descendants just looked like a sabertooth tiger... mated with a velociraptor. And another one just ended up looking like a hippo... mated with a triceratops. Very strange animals. That's why I just say... they looked like mammals... in dinosaur pajamas.

And their official name was the "synapsids," but I just like to call 'm the dino-mammals. ... even though they weren't even remotely related to the dinos. Their last common ancestor was Abednego and Meshach - tens of millions of years earlier. ... But you've gotta give 'm credit though on their follow through. They were old school. A promise re-ally meant something back then.

And follow through they did. Meshach's descendants looked weird. One of 'm just looked like a little dino dog... pretty much the size of a little ground squirrel. So, it basically just had this like little Pomeranian's face, but like with a ground squirrel's body, and no hair yet. ... - just this sort of smooth dino skin. ... like some kind'a pet you would expect to see in a scientifically inaccurate stone-age family.

Not to worry though... the little dino-squirrel-Pomeranian was slowly getting her hair. ... - slowly but surely. ... So, for a while there, she just looked like a sort of hairy little dino squirrel dog.

It still blows my freakin' mind though, that they literally had a little *dog's* face. ... ...

I mean, freakin' seriously... ... ... what are the odds?! ...

I can't believe that these little doggy faces are actually... hundreds... ... of *millions*... ... of years old.

Woe...

Just wait for your buddy to get really high and then start explaining that to 'im.

Pfff... forget the high part. ... Just start explaining it normally. ... Freakin' science is crazy enough.

Alright... what were we talking about? ... Oh yeah... our little Pomeranian dino squirrel. ... Alright, so this little gal was pretty humble. She kept a low profile. She just nested in a burrow in the soil. And her little diet was pretty simple also. ... - basically like a squirrel's diet.

And even though she had a bunch of massive synapsid cousins, and they were smashing up the world all around her, she just kept on being humble and keeping her head down.

And then a major mass extinction came and got 'm. The only ones who made it were just these little ones. And then a whole new class of monsters rose to power - it was the actual dinos this time. And these guys were worse - way worse. They got bigger and badder and louder.

But she just kept on being humble. And as the dinos shook the landscape all around her, she just slowly kept on changing and becoming more mammal-like. She got hairy because it helped her to stay insulated. She got warm blooded so she could become more active. She got breast milk so that her babies could grow up stronger. And after enough of these upgrades had happened, she eventually became what we would recognize as a modern mammal.

And then somewhere along the line, this little dog faced squirrel girl started playing around in the tree tops. And then another mass extinction came and scorched everything. It obliterated pretty much all of the dinosaurs. ... - all except a few - the fluffy, warm-blooded ones we call birds now.

But, just like the previous mass extinctions, it was mainly just the little things that made it. ... - at least on land anyway. So, now that's TWICE that our little squirrel girl survived a mass extinction. And now the dinos were all gone, and it was finally her turn.

And this where her descendants just started diversifying. They went in all sorts of different directions. And once again, the descendants of Meshach would rule the planet. ... but this time they were doing it as modern mammals. They became the hippos and the grazing wombats and the saber tooth tigers again, but this time they were finally doing it without the dino skin.

And of course some of them just remained within the tree tops. Then their paws got better at grasping onto branches. Then they developed some opposable thumbs, with some really long skinny little monkey fingers. And then they slowly started shrinking down their snouts... so that they could eat a diet that mostly fruit based.

And then time went on, and eventually some of these little monkey critters turned into the ape family. And then one of these larger species discovered the beach one day. Then they started eating shell-fish and wading around upright. And so, they ended up with stork legs and a naked hippo body.

And then these foragers started joining up in mating pairs – that competition drove their brains to a whole new level. Then these brains started helping their families become specialized. And so, they traded their marketable skills in a symbiotic relationship.

So, these mutually beneficial relationships had thus become their new normal. It was no longer about the law of the jungle. There were no more alphas with the authority to just command people. I you wanted to get something done, then it had to be reciprocal. ... – you had to make a deal.

And so, that's basically the story of us - from amoeba to a beach bum.

And the only reason that I mention this history here is just to point out that there's a pattern in it. ... See if you can spot it.

First things first, we had Otis and Milo teaming up together. Then we had Siam and Misa sticking it out together. Then their tadpoles started forming into row boats. Then these row boats started forming into tube structures. Then the tubes just started swimming and got a lot bigger. Then these swimmers became a fish, and one of these fish learned how to breathe oxygen. Then these breathers started walking, and became a salamander. Then these walkers got a little tiny dog face. Then these dino dogs started getting a lot more fuzzy. Then these squirrel dogs started climbing around on branches. And then these climbers started swinging and eating fruits a lot. Then these monkeys got a lot bigger to dissuade the jaguars. Then these big ones became the apes, and one of'm found a beach to live on. Then these beach ones started forming into pair bonds. Then the bonders became much more intelligent and started developing mutually beneficial relationships. ... ... And Bob's your uncle.

Did you see the trends in there?

We were never the sharks or the squids. We were the minnows.

We were never the freakishly massive amphibians, the size of a water buffalo. We were the little salamanders.

We were never the lumbering, violent synapsids. We were the little dino dog.

We were never the towering and tyrannical dinosaurs. We were *still* the little dog squirrel.

We were never the lions and the tigers and the bears oh my. We were the monkeys and the apes and the forager guys.

Do you see the trends?

We've... been... humble...

We've almost always been a little tiny creature.

And we've often worked together and created something greater than ourselves.

And our diet has almost always been quite simple.

We've outlasted and outperformed nearly everyone by simply manifesting existential humility.

There's an old saying that the humble will inherit the Earth. ... - not the fearful or the obsequious... - just the humble.

And just because something's an old saying doesn't necessarily make it prophetic. It could just as easily be some old bullsh** for all we know. ... But in this case, it's really pretty obvious. The trends are very clear and undeniable. Fortune often favors the bold, and the future undoubtedly belongs to those who also know humility.

Now, there's just one more thing I have to address here. I know that some people out there might not take too kindly 'roun' here to me using these religious names for some of my evolution characters. But here's the thing... ... I'm not using anything - those are their actual names.

And I know you might say that a lot of this is blasphemy... ... but all I have to say to that is... ... I know you are but what am I?

I see your blasphemy... and raise you... one blasphemy. ... ... IIIIIIII say it's blasphemy... to claim that what IIIIIIIII'm saying is blasphemy. ... ... IIIIIIIII say it's blasphemy... to presume that some super epic genius god beings, like Doctor Manhattan, are just sitting around all day trying to control everything.

Because if there's one thing you can say about intelligence... it's that intelligence... loves... novelty. It likes learning and science and music and drama and a million other things. It's not intimidated by new stuff - it freakin' loves it. It loves to discover. It loves to investigate. It loves to explore things. How can you ever explore things if you never let nature run its own course? ... if you just constantly try to force into this static, unchanging freeze frame? ... That doesn't sound very interesting.

And mankind gets a free will, so what about nature? Doesn't nature get to pursue a destiny? Doesn't nature get to adapt its convictions?

And besides, how humble is it to just automatically assume that you're superior to everything? ... that you're just better than everything else in nature... just because you've inherited a form of intelligence that's been modified for wanting to manipulate its environment? ... That's not very humble. ... It's precisely the opposite, in fact. And different doesn't equal better - it just equals different.

And just because we've been given more power doesn't make us better. ... It just means we've been given a greater responsibility to be humble.

And besides... what do you care if your human body evolved? ... It's just your Earthly vessel... It's not your truest immortal essence. ... It's like, seriously... who even cares? ... It's just this thing you're briefly using for a moment - for like one little blink of eternity. ... So, who the heckfire even cares?

Am'a right?

Or am'a right? ... Or am'a right? ... Or am'a right?

Or'am'a'right? Or'am'a'right?

Or'am'a'right'or'am'a'right'or'am'a'right?!

Or'am'a'right?

"Ned Ryerson?! ... ... I have missed you... ... soooo much."

That's just a quote from our new, Punxsutawney holiday coming up.

But people aren't really worried about these arguments though. They just feel like evolution somehow negates the possibility of a spiritual dimension. ... as if those two things are just automatically mutually exclusive. ... But that's like saying... that just because there's ripples on the ocean, then there can't be any waves or any tidal rhythms.

Uuuuuuuuuuuuh... ... no. ... They're just waves. ... They can easily coexist in the same exact place at the same exact time. And they can break. And they can bend. ... - albeit just briefly. But sometimes briefly is all you need. ... But never mind all that.

Acknowledging science isn't a negation of spirituality - it's actually allowing for it. The more we learn about the fabric of our reality, the more it opens up the possibility of these other dimensions. ... Like I said... spirituality isn't anti science - it's actually the ultimate science. ... If anything, it should be the strictly materialist atheez trying to run around and block science.

I've been studying just about every subject imaginable since I was a little kid. I just instinctively knew that everything was connected. ... and
if you wanna get some clues about your favorite subjects, then there's a
really good chance you can find some in another one. So, I just always
studied everything. ... even the paranormal. And I learned all kinds of
stuff. And I became like a little magic girl who always knows all the
answers in magic class... and just keeps raising her hand constantly.
"Pick me!! ... Pick me!! ... I'm a little smarty pants!!"

... except I'm not a girl.

... and I didn't have a fro.

But I did study the paranormal. ... among other things. ... for decades
actually. And my friends and I found all sorts of stuff. And there is
definitely more to this reality than meets the eye.

But there's also a lot of bad arguments too. There's been a maaaassive
amount of stuff that people claaaiiim is some sort of proof... but actually isn't. I could go on for days. ... about all the false evidence out
there.

One of my favorites, though, is when people say this one - "uhhh, you
know that consciousness continues, because you can't destroy energy."

Ugh... ... It's painful for me to even repeat that. It's such an outra-
geously erroneous thing to say. ... Can we just please stop saying that?
... Consciousness is STRUCTURED energy. It's not just "energy."

If I take your structured energy consciousness and I put it in a blender,
I will destroy it. The structure will be gone. ... - not the energy... ... -

the structure. ... "Energy" doesn't mean "holographic ghost body."

So, if you wanna say something different like, "uuuh... we know consciousness can still exist, because, like, molecules are like, cymatic cavitation fields, or like aether or something... " then fine! At least that actually sounds plausible. ... But can we PLEASE stop parroting this "energy" thing?! ... It's a mEANINGLess statement.

It's like using the word "literally" wrong all the time.

Oh... and another one that drives me nuts are the ghost hunters.

I used to be a ghost hunter. That's how I know that most of it is a bunch uh gobbledygook. My friends and I used to go around and film all sorts of haunted places. We went to some of the most horrifying places you can imagine. ... - places where some reeeaaally bad stuff had happened. I'm talking like, murder sites, abandoned insane asylums, haunted old historical buildings, ghost towns, old prisons, old hospitals, old places where they used to hang people and execute them. ... - all sorts of crazy stuff. We wanted pROOF!!!

Probably the worst one, though, was this old abandoned insane asylum we used to go to. It was all old as hell, with like all this creepily antiquated medical shit... with like straps on it for holding people down and stuff.

And of course you had to go there at night. ... because the whole entire compound was gated off. ... and you had to sneak in. So, you'd just be going around through the corridors in total blackness, just hoping to God your flashlight didn't fail. ... because you'd be completely f***ed without it. You would never find your way out. ... especially down in the basement. ... That basement was fu**ed. It was so dark, when you shined your light down a corridor or something, you couldn't even see to the end of it. Your light just ended in blackness.

My friend Rob always wanted to go deeper into it. I was like, "f*** that." ...

He always wanted to go back where it was all flooded out with water... and just go wading around in it. I was like, "f*** that. ... I'm not going for a fu**in swim down here. ... ... in the fuckin pitch blackness. ... with like some eighty year old demon water."

... fuck outta here.

Yeah, that place was a freakin' safety hazard. ... ... hence the gates around it. ... - the ones that we so eloquently bypassed. That was actually half of the terror. The whole roof was all caving in all over the place. And you never knew if you would just like, take one wrong step and go crushing down through like several floors of the building.

GJJJJ!!! ... GJJJJ!!! ... GJJJJ!!! ... GJJJJ!!! ... GJJJJ!!! ... GJJJJ!!! ...

... still end up in Rob's water anyway.

"SomebODY HELP!! ... ... SomebODY HELP!!"

The worst part though was having to walk through those creepily tiled corridors. And you had to walk past every single open doorway. And of course every single room was just pure blackness. And you were always terrified to shine your light in. Because no matter what was inside it... it was always gonna be something messed up. ... like, no matter what it was... just like a fuckin wheelchair sitting there all by itself. And you were like "what the fUCK?!" ... "why is there a fuckin wheelchair by itself?!?!"

And like, an old timey, weird one too. ... with like, old steam punk contraptions.

Ububububububub... still creeps me out.

But anyway. In all our years of taking our girlfriends to those horrifying places, we never once observed anything metaphysical. Like, we never saw like a wheelchair just come wheeling out on its own, and then go wheeling around by itself. Nothing like that.

That would be messed up if it did though. ... - just wheels itself out all incrementally through the darkness... as if someone's feebly pushing the wheels forward. ... - r'e'e'... ... r'e'e'... ... r'e'e'... ... r'e'e'... ... - slowly turns to face us. ... - i'i'i'iiii'i'i'i'i'iiii'i'i'i'i'iiii'i'i'i'i'iiiiiii...

Fuck that. ... ... I'd be outta there so fast. ... ... as soon as I finished recording my video evidence like a good white person... ... I'd be ouTTA THERE!! ... ...

... soooooo fast!

... as soon as I got done recording!

You don't even knOW! ... once I finish recording!

Nah, that's more of a mad scientist thing. ... - just willing to do whatever for science. ... And that's why we did it... ... - for scIENCE!!

... and for the effect it had on our girLFRIENDS!

... but mostly fOR SCIENCE!!

And I can prove it... because sometimes I just went by myself. ... - not to the insane asylums. Fuck that. But to some of the more comfortable horrifying places.

... fOR SCIENCE!!

... and cOMFORT!

One time I went to this old execution site that had this really old hotel nearby. And it was still in use actually. ... - the hotel I mean. ... It was so old it had like this pre-Victorian, early colonial look to it. It looked like something you would see when New York was still becoming urbanized. ... - back when biggest things around were just a few stories tall. ... and you could still get a world famous New York oyster for a nickel.

So, that's what it looked like. And the proportions were all weird and warped from being so old. Like, if you were standing in one of the hallways, you could see how it sort of bent and curved a little. And the lighting fixtures were all ill-conceived. So, there were just all these dark spots up in the corners by the ceiling everywhere. And it was weird too - it was almost like the darkness was pushing back against the light. ... - like, almost squeezing it and confining it to just these little areas.

One time I was up late at night during the witching hour, around three in the morrow, just sitting in this dark unoccupied room on the second floor. It was like an old room where they used to play poker, and then lose their sh** and start gunning each other down.

And I was just sitting in there.

... in the darkness. ... waiting for something to happen. ... And I'de be lying if I said I wasn't terrified. ... ... But I'm also pretty agro. ... So, I persisted.

I was just waiting there for something to go like, flying off the wall or something. ... And I ended up sitting there for what felt like forever... ... until I finally started to notice this subtle whispering. And then I realized that if you just listened really really closely, it actually sounded like someone was trying to say something. ... And not someone in another room or something - it actually sounded like they were in the air, all around you... kinda like moving around a little. ... and then getting really really close to you. ... like right up whispering in your ear sometimes.

Yeah.. ... ... all of my hair was standing straight up. ... ... I was ready to flip that card table and make a run for it. ... But I just did that thing you do when you're going over the edge of giant roller coaster and you just brace yourself. ... ... But in my mind I was like, "these mother f***ers are whispering in mY EAR NOW!!!"

Ubububububububub... ... still creepy...

But I just braced myself and waited. And I just kept listening and listening. I was trying to make out what they were saying. ... trying to catch some words here or there. ... ...

But then after I had sat there a while... I finally realized what the whispering was... ... It was the freakin' aIR vent!!! ... It was that little thing up on the roof that starts to spin when there's a little gust of air rustling past it. ... There was NO mistake. It was that exact little tinny spinny sound. ... - clear as a bell. ... and with that exact same little periodic, brush motor effect. ... and even synchronized to when the trees would start to shimmy outside.

I was like, "oh man, I'm a freakin' *idiot* for being afraid of that."

But that's the sort of thing that a lot of these ghost hunters would just record on their little cassette players and then assume was an apparition voice.

See? ... You gotta get to the bottom of stuff.

But in my quest to either prove or disprove this haunting I did accidentally created another ghost story. There was this short little moderately heavy set girl who had gotten up to go use the restroom. And since the hotel was all old-timey, you had to go all the way down the hall to the very end to use the common restroom. And, of course, the place was so wooden and creaky that you could hear every little footstep - i'i'h... i'i'h... i'i'h... i'i'h... i'i'h... i'i'h... i'i'h...

But then on her way back, she caught a glimpse of my silhouette in the faint little moonlight, just sitting there in the darkness completely motionless, and she sort of tried to play it cool for a couple uh steps, but then her footsteps just shot up by like a factor of five - ii'ii'ii'ii'ii'ii'ii'ii'ii'ii'ii'ii'ii'ii'iih...

I just became a legend in her family folklore. ... - the time she saw a ghost sitting in the poker room. ... "Why would any nORMal human being just be sitting in there like that?!?!?!"

I even interviewed the security guard down in the lobby. I got bored of all these fake ghost whispers. And I thought I better get outta there before I create another fake ghost story. So, I thought ol' Billy or Bobby or whatever might have something interesting for me.

He'd been working there for a number years at that point. And he was still terrified of the place. But upon questioning him, I found out that he had never actually witnessed anything.

And he was this funny sort of guy too, sort of like a big simple farm-hand type. And he only had this one scary story from all of his years there.

One time he thought he saw an old timey man just hovering there staring at him, in the corner of his eye. So, he just froze there immediately, right in mid stride. And then he just spent an entire hour of his life like that... - just standing there completely immobilized. And then after about an hour of absolute terror, he slooooooooooowlyyyyyyyyy tuuuurned to see what the thing was. ... And then he finally laid eyes upon the insidious specter.

It was a coat rack.

... - the same exact coat rack that had always been there. ... ... as he told me.

But he was already so terrified at that point... he just assumed it was a haunted coat rack.

So, he basically just waisted one entire hour of his life like that, just standing there perfectly frozen. ... like that famous picture of the

Sasquatch with his arms out.

... for an hour.

Too bad it didn't catch him right as he was lifting one foot off the ground. ... ... - just freezes him there like that.

One legged Sasquatch.

At least the guy was honest though. He didn't try to make up some fake, mumbo jumbo, ghost story. ... ... Touché, my good man. Touché.

He was too terrified to make up a story.

Alright!!! Anyway. To wrap that all up... I had one final prospect to find a haunting there. ... - in the "hell room." ... as they never called it... but they should've... cuz that would've been a good name for it.

There was this little old boarding room up on the top floor, which, as they attested, was the single most haunted and hell-raising location on the entire premises.

The room itself was completely boarded off. Supposedly it had become too dangerous.

It used to be the favorite room of this well-known outlaw guy. He would come and sort of lie low there for a while. And I guess he had a good reputation for being a douchebag. He would always gamble and get belligerent.

One night he was playing against the owner, of all people, and apparently he was on winning streak. The owner just kept on losing and trying to get his money back. So, finally, the owner just bet the entire hotel. And then he lost it. And then this outlaw guy just meandered up to his room all giddy and drunk. And then the owner just had some guys come up and execute him as he entered his doorway.

And he was neither the first nor the last to be murdered in that area... but he was definitely the angriest. He had been cheated out of an entire fortune.

And actually, now that I think about it, I wonder if that was a scam that the owner was pulling - you just play these unsavory types until you beat'm finally... and if you accidentally lose too badly... you just kill their asses. ... ... Yeah, I guess that would work. ... ... until they start haunting the shit out of you.

But, yeah... I guess that room had just become too dangerous after a while. Too many people were getting attacked in there by something. So, they finally just fixed it all up how he liked it and stopped letting people in there.

... according to legend.

And I know what you skeptics are thinking. You're thinking that obviously it's a gimmick. That's why it's only a little tiny room, way up on the top floor that they're boarding off, and not some big fancy suite at ground level. And that's exactly what I was thinking... ... until I saw my room.

It turns out, in fact, that half the freakin' rooms in there were just these little tiny boarding rooms. That was place was old school. ... My little ass room was even smaller than the hell room. ... You could just reach out and touch like everything from the little bed where you were lying. ... Lucky for me, I was already used to these small conditions.

... and the sounds of people going up and down the stairs above me.

And I was actually right down the hall from the hell room. I was trying to get as close as possible. ... even though they told us, under no uncertain terms, to *always* stay away from his doorway. And then they told us very gravely, no matter what you do, neeeever try to talk to him.

... especially in a rude way.

So, I went ahead and did all those things. And I did it at night when everyone was sleeping... and when everything was even darker than usual. ... especially that area.

The hell room was sort of further back towards the end of a long corridor. And as you looked right down its maw in that direction, the walls just sort of faded into blackness. And that was kind of a weird thing too, because there were other rooms back there with guests in them. So, it was like, "why is it all dark as hell back there? That's like a safety hazard."

But, yeah, I was pretty terror stricken at this point. And none of my earlier successes had made me feel any better about it. But I was like, "f*** it," and I just walked back into the darkness.

But once I got back in there I could sort of see a little better. There were these little like, old school windows above every doorway. And there was a little bit of moonlight coming in through old man Jangles's window. ... - or whatever his name was. And I guess like everyone back then had malnutrition, because all the doorways were really short, and I'm a little on the tall side, so I could juuuuuuuuuust see in through that little window above the doorway.

But as soon as I did, it was freakin' bone chilling. You could just see his weird little bed where he used to sleep. And there was this like miniature little night stand next to it. And you could just make out the shape of an old oil lamp. But everything was like tilted and broken and warped. And like one side of the room was faintly lit because of the window, but then the other side was pure blackness. You couldn't even see into it... ... ... ubububub... still makes my spine tingle...

But I kept at it. ... ... And I was even thinking about shining a flash-light back in there. ... But then I was like, nah, dude, I'm probably not gonna like what I see in there... There's probably like some old creepy sh** just sitting there staring at me. ... ... probably like a wheelchair in there.

So, I decided against it. And instead I just braced myself and started breaking the next big rule - the next big mortal-epic-sin they tell you not to do. ... ... I started talking to him... ... ... directly.

I did one of those loud whisper things - and I forget his exact name - but I was like, "Hey!! ... ... Old man Jangles. ... You in there, boy? ... Hey, how you doin'? ... ... You wanna come out and say hi?"

Oh, man, I was freakin' horrified. In my mind I was like, "Oh my f***ing god... this dude's about to come out here, now that I invoked him!!!" ... ... And I was just waiting for like, some sort of icy chill to just go walking right through my body or something...

... or like some crippled old lady comes out of nowhere and knifes you in the back.

"A'A'A'A'A'A'A'A'A'A'A'HHHH!!!!!!!!!!!!!!!!!!!"

"SomebODY HELP!! ... ... SomebODY HELP!!"

But nothing happened though. ... not yet anyway.

It was quiet. ... ... And I was just way too afraid to look in through that little window again. ... probably get knifed in the eyes or something.

So, then I finally just worked myself up for a final battle... and I did the absolute last thing they told us not to. ... ... I got rude.

"Hey, old man Jangles! Why you ain't comin' out? ... You some kind'a bitch? ... ... Why you hidin' in there, boy? Why don't you come out here, you old bitch?"

And you wouldn't even believe me what happened next. ... ... after I finally took things to that absolute most unconscionable of levels - the ONE they told us not to... ... ... I almost ended up dying that night...
... ...

... - almost dying from boredom.

I guess old man Jangles tries to bore you to death.

That son of a bITCH.

18

9 ¾ 7 -- YOU'RE SITTING ON THAT BROOMSTICK WRONG

Luke... I am your father... - not Gary. He's just a friend of your mother. You don't have to do anything Gary tells you, Luke.

I pitched that idea to my old boss. But he just got all butt hurt and started choking me out with his mind.

A lot of my ideas get rejected actually. I'm not just this perfectly flawless human that you think I am.

I wanted to open a gym for women, and have like a realistic movie set in back, where it looks like a child is being trapped under a car, and then the woman's instincts are triggered and she just lifts the whole car off of him and gets like super buff.

I just want to help women realize their full potential. ... But then the stupid lawyers were just complaining about everything. "Myeh, myeh, myeh, myeh, I'm a stupid lawyer."

I just wanna help people.

I'm working on a plan right now to go out to some of the most starving parts of the world and bring those people candy.

I just wanna bring candy to starving people.

"What the fuuuuck is this? I don't want fuckin' candy. This is a billion times too sweet."

For just three cents a day, you can join me on this quest and become a sponsor.

And while I'm feeding the world with your money, I'll also be helping them to learn important life skills - skills that they'll need to survive in our modern world. ... like how to speak Australian.

"Oy rang moy mates to gay dayn to the baych, because Oy wuan'ed to gay 'av a bid oh vuh swim..."

"Auoo'strayans loyk to *gay* places."

Australians are a special people. They're like Italians on the outside and Canadians on the inside. It's basically like if you took a stromboli pizza pocket and then slit it open on one end and then forced it inside out and then put one of your hands in it like a pizza puppet. That's basically what you should think of when you think of Australia.

That's important too, because learning a new language isn't just about learning the vocabulary - it's about learning the culture. You see, it's a well known fact... that if you wanna get a hundred Canadians out of a swimming pool, all you gotta do is say, "hey, would you all mind getting out of the swimmin' pool, eh?" and then they'll be like, "yeah, sure, no problem, one sec. Le'me just grab my maple syrup and I'll be out in a jiff."

And it's a further well known fact... that if you wanna get a hundred Italians out of a swimming pool, all you gotta do is say, "hey, would you all mind getting out of the swimming pool?" and then you'll be met with a barrage of HEYs and noncompliance.

However, if you wanna get a hundred Australians out of a swimming pool, you just gotta say, "hey, would you all mind getting out of the swimming pool... you fuckin' cunts?" And then they're like "alright, al-right... we're gayin', ... hold yuh kangaroos, yuh bloody cunt."

You see?

It's a different culture.

Or, if you wanna translate that into the British language - "Eet's uh different cauwchuh."

Oh, and since we're teaching people how to speak Australian, we might as well teach'm a little English while we're at it. Apparently if you put a gun in people's face they start to learn English REAL quick.

That's gonna be a weird classroom experience though. ... I don't know. ... Maybe we'll hold off on the gun teaching for now.

"Your word is 'CAT', mother fucker. *SAY IT!!*"

Yeah, maybe not. Maybe I'll just stick to my t-shirt cannon idea. I really love helping poor people. But, like, I don't wanna get like physically close to them or like physically touch them in any way or whatever. ... So, like, I just shoot all my donations at'm with a t-shirt cannon.

"You're damn right I can spare some change. Where do you want it? The chest? The leg?" ... ... !!!!BOOF!!!!

Not really. I just give'm food. I set it down at a distance so I don't have to get close. I put it down like a scene in a movie where there's a stand-off and one person has to set their gun down all nice and easy. "Hey, I just brought you some food, EASY... EASY... I'm setting it down. Everyone be cool. I'm backing away. Woe woe woe I said *STAY the fuck aWAY from me!!!*"

I just like helping people.

Sometimes I do volunteer work. I go stand on the street corner with a sign that says, "Will hand your donation over to this homeless guy so you don't have to get close to him."

Alright, alright, alright... Just relax. ... Just take a deep breath. ... NO-BODY'S trying to dehumanize homeless people! ... Just take it easy. ... We're just brainstorming here.

We're just trying to make the world a better place, you bloody c...

Oh, hey... before I forget... I forgot to teach you the other one. How to get a hundred jews out of a swimming pool'WOOOOEEEE HEEEEEEY WOOOOOOEEEEEEEEEE... WHAT'S GOIN' ON HEEEEEYUH?!!!! ... ... WHAT IIIS DIIIIIIIIIIIS?!!! ... WHAT KINDA BOOK IIIS DIIIIIIIIIIIS?!!! ... WHEH DID ALL THIS HATRED COME FROM ALL DUH SUUUUDDEEEEEEEEN?!!!!

Woe!! Easy!! Jeez!! Alright!! Take it easy!! I haven't even said anything yet. ... ... Jeez! ... ... A lot like the Italians, these guys... ... ... - ancient rivals... ... but basically identical.

That was back in the Roman days. ... - back when they used to get physical with people. ... ... No, no, I'm not makin' that up. ... ... They used to make it physical.

True story.

But then they learned from their mistake.

WOOOOOEEEEEEEE HEEEEEEEEEEEY... ... WHAT IIIS DIIIIIII-IIIIIIS?!!!

Alright, alright, alright... just relax!! ... Everyone just relax!!

Okay, I'll finish it. But I'm gonna say it really fast. How'do'you'get'a'hundred'jews'out'of'a'swimming'pool??? You'say"hey,guys,did'you'know'that'swimming'is'a'really'good   ...exercise...?"

Boom!

Out in a flash...

WOOOOOOOOEEEEEEEEE   HEEEEEEEEY
WOOOOOOOEEEEEEEEE... I JUST GOT DAT... ... WHAT IIIIS
DIIIIIIIS?!!! ... ... We come here in good faith, AND NOW DIIIIIIS?!!!

Alright, alright, alright, let's get the hell outta here. ... This book is
about HELPING people. ... ... - not riling up the Mediterraneans.

And it's NOT true. ... Those are just hateful stereotypes. ... The sports
fields in the jewish community centers are NOT the most unused facilities I've ever seen. ... That is NOT an observation which I have
made.

Okay, okay, okay... It's time to save the world. ... And we can do it. ...
We can make the world healthier, wealthier, and wiser.

We can make it more humble and more ethical and more transcendent.

And that's not just platitudes and hyperbole. We can *actually* do those
things.

We can simultaneously make our lives better AND the planet better. We don't have to pick and choose. Everyone can win.

We can have our cake and eat it without getting crumbs all over. We can be like a rocketeer who doesn't barbecue his own legs off.

We can barbecue other things instead. We can barbecue a real natural meat product which is so incredibly ethical that even a lot of vegans are actually eating it. ... Seriously. ... - a real natural meat source that even a bunch of vegans are completely okay with. Not all of'm, but still a lot of'm.

And what IS this barbecue meat product?

Well, it's mussels and oysters. The single most nutritious and power packed meats on the entire planet. Both of these mind blowing bishells are actually nonsentient. Yeah! Seriously! They have no awareness. They have no consciousness. They have no need for one. They're just the animal world's answer to a vegetable.

WhaaAAaaaAaaAAAaaaaaAat?!

Yeah! They're like a plant. So, their tissue can still detect perturbations, but there's nothing actually hooked up to it. It's not connected to any consciousness. There's no awareness there. So, there's never any actual "suffering" taking place. ... ... just like a plant.

And that's still not good enough for a lot of vegans out there, and that's totally cool, but really though, it's not anything to worry about. If you're going to the dentist and they put you under, is there any "suffering" going on, inside of your mouth? Is the dentist do anything unethical to your mouth tissue? ... ... No, they're not. Suffering is an artifact of consciousness. ... So... no consciousness - no suffering.

Nature doesn't just arbitrarily dole out the power of pain perception. The ability to perceive pain is exclusively a feature to facilitate damage avoidance. And if you have no actual means of subverting this damage, then there is simply no justification for ever wasting precious resources on this. Otherwise, the act of simply mowing down your lawn grass would be an absolute nightmarish horror show. And nature would never do that. That would be insanely stressful. And stress is a killer. So, nature would never do something that preposterous. Only a human imagination would even conceive of that.

"Okay, well, then, like, what about the spirit of the grass or whatever? Wouldn't it be, like, electrically suffering or something?"

Uuuuuuuuuuuuuhhhhhhhhhh... Why in the world would the spirit world do that to an innocent little grass plant? ... That would be insane. ... If anything, the spirit world would be PROTECTING consciousness, not subjecting it to a pointless bombardment of torment. ... No... ... No... ... There's no "electrical etheric" suffering going on. Maybe a little bit of sadness... possibly... but certainly no writhing around in agony. ... No... ... No'no'no... ... Those little blades of grass don't give a eff. ... They just keep on growin' steady. It's one big grass concert. They're down there rocking out with their stocks out.

This rumor actually got started because of this one guy back in the day. He did this rudimentary experiment where he sliced through a tomato, and then he measured a little electrical signal, and then people just automatically jumped to the conclusion that the tomato was "suffering."

Uuuuuuuuuuuhhhhhhhhhhh... No... This is just us projecting our feelings. We live in an electrical universe. A trillion different things can produce an electrical signal. Go stick a couple wires in a potato and you can power a small light bulb.

And besides. A tomato is a fruit. It's MEANT to be eaten. That is precisely why the plant makes it in the first place. Its whole entire reason for existence is to be eaten. It will have failed its intended purpose if it is NOT eaten... ... So, with all due respect to the people who've been keeping this rumor alive over the years... the idea that a tomato is experiencing pain is fffucken ret***ed.

It's okay though. That's how we learn. We try stuff out and sometimes we make mistakes. It's just the price of admission. ... It's just the catalyst of progress.

But grass doesn't give an eff; plants don't give an eff; oysters don't give an eff; and mussels don't give an eff. These bishells are essentially just meat plants. ... - very very *magical* meat plants, yes... but still meat plants.

These foods are not just perfect for our hominin human bodies, they're also perfect for our humanity.

These foods are simpatico with our natural human psychology... barring any aforementioned schadenfreude.

They are perfect for our body, mind, and soul... - these are the foods we came up on. These are the foods that made *us*... into *us*. We literally owe our existence to these miraculous little bishells.

And as far as the others are concerned, the scallops and the clams aren't exactly vegan, but they're awfully darn close. Their level of "awareness" is essentially nothing. And their harvesting is easily instantaneous. So, on the scale of ethical food choices, scallops and clams are pragmatically perfect.

And clam chowder is the food of the gods. Let's not even pretend.

And fried scallops are orgasmic. Let's just be real about this.

And throw some tartar sauce in there. ... Good LOWD!!!

But either way, even the most discerning of ethical consumers can still enjoy mussels and oysters.

How insane is that though? ... the fact that two of the most perfect food sources for the human body are ALSO flawlessly ethical? ... ... *Seriously*. ... That is *insane*. ... It really makes you wonder about that old saying that the humble will inherit the Earth.

Oh, and let me just address something right up front. I know that a lot of people have unfortunately been given a very wrong idea about oysters. Like, they think that the texture is more like mangos or yogurt, but it's not. It's nothing like that. It's more like a perfect little provolone ravioli with a crème brûlée finish.

Oysters are one of nature's most convenient snack foods. They come sealed up hermetically between their half shells. Their shells have a sort of flinty, rock candy, camouflage appearance. One side is really flat, just like suction lid. And then the other side is a lot more rounded like a little bitty serving bowl.

If you get a fresh oyster and shuck it, the seal just pops open like a classic bottle top. It gives you that nice little hhhhwwwffffff'pOp... sound.

Then you lift off the lid by holding it from the cup side, and the interior is like a gleaming little pearlescent jewelry box. The entire circumference is just this brilliantly white iridescent nacre. It seriously looks like some sort of absurdly decadent food that only the wealthiest, spice-trading, space families could afford. You're just like eating off of literal jewelry nacre.

And then situated right at the center of this culinary masterpiece, immersed in a purified liquor of the most immaculate salt water, you've got your smoky vanilla oyster wedge, like a glistening leaflet.

And it's actually that salt water which gives them their slippery sensation. Salt water is inherently very slick. That's why they call it "soft" water. So, people will often *mistake* this for a type of squishy softness. But in reality, oysters are a very firm and tender truffle.

And if you wanna just consume it like an emperor, then just slide it right on back and then chew it a couple of times and savor. ... Or you can make about a billion different recipes out of'm. ... Either way.

Their flavors are extremely diverse, and you can customize them just like a wine vintner. They can be like cheeses or like meats or even olives. And you can cultivate their flavors on a level that's completely unheard of with any other meat variety.

And I don't wanna make this whole entire thing about exclusively just oysters, but they are *kind of* a big deal. We've been enjoying them on every single continent throughout all of human history. From London to Paris, and from the Orient to the New World, they are one of very few foods that both the modest and the wealthy have enjoyed simultaneously.

New York was a great example. Long before there was any such thing as a Coney Island hot dog, New Yorkers were already dotty about their oysters. Oysters on a half shell were everywhere. They were famous in the finest upscale restaurants, and you could always find them featured along the thoroughfares.

Yes sir. The Big Apple was once The Big Oyster... and for *centuries*, they were crackers about their clackers.

It was the same thing everywhere, really. ... - everywhere within wagon distance of a decent inlet. And it wasn't just because of their flavors. ... No... There was also something else. Oysters have had always had a sort of magical effect on us. They get people all "fired up." They get us all... "randy." They get a person all... "hot and bothered." ... if yuh know what I mean.

They get the ladies all fired up in the ol'... "Netherlands." ... - in the ol'... "Mount Pleasant." ... - in the ol'... "Misses Fubbs' parlor," ... as it was known back then.

Or... if you were a man... that would be the ol'... bits and bobs conductor. ... - the ol'... "Master John Goodfellow." ... - the ol'... "silent flute," as it was once known.

Or... as it was once more festively known at one point, the oooooool' "gaying instrument."

... ... You know... ... as in, the ol' bringer of good times. ... - the ol' maestro of merrymaking. ... - the ol' Tiny Tim of tallywhacking.

Okay, maybe not that last one. But it's just like the name implies... - the oooooool' gaying instrument.

And people really like that. They like having a merry ol' time. They like going out and just gaying it up every once in while. And so, that's why you could almost always find oysters around places of... "high spirits." ... - around districts of... "congenial congress." ... - around establishments of... basically just being a hole in the wall where you could go on down and "get your gob sluiced."

Oh, sorry... that just means going and "having a nice hardy drink." ... - just havin' a nice tasty beverage. ... - just havin' a nice refreshing libation. ... - just goin' on down, to your favorite old-timey eatery, and gettin' the ol' pie-hole, sauce-box... gob sluiced. ... ...

... Why, what did you think it meant? ... ... Oh... That's what you thought it meant?! ... ... HA!! ... That's funny... ... ... No, that's not what it means. ... ... not unless that's your partner's idea of having "a nice hardy drink."

But then in that case, that would be *her* getting HER... ... ... Alright, never mind. Let's keep this thing movin'. The point of all that is just this, oysters get you all jazzed up. And who doesn't wanna be jazzed up.

... ... Nobody... That's who...

That's why people in the last several decades have been rediscovering oysters. We've been getting back in touch with our roots and rediscovering what it means to be human.

So, now, all around the world you can find these massive oyster festivals. It's just a bunch of people partying and making merry. And there's a whole lot of gob sluicing going on. ... right out in the open.

But if you fancy yourself more of the genteel type, there's also a lot of oyster tasting around the world. And it's no less like-a-sir than any wine tasting.

But, like I said, I don't wanna make this whole entire thing about exclusively just oysters. Every single bishell is miraculous - oysters, mussels, scallops, salt water clams, fresh water clams, all of it.

... Oh, and by the way, "bishells" is just a term I came up with. It's better than their official name of "bivalves." ... - that one just sounds like a car part to me. So, I just call'm "bishells." That's a way cooler name. And it actually sounds intuitive. ... - bi (two), shells (shells) - bishells.

But these bishells can be used for making anything - any type of meat that you're accustomed to - bishells can do it all - pepperoni, sausage, burgers, bacon, deli meat, hotdogs, white meat, red meat, salami, steaks, sushi... any of it... They can do every single flavor you've ever tried, and a million that you haven't.

Mussels will give you those rich hardy base notes that you get from a red meat variety. And the oysters and the clams and the scallops will give you any other flavor. You can do anything.

You can cultivate their flavors just like a wine connoisseur. So, if you thought that the classic array of land meats was giving you options, you ain't seen *nuttin'*.

And I know that the word "epic" is a really overused one these days, but seriously, how freakin' epic is this? We can make all of the classic varieties in a way that is infinitely more healthy and ethical?! ... ... Seriously... ... How freakin' Earth shattering is that?! ... We don't have make these terrible impossible decisions anymore... We don't have to choose between our health and our humanity. We can have our cake and eat it, like a rocketeer with fireproof pants on.

Everyone can win!

And it's not like this is just some brand new, experimental new, science meat either. This is perfectly pure and natural, the way that God and Mother Nature intended it.

And anyone can raise these bishells on their own if they want to. You can have your own little aquaculture ecosystem. Tons of people do this already. Human beings have been cultivating bishells since the Neolithic era, and surely much much longer than that.

For example, the Native Americans of the Pacific Northwest, from Alaska down to California, used to be experts in cultivating bishell gardens. ... ... They used to be experts in all sORTS of things - *lots* of things that people still don't know about. ... ... wink, wink. ... ... tee-hee... ...

But we don't have to be keeping this dirty little secret anymore of how the sausages are getting made. We don't have to be putting a bunch of food on our plate that just freshly came out of a real-life horror show. It can be food that actually came from some place awesome.

When I was a wee little child laddie, they used to take us to all sorts of different places on field trips. They used to take us to like, coal mines and dairy farms and like old archeological dig sites and stuff. And it was actually pretty awesome. We got to see how things were managed. ... But they *definitely* never took us to a slaughterhouse.

Well, how cool would it be... if we were living in a world where that was different?

And our meat doesn't have to be this wimpy little, empty protein, flabbiness anymore. It can ACTUALLY be nutritious now. It can ACTUALLY be brain food. It can ACTUALLY have something more than just empty protein in it.

You could be munchin' on a hotdog at a sports game, and it could finally be a hotdog that's actually good for you. It could finally be a hotdog you don't have to feel guilty about. ... not on ANY level. ... ... except for maybe eating too many. ... ... but that's it!

Now tell me that's not fU**in' awesome. It's a win-win-win-win win-win-win win-win-win-win scenario.

... win'win'win...

And I'll tell you what else is a wining scenario... EFFICIENCY! ... These bishells are super efficient. ... They don't have to waste all kinds of resources on fighting gravity all the time. Nor do they have waste a bunch of resources on thermally regulating as much. ... "unduh dee watuh... ... down wheh eet's hottuh... ... unduh duh seEeEEeeEaAAaaAa!!!"

And bishells don't even have to swim around like a fish does. They just sit there like a plant and keep getting fatter. And what that means, is you get WAY more food output for the amount of food you have to give to them.

They're basically storing all of that food up instead of burning it.

And all you really need to keep them happy is just some nice clean water and some tasty algae. So, if you wanna raise these bishells up on the mainland, then all you gotta do is grow some algae.

You can grow it in big beautiful sun pools like they do in Hawaii. Or you can also just grow it in giant sun tanks. But the tanks are also translucent, so they look like giant green glow sticks. And that'll make you look like a mad scientist. So, it all just depends on what look you're going for. ... - which one of these food production facilities will *look* the sexiest. ... ... *That's* what really matters here.

But as far as the algae goes, they just need some sunlight and some minerals. They're a pretty cheap date. Just ship in the minerals and add it to the water. Not a big deal. Not a big expense. And then the sunlight and the algae do all the work for you.

... Talk about solar power. There's your freakin' solar power. ...

And if you live up in the latitudes, you can always just use a grow house. The Dutch have already been perfecting these giant, glass, grow houses for large scale food production for a while now. It's been *enormously* effective. They've been the world's *second* largest exporter of agricultural products. And this is all despite living on a tiny little postage stamp of real estate. ... ... *Tell* me that's not *frickin'* mind bending.

But you can't exactly stick a cow in a grow house. So, this efficiency doesn't really translate into the world of meat production. ... - not unless we had some sort of meat product that tended to grow like a plant does.

... Hmmmmmm... Now, where might we find such a thing?

Boom... ... ... There you go. ... ... You just use grow houses for algae and bishells. ... Now you have that ultra efficiency in the world of natural meat production.

You can do this in any climate, and it's enormously cost-effective. And it's extremely energy efficient. You can run this whole entire growth process off of just the power of the sun.

And as far as the whole *harvesting* process goes, it's roughly about the same as the other meat sources. With the other meats, you have to do all sorts of little detail work. You have to separate out the bones and all sorts of other stuff. It's a whole lot of tedious dissection. I used to have that job. It was freakin' gnar gnar.

But with the bishells, you don't have to worry about all that. All the work is just in popping open those hinges. And once you get those shell hinges open, all the meat is just perfectly ready, and there's nothing else to worry about.

And, of course, we can make machines that are extremely efficient at doing stuff like this.

So... as it currently stands... the actual harvesting of the bishells is still about the same as any other meat. ... ... HOWEVER... ... we can *also* make *this* process more efficient!!! ...

And this is another beauty of these aquatic lifeforms. These creatures don't have to support their own body weight. So, we can breed them up as big we want to, within a relatively short time frame. And if you think they were efficient before, just wait until they're gigantic and you can harvest one in under ten minutes.

And who knows... ... maybe in the years to come, we can have some like the super massive Venus clam. ...

But it won't take a Venus clam to outcompete the harvesting efficiency of other meats. All we need is some varieties of bishells about the size of a dinner plate. ... - basically how they were before we over-harvested them back in the day.

These creatures really are a gift. Even if there was nothing else special about 'm, we could *still* use them to outcompete the efficiency of every other meat sources. These lifeforms aren't just a gift - they're a freakin' miracle.

And, it's like I said, we can do this just about anywhere. We can do it way out in the country just as easily. We don't have to be near the ocean to do this.

And we don't need some super advanced technology either. This is almost child's play compared to what modern-day farming is already doing.

And we can do this on the areas of land that aren't really suited for growing plant crops. We can have fresh local super-meat in literally any community. ... - brainfood meat!!! ... - longevity meat!!! ... - meat that all the longest lived people on the planet are eating!!!

So, why we brewin' up some moonshine to get ourselves stupified, we can also be brewin' up some brainshine to get ourselves unstupified! ... ...

YEEEEEEEEEEEEEEHAAAAAAAAAWWW!!!

Country music ain't gotta be sAD NO MOE!!! ... ...

We can have more of those really fun country music songs... that sound suspiciously like a paid advertisement. ...

"Now coooome oooooon doooown...
It's time to paaarty dooooown...
We're just some regular folks...
that like to get oooon dooown.

But I aaaaaalways driiiiiiiiiiiiiiive respooooooonsiblyyyyyyyyyyyy... ...
...
when I'm sippin' on Milwaaaaaaaukee's fiiiiiiiiiiineeeeeest... "

... keeps on going on like that for like three minutes straight. ... ...
"What'ufff..."

"Was that the world's longest freakin' commercial... ... ... "

"... or is this dude A SHILL?!!!!!!!"

Speaking of shilling for stuff... I've got some of my own music f'r yuh...

Now cooome oooooon doooown...
We like tuh party on down ehhhhhh fuck it, I'll just tell you about it. ...
There's this stuff called kava beer, and it's a pretty fine libation in its own right. ... And I couldn't care less where you get it from - just gET IT! ... ... Not really. But it IS worth knowing about.
...

It's this beer that they've been brewing in the south Pacific for like several thousand years now. And it gets you all drunk and tipsy and stuff, but it never just annihilates you or destroys your livelihood. You just get all jovial and start chitty-chitty chat-chatting. ... and then doing an Irish jig that no one ever taught you how to do.

And you never get all sick and just start puking or anything. And you never get all trashed and then feel like garbage the next day. And there's never any negative impact on your organs either. ... as long as you're doing it traditionally like they do it down in sopac - with just like the roots and not the other stuff.

And it never gets you all sketchy and wanting to fight people either. It's exactly the opposite. It just gets you all smug and dopey. And it's even therapeutic, believe it or not. It actually calms your central nervous system a little. ... ... So, technically it's actually good for you.

And the reason that this beer is so different is mainly because it uses a different mechanism. There's no ethanol in it. So, then it also doesn't have all of those extra calories in it.

... ... So, yeah, it's basically like an old timey banjo song paradise. ... - all these sweet benefits that allow you to keep on being a drunk-ard.

But it's not cheap though. That's the catch. And there aren't any big lakes of it out there, where you can just row around and drink it for free. It basically costs what a regular drink costs in a big night club.

But that could all change in the future. If more people started getting into it, then the market would just drive the price down, and it would be just like any other beer. ... as far as the price is concerned.

Oh yeah, and it's not addictive. ... - not chemically anyway... not like alcohol. ... I'm sure it could become habitually addictive. ... Just about anything can become habitually addictive. ... But it doesn't have that chemically addictive property to it.

So, yeah, it's some pretty darn fascinating stuff. ... And I'm not really a fan of just disrupting our natural state of consciousness like this, but if you're ever in the mood to just geeeet oooon dooooown... with some regular folks that like to geeeet oooon dooooown... ... then you might as well do it legitlyyyYyyYYyYyYYYyyyy...

And I find it pretty darn fascinating, honestly, that you could just be sitting around with an ice cold beer in your hand, while you were chowin' down on a bratwurst and getting tipsy, and *neither* of those two things would even have to be bad for you. ...

That's pretty insane.

Creating this big, awesome future doesn't even have to look different. It's just a tiny little tweak from what we've been doing.

And this isn't like some hardcore, futuristic, space technology either. This is ancient freakin' food craft from the Neolithic and Bronze Age. ... - kava beer and mussel brawts. ... - literally what I'm already doing at social gatherings.

And the whole wORLD can be doing this. We just have to *do...* it...

And this isn't just the past – it's also the future. This is how we, as history's actors, are going to be questing across the galaxy. We'll have ships with a bunch of big ass tropical food forests inside. And those food forests will be punctuated with all sorts of beautiful aquatic systems.

Plants don't care if they're out there flying around the cosmos. Bishells don't give a hoot if they're out there floating around in zero gravity. These lifeforms kuh care less. All they want is some nice warm lighting, some nice fresh air, some nice tasty nutrients, and some nice clean water. That's it.

These lifeforms are already perfect for space travel. And then this way you can have these beautiful tropical food forests all throughout your ship. And shipmates looooove that sort of thing. ... especially crewman number six."~

That's better than having a bunch of slaughterhouse on your spaceship. ... or just like, keeping a bunch of animals in little cages. ... Imagine meeting another advanced civilization and then trying to convince them that we come in peace when we've got that shit in the back.

But lucky for us, though, we dON'T HAVE to!! ... It's cOMPLETEly unnecessary.

And that's another thing that's so cool about these aquaculture systems - they're a perfect reciprocal partner for plant life. You just take the old deposition on the bottom of the growth pools, and then you give it to the plants, and it totally supercharges them. ...

The plants provide us food and supply our oxygen, and then all we have to do in return is give them nutrients and growth lights. ... and maybe a little classical music. ... It's a perfect symbiotic relationship.

Bishells are truly miraculous. Wild populations of bishells can very efficiently keep our waters clear, and then captive populations of bishells are the most universally perfected food source.

And they don't even know that they're "captive." All they "know" is that we're supplying them with algae. So, we can cultivate them in extremely high densities. And they don't even care, as long as you're feeding them. ... If you tried that with any other animal, it would be cruel and inhumane. But bishells couldn't care less.

So, that's just one more dimension in which these bishells are A LOT more efficient. Just give'm that tasty algae and the world is their oyster. They'll be as happy as a clam with hired mussel.

It's all about that tasty algae, baby. As long as you're supplying them with that high quality micro plankton, then the ultimate consumable product will be absolutely impeccable.

The same goes for any animal. A lot of these factory farmed land animals are being fed terrible feed meal. Like, I'm not even gonna talk about it, because, just, you don't wanna know... but all of that stuff is going straight into their bodies. And then it bioaccumulates there at exponentially high levels. And then you come along and buy it in the grocery store and it goes straight into you. ... ... *Yikes.*

And it doesn't even matter where the animal is. Even the wild ones are still stacking up these toxins. There are dozens of little shrubs that are extremely poisonous to us. And these bushes can pop up anywhere and then get eaten by these animals.

But these animals, like many birds and many ruminants, are well adapted to consume this stuff without it hurting them. For them it's just like an herb or a tangy little spice plant.

However, just because these animals are eating this stuff, it doesn't mean that these toxins are necessarily being neutralized. A lot of this stuff actually ends up in their organs and tissues. ...

And sometimes ranchers will actually give their animals certain food items in order to "marinate" them before they slaughter them. ...
... Well, they can also be getting "marinated" with some *other* stuff too.

Many animals actually do this on purpose. They've been specifically adapted to seek out certain plant toxins. So, they just accumulate these organic compounds within their bodies. And when a predator attempts to consume one, they get a nice little dose of that poison. ...
... They like to get all nice and "marinated" for the predators.

So, if your animals are just consuming a bunch of random stuff, like they're just grazing in the open pastures or consuming some random feed meal, then you really have no idea, honestly, whether they're actually "clean" or not. ... You're just operating on probabilities, not guarantees.

However, if you DO control the quality of their food intake, then just about any type of animal product can be certified as copacetic. ... And bishells are one of the most straight forward and simplistic things to do this with. All you have to do is control the inputs. This is just basic, run-of-the-mill, Bronze Age, Neolithic aquaculture. ... And it's a lot more copacetic than honey - that's for sure.

But, again, if you're not actually controlling all of these inputs, then you really have no idea what condition an animal's tissues are in.

Take dairy for example. It's one of the most notorious and effective natural vectors for being exposed to environmental contaminants. ...
...

... - sadly.

I freakin' love dairy. ... or cheese, to be more specific. I usually use like coconut products for all of the other dairy applications. But cheese is just freakin' divine – even the nut milk cheeses. Most of the planet is completely in love with it. And believe it or not, we still don't exactly have a theory for why so many humans are so addicted to it. ... ... It's a mystery...

They've looked for any would-be addictive substances in it; and they've done tests to see if its various components would still be just as addictive in another context - namely the salt and the fat and the umami flavors. And they sort of are... but alas... nothing has *yet* quite explained why the humble cheeses are so addictive.

... or has it??????

If there's one thing that you'll definitely notice about the bishell community, it's that the people who are thusly initiated tend to get pretty bañañas about it. And then they do things like put on these festivals where everyone just starts going all Elvis-girl and taking oyster shots.

... And then like every time a person does an oyster shot, their eyes just start rolling back in their heads like they're having an oystergasm. ... ...

Does that reaction to a simple food item sort of remind you of something?

... any foods out there that aren't already loaded with a bunch of psychoactive chocolate or sugar?

... ring any bells???

... Ehhh, ... Pacha?!?!"~

Well, how 'bout this one? ... What if I told you that not every single culture is in love with cheeses? ... ... HUUUUUUUUHHHH?!?!

I'know'I'know... ... I know... ... It's crazy... ... It's a crazy world. ... But not every single culture is in love with cheeses the way we are. ... ... A few of'm got some kinda pROBLEM!!

But they also have another thing in common. ... ... - lots and lots and lots and lots of sushi.

These cultures never needed our cheeses... ... ... because they never lost the original addiction.

So, is it really any wonder why cheese is so addictive for us?

We love its saltiness and its tangy aromas and its delicate consistency.

And we love this... because that's exactly what we're built for.

We love this... because that's exactly how a *bishell* is.

Boom...

The future can be lit... ... ... if we let it.

And it will be... ... ... eventually. ...

I have seeeeeeeeen the promised land... ... and it... is... maaAAG' ... 'nIFICent!

Lemmeh tell yuh, brothuhs and sistuhs... ladies and gentlemen... and gentle-boss-modes... and all you othuh folks out there.

I have been. To. The. Promised. Land. And. It. Is. Mag. Ni. Fi. SUNT!!

Lemmeh tell yuh...

But first we gottuh problem, ladies and gentlemen. ... ... and all the rest. ... ... Yuh see... ... these anthropologists out heeyuh... they think they own us. ... ... They think we ain't got no say in how evolution goen be.

They think they just goen sit up there on they castle, and dictate us. ... tellin' us how evrehthang goen be down heeyuh. ... ...

Well, let meh tell yuh, ladies and gentlemen... thangs ain't goen be like that. ... 'Cuz we goen break 'm. ... We goen breeeeeeaaaaaaaak these chains... ... and we goen break these anthropologeez...

We gotta get'm in the ring... and we gotta break they body. ... We gotta put'm in the nursin' home. ... Then we gotta visit dem in the nursin' home. Then we gotta read'm their favorite stories and feed'm with a spoon. Then, when the doctuh finally says they body brace can come off, we gotta push'm 'roun' that courtyard to see the flowuhs. ... We gotta push'm in the wheel cheh, ladies and gentlemen. ... But first we gotta break'm.

... allegorically, yuh understan'? ... We goen break they a**es. ...

... metaphorically... ... of course. ...

That's what Apollo would say. That's not me sayin' that. That's just Apollo.

I would never say such a thing. ... But what I wOULd say... ... is that we all need to come together now. ... because we gotta fight these anthropologists. ... They're too cocky and too arrogant and too conceited. And they think they're invincible. ...

They're just sittin' up there in their ivory towers makin' jokes about us. ... callin' us aquatic apes and shit. ... makin' up swamp gorilla memes... like they think that shit's funny.

Well, I've got a little surprise for these cocky ba***rds. While they've been making their little giggly snickers... we've been gettin' stronger. ... We're a lean mean fightin' machine, yuh dig? ... We're like Apollo and Rocky and Drago now. And these anthropologists are like that one bloated Harkonnen guy from Dune.

... - just floating around all sloppy.

... makin' their jokes.

... swimmin' around in their bacon grease.

Vell... vee muzz break zem... comrades.

Zeez anthropologeez muzz be broken. ... ...

... metaphxorically, of course. ... Vee don't vant vun leettle hair on zeir precious head to be tushing... ...

... vhen vee !!!bREAk!!! zem!!!

Don't worry though. ... This is all for their own good. ... They must be humbled... so they can inherit the Earth with us.

It's the only way. We have to get better. We have to advance. We have to rise above our non-magical, mugg-blood parents. And we have to become like a little magical girl who always raises her hand in magic class like a little smarty pants. We have to learn how to be a little humble - like a little magical kid who had to grow up under the stairs.

And we have to learn how to ride these broom sticks properly. We've been sittin' on'm all wrong. ... especially these anthropologists. What they've been doing just looks painful. ... No wonder they can't fly straight. ...

We're supposed to be sitting on the big cushiony end. ... - you know, the side with all the fluffy bristles on it? ...

That's how you do it. ...

And as soon as we remember how to sit right, we'll all be flyin' like the dickens and catchin' snitches.

And as soon as we remember how to treat our bodies right, we'll all be flyin' around the landscape and learning tricks we never even dreamed of.

We'll all be happier and healthier and smarter.

We'll all be wealthier and longer living and more transcendent.

We'll all be protecting our little children and our little sheeps and stuff.

We can do this.

We can build a better world.

We can work wonders.

And that ain't even the half of it yet. ... ... Just you wait. ... ... tee-heehee...

The

End

... of this chapter.

Alright. We're done. Chapter's over. Get the fuck out of here. We're cleaning up for the next chapter now. ... One of you bastards spilled popcorn everywhere. ... And now I gotta clean this sh...

Wait... ... are you serious? ... You want more?! ... After all that?! ... You still want more?! ... ... Man, I already gave you a show. Just wait 'til the next chapter.

Get the fuck outta here.

What? ... ... You want more of thIS chapter?!

What the hell, I already gave you most of the highlights. ... What else could I even... ... ... wait... ... ... hang on a minute. ... ... Maybe we can help each other out here.

I'll tell you what. If every single one of you in this auditorium tonight signs my petition, then I'll give you ONE more performance... ... ... Alright? ... ... Okay, here. ... Yeah, you just take it and sign it and pass it on... ... ... No... ... no ... ... no, you gotta sign it first and then pass it... That's the whole point... ...

The pen? I just gave you the pen. You lost the pEN alrEADY?! ... on the third person?! Man, what the f... ... Alright, here's the last pen. Don't nobody lose this pen. I'll freakin' buy everyone in this auditorium their very own personal pen... if you just please don't lose this pen... ... ... What's that? Can you use your own pen? Of course you can use your own pen. Just take the common pen and pass it on then. ... ... Please... yes... thank you...

Why? Uhhhhhh, I already said why. ... ... Oh, I didn't? Uhhhhh, wait... yeah, that's right ... ... I didn't... uuuuhhh this petition is basically to demand that the Apollo Theater in Harlem lets me back in. ... ... I'm not allowed within five hundred yards of the premises...

Why? ... Uhhhh... Well, there were various attributions for which it pertains to notwithstanding the grievances of the counsel, forthstanding outright amidst the, as it were, remedial connotations pursuant to... ... yeah, exactly, yeah, it was complicated. Yeah, yeah. Exactly.

But basically, I went in there and started layin' it down with some freestyle beats like they ain't nevuh seen b'fore. ... And I'm'a tell you straight up... foos got jealous.

I went in there and started layin' it down like...

"Hey y'all! ... What up?! ... Yeah, I know you ain't been seein' nobody like me up in this beeyatch b'fore... but I'm fitsin' to set this shit off!

Alright, drop that beat...

mmm, mmm, mmm... mmmmm,
mmm, mmm, mmm... mmmmm...

I'm straight up God's gift to the world,
when I roll up in this club.
I'm straight up God's gift to these girls,

cuz they wanna know a thug.

They call me the illest cop-a-feel dis,
but I'm not sure what that means,
That better not be some joke shit,
like secretly making fun of me.

If I hear dat dat's some joke shit,
... like I'm some punk ass chump,
then I'm'a come back and do a drive by,
and shoot this whole club up.

Just playin', y'all, I ain't like that.
I would *never* do that shit.
I ain't no weak ass pu***...
... runnin' outta here like no bitch.

But for real doh, if y'all's makin' fun uh me,
I'm'a run outta here like a bitch.
Then I promise I will go to the authorities. ...
... and start squealin' like a snitch.

PSHHHHHHHH, like I would do dat!
Y'all should see y'all's trippin' faces!
All worried like y'all cayn't trust me,
like I'm'a talk about Marcus's hideout places.

Alright, now y'all bettuh back up off me,
'fo' I start swangin' BLOWS!
Y'all be makin' me incumftuhble at this point!
Y'all bettuh stop ackin' like hoes!!

Yannick, I know you dealin',

... and Essex, I know you strapped.

If ya'll goen keep makin' fun uh me,

I'm'a tell the cops all *dat*!!!

19

X -- TAKE THE CANDY BUT DON'T GET IN THE SAVANNAH

C'mon, baby, pull my hair and spank me... Mmmm, there we go... Now choke me and pour this bag of potato chips all over me and say "yeah you love it you fat bitch..."

AAAHH YEEESSS!

Okay now make it really dangerously cold outside for half the year and almost impossible to find any food to survive, you dirty... ooOOOOWWWAAAAAAHHHHHHhhhh

Oh yeah. There it is.

She's always like "But, you're not even fat." I'm like "Just do it."

Some of us are freaks and we LOVE weird environments like cold and snow and mountains.

Part of the reason for this is because the people in those areas that *didn't* like that kind of stuff either got so depressed and melancholy that they stopped passing on their genes, or they just packed up and left, leaving behind only us mixed nuts and masochists.

Now, this sort of thing may actually apply to you personally and you might be into some freaky pain stuff in your own private dance party time, but that doesn't mean that this activity is appropriate all day long, like when you're trying to get some coffee in the break room and someone sneaks up from the side and cracks a leather whip across your balls.

Just because we became total kinks and gluttons for punishment doesn't mean it was always that way or that this was our original default inclination. The deadly cold environments are not natural for us. That's why we have to wear thick clothes. Obvi!

We psychologically adapted to enjoy those environments. But despite this recent human adaptation, however, we will always have the monumental chapter of history that created us, and it will forever be written on our souls.

What is the first thing we do with our year the moment we're able? We go to the damn beach. Everyone does it! The people in godforsaken Siberia do it. The people who already live their whole lives next to the beach go off and find an even better beach. It's in our bones. We also love swimming in lakes and rivers. We're drawn to them. But there's nothing like a beach.

Most land animals try to avoid bodies of water. We seek it out.

What does every kid want to do? They want to climb things and swim. Well, that's what they used to want to do before video games and being trapped inside all day.

That was back before you suspected every other person in society of the most atrocious things and watched the mailman drive away like "Yeah, you better."

Then you went back to sipping your morning coffee and gazing out the window at your wonderful yard and wondering what it would be like if a futuristic hovercraft from the future came and hovered over your lawn and there were some police lights on it and then these robocop looking guys repelled down on ropes then ran over and kicked your door in and threw you on the ground and put their boot on your neck and said "You're under arrest" and you were like "AAAAAAAH! WHAT THE FUCK?! MY NECK! I DIDN'T DO ANYTHING!" and the robocop was like "Not yet, you piece of shit."

So back to the point. You think we spent all day on the hot a** sh**** Savannah staring at nothing in the suffocating heat with flies all over our mouth? And we did this despite having access to endless amounts of beautiful coastline whyyyyyyyyy exactly?

Why aren't we deeply and instinctively psychologically attracted to the scorching grasslands and plains then?

Like we talked about, you WILL be selected to love your environment. It will feel like home to you down in your bones. And you will love it, or nature will MAKE YOU love it. DO YOU UNDERSTAND ME?

Oooooor does this sound more familiar? Swim all day on the beach collecting clams and fish and lobster, then have a big feast with all your friends and family watching the sunset and listening to the waves, then sneak off and have special time with your favorite monkey "friend," then roll over out of breath and pass out in your tree house nest, then wake up in the morning and head out for a new beach adventure.

Does that sound like a vacation to you? Does it sound like paradise? You think it's purely coincidence? Nature didn't design you to sit around suffering all the time. People often gravitate toward suffering because there was so much of it during their childhood that it psychologically imprinted on them that suffering is normal. It's not normal.

A little trouble in paradise is normal. I bit of angst or anxiety is normal and healthy from time to time, but not straight up existential torture.

Returning to the Savannah, do you really think nature made us to desperately need to try to get our hands on as many thick ruminant skulls as possible and slowly fish the brains out through the nose before it rots? Notwithstanding the fact that you still can't get anywhere near the proper amounts of omega-3s or trace ocean minerals needed to support the brains of your whole entire troop in this manner, do you know how crazy fast brain matter rots? Do you know how fast it would rot inside the insulated oven of a ruminant's skull on the sweltering Savannah? Your monkey ass would have to chimp waddle over there to the kill leftovers long enough after the lions have left that they don't notice your slow butt and run you down for dessert? And this whole time of sitting around nervously waiting until the heat is off, that fragile brain tissue is just sitting out there inching closer and closer to a state of rancid.

And the vultures don't have to wait like you. They have wings; you don't. If the high-speed killing machines start coming back, the vultures can just flap their merry wings out of harm's way and say toodles. But if YOUR silly butt is still out there when those free-range organic *ass fed* serial killers show back up, then shit just got real and you just became slow seconds. So, you better make AWFULLY damn sure the coast is clear before you start sneakin' and creepin'. This ain't no children's film about two animal friends. This is the realness.

That being said, the vultures are going to have their fill long before you and your sad monkey self out in the Christmas cold looking through the warm toy store window is going to be able to get out there and get some. What's going to be left after them?

And you know what usually comes and chases the vultures away before any hypothetical jungle monkey could show up looking for flavor scraps to lick?

Hyena.

Yep, those sweet laughable characters that would probably share a giggle and a snicker with you while you clinked your imaginary wine glasses together and pretended to be proper ladies.

Except they won't. Hyenas will fuckin' kill any Pit-bull or Rottweiler you put them up against. They are massively powerful creatures. The documentary cameras can never do them justice. And you see them right after they show you the huge lions. So, by comparison, the hyena look small...

They're not.

The jaws of hyena are built for crushing — something like biological wood chippers. They're not so in to picking meat off bone. They'd rather just bite straight into everything, and you can hear and feel the insane crunch of solid bone if you're standing too close.

They're jokers, alright. Think joker like a raspy whisper asking you if you know how they got those scars — not joker like children's birthday party.

What's real fun is to be in a safari group where the hyena can see you all and then be a little bit of a silly joker yourself and go walk out away from the group and start hobbling with a limp so the hyena think you're the weak one, then watch how they all pop their heads up super alert and point their ears at you and zero in on your every movement and NEVER take their eyes off you or blink for a second. Even when you stop the limping act and go back to the group chuckling to yourself and stand in the very back behind everyone, they'll still be watching you. They will always be watching.

What you didn't realize when you did your silly joke, though, is that you actually put a permanent tracking beacon on your ass and revealed that you are the weak target. They are never going to forget.

They won't break their laser focus until they get an opportunity to run you down and kill you or until you get back in the jeep and says "C'mon, guys. I think we should go see some different hyenas. Haha" trying to play it cool and fake laugh like there's something funny, even though there's nothing funny at all, because the current observation point took hours to establish. So, no one else is laughing except you as you climb back into the jeep and close the doors behind you with the electric windows still up and the motor off and no AC, out in the blistering Savannah heat.

Then the temperature is so intense with the trapped air inside the vehicle that you are gushing sweat within one minute, and it looks like someone is having sex in the jeep even though it's just you in there alone by yourself. You would crack the door for some relief but it makes the doorbell dinging sound and drains the battery and you don't want anyone to suspect you for hiding. So, you think it's fine because you can just tough it out, and you can still see the hyena staring straight at you with a death glare through the foggy windows. So, you stay inside pretending to be chatting to an old friend and laughing at old jokes on the satellite phone for emergencies only.

Time passes and then finally the group comes back later to get back in the jeep and open the doors and then all the steam pours out and you collapse out onto the ground soaking wet and unconscious and they have to get you medevaced by heli immediately.

Safety is our number one priority while on safari, and safety is no laughing matter for the hard working men and women of the safari teams and medevac crews that proudly serve throughout the greater Savannah territories. When there's a weak link in the safety chain, it affects all of us.

I don't personally know who would ever do such a thing, but hyena are serious wild animals that deserve our respect. Don't joke around with them or the joke will be on you.

Now, where in blazing asses we're we? It was something with, OH, that's right. How the actual fuck are you supposed to sneak in between the lion-vulture time and the hyena time and access the leftovers of a kill? I mean, if you tried to shew off the buzzards then they would probably raise a ruckus and a fuss and develop some special call to draw in the hyena early and spitefully smite you, thus ridding themselves of their monkey problem. The Savannah theory is such flaming horse sh** on so many levels.

So, where exactly between the buzzards and the vice jaws of death are you supposed to slip in there and have a little taste? Can you imagine a chimp running in there and snatching a single bone and then trying to get the eff out and do its weird sideways dumb gallop run on all fours and trying to carry a bone at the same time? And you're not going to outrun a hyena. They gallop too, but more like a horse — you know, that big powerful animal that humans get on to go fast.

Never mind the fact that as a slow defenseless chimp you would never want to be caught anywhere NEAR the open Savannah. There is nowhere to run. There is nowhere to hide. You think you would just scurry up into a tree for safety?! Guess what else can scurry up into a tree — CATS! You think a tree is going to save you from a big cat?!

Sure, a tree may save you from hyena or a raging buffalo, but that's about all that a tree works for. Everything else can either shoot up the tree faster than you or just reach right up from the ground and snatch you out or just straight up crush the entire tree down with you in it.

If you were a chimp on the Savannah you would be the easiest snack in the world for the lions and cheetah. Why would they even bother at all trying to take on powerful buffalo or run down a turbo jet gazelle when they can just go snatch your monkey butt conveniently on display up in the tree like a hot Kalahari rotisserie.

Nestled deep in the heart of the jungle the chimps don't have to worry about the plains hunters, and they are physically large enough that they would pose logistical challenges to the occasional jaguar that may want to make off with them. But your size and monkey ferocity would not bother the big pack hunters of the plains in the slightest. For them it would just be more to love.

We've been discussing the idea of plains apes trying to scavenge like vultures, but what about the idea of them persistence hunting?

Can you even fathom a defenseless early hominid trying to run around after a gazelle without getting gazelled himself? Can you picture him running around out in the open far from any protection, and at that, running remarkably slow for a plains animal? Do you know how fast they would snatch him?!

Let's just wave a magic wand then and say he's able to slowly persistence hunt and successfully track this tricky gazelle using his simple hominid brain.

Alright. Now how is he going to keep his catch from the other predators? Other animals aren't that dumb. They would learn that the persistence monkeys run gazelles to death and then you can just go steal it from them. You could even get them TOO and get a two for one deal.

Or let's say persistence monkey is going to slowly carry back the heavy catch with a couple of his mates? If you thought they were slow before, just wait until they're carrying a whole other animal with them.

Or would they just eat it all right there and wobble back to the troop with huge bloated stomachs of bloody meat and then get back and throw it all up on the ground for everyone else?

Birdie style.

And what happens when the hunters start tearing into the flesh and release the blood smell onto the wind? Other animals have full on super power smell compared to us. They even have way better smell than chimps, because chimps are more visually oriented for spotting bright colored fruits and distinctly shaped nuts.

And as bad as a chimp's sense of smell is, ours is WAY WORSE! Why would we have such a flattened off snout and even way less power to smell than a chimp?! Could it be because the ability to smell doesn't help us gather beach food and because it does absolutely nothing for us underwater?

Everything in the Savannah has either speed or mass to protect them. Apes have neither. And we will see in the next chapter how fantastically delirious it is to suggest that apes were using spears or intimidation tactics to defend themselves.

So, let's set the finishing touch with this and just pretend for a moment that there actually were apes somehow beginning to become more advanced on the Savannah, and by some bizarre miracle they had available niches that defied the odds and that would be available uninterrupted for hundreds of thousands of years. Okay. Totally insane. But okay.

Because of the sheer difficult brutality of trying to get super scarce nutrients in the Savannah, if any simian competition, at all, in the form of EVEN A SINGLE OTHER GROUP OF APES made it out to the coasts, they would be showered with such plentiful attainable abundance that they would sky rocket straight past the Savannah apes like they were standing still.

Coastal apes would already have been all revved up on Halloween candy and rearin' to go while Savannah apes were still just screaming in front of their TVs for hours flapping their arms like an airplane.

If some professionals are not thrilled to hear the Savannah theory being demolished, I would just say that our time spent studying it was not wasted. Barking up the wrong tree helps us understand what answers are *not* correct, and there is just as much value in that as finding answers that *are* correct.

Plus, life is tough and shit happens. If you want to be in a profession where you will never fail and you will always be the greatest of all time, then you are going to have to go become a gangsta rapper.

That's the only place where everyone is the number one illest and everybody always gets first place, photo finish. They all IS the best up in this game. You think they would just be spoutin' off, cuzz?! HUH?!

You could bring something fresh to the street...

Mmm, yayuh! Drop that beat...

This is MC Anthro comin' right at YUH!

I'm at least top thirty.
This is where you heard it.
You ain't takin' my diploma.
I did just enough to earn it.
I was there with all the rest,
at the graduation fest.
I ain't tryn'uh be no best.
I was FORCED to leave the nest. WHAT?!

Aaaaah hell yeah, Anthro!
Tag me in, SON!

My name is DJ paleontology.
And I owe you all an apology.

Cuz I been spittin' so much bullsh**
About all dis ancient biology.
My only goal was to get laid
And impress these ladies with my brains.
But now I got a wife named Amber
So none of this BULLSH** even MATTERS!

Yo yo yo
Yo yo, yo yo
yo YO yo YO, yo YO yo YOOOH
You blew it UP, PALEOOOON!!!
Now le'me at'm

My name is lil' archeology.
My raptor claw will slit your belly.
I'm in this game cuz I saw a movie,
but did not realize you don't make money.
All I wanted was to be a consultant
On a secret dinosaur cloning project
And get my chance to dramatically say
Mothuh fu**** LIFE WILL FIND A WAY

20

XI -- DON'T SPEAR THE REAPER

Chimpanzees would all be slaughtered within the first five minutes of setting foot on the Savannah.

But you're telling me otherwise. You're telling me they switched up the whole game by making pointy sticks.

You mean to tell me, sitting there, looking me dead in the eyes with a straight face and no alcohol on your breath that some fruit creature climbed down from a tree, got a stick and poked a perfectly designed killing machine about five times its size, and this ended well for the fruit animal?

Are you having a go? Are there cameras? Is this one of those secret camera shows they're doing now? Haha. What fun!

Has this been some distastefully long hideaway camera program my entire life, exceeding all moral and ethical boundaries and standards of decency so the Illuminati can have a cheeky laugh about me in their meetings? Haha. How delightful!

Well, as it were, some people would presume to have you believe that apes were able to survive in open grasslands because of using spears and intimidation tactics. Now, this may all *seem* like a ret***** idea at first, but under closer scrutiny and examination we will find that it is actually a *really* ret***** idea.

Do you know where the safe space is for apes in the open grasslands and sparse trees of the Savannah? Do you know where they can go to be protected from harm? The answer is nowhere. The safe space is nowhere.

Apes aren't super fast runners that sleep on their feet and have strength in enormous numbers for monitoring their surroundings. They aren't extremely massive animals with huge hippo teeth or buffalo horns for protection. They aren't birds that can fly away, or dangerous hyena or crocs capable of defending themselves. They don't have even one single necessary quality for surviving on the open planes.

Apes are not fat, fast, flappy or ravenous.

They like to eat fruit and sleep in nests in trees at night, deep in the jungle, far far away from the open grasslands, because guess who else loves to sleep in trees. Cats. Big ones. Leopards, cheetah, and lions of the Savannah.

Apes like to use the night for sleep, while big cats like to use it to hunt. You would literally need posted armed guards on the lower branches where they could funnel the attackers down to one point and be ready to go all 300 Spartans on them if any lions tried to get up the trunk. This would be tricky too since the big cats have night vision, which the apes do not, and they are masters of stealth, which the apes are not, and they are proficiently armed with an array of deadly weapons, which the apes are not, and they can easily leap straight up as high as the roof of a jeep, which the apes cannot.

The big cat hunters are basically Navy Seals that would swim through the darkness up beneath them, slowly rising behind them in absolute ghostly silence and garrote them around the neck with a wire while they were still there thinking they were doing a good job with their Soviet weaponry and third world country clothes and ill-conceived battlefield tactics. The big cat angels of death could be up that tree so fast the guards wouldn't even know what hit them until they all woke up in the morning dead.

So, the guards would just have to pack a cluster of spears splayed downward around the upper trunk like some kind of savage medieval siege defense just to get a good night's rest.

If instead they wanted to sleep on the ground, they would have to surround themselves with a huge fan of packed spears as long as those of the phalanxes of Alexander the Great -- huge spears everywhere. Cats can jump.

Is this starting to strain credulity yet? Anyone? Have we wacked Occam's razor completely dull at this point? No? Good. There's a lot more.

And that's just sleeping. What about everyday life? Would they have to march around everywhere like a Roman legion? Well, they better, because if they didn't, then a hunting party of lionesses could just waltz right up to them and snatch one and walk off with him and rip him to shreds. The slowest and oldest chimps probably wouldn't even bother running because they know it's just going to be them. They'd just be like "Fuck it."

In fact, the predator groups would constantly follow the apes around everywhere knowing it was the most convenient form of food in the world; you just walk right up and snatch one at your convenience, like some sort of Savannah refrigerator. "Hey, I'm goin' to the bonobos. You want me to grab yuh one? Yeah, they don't even have spears to protect them. Haha. I feel bad. Poor guys. Sure you don't want one?"

They would HAVE TO walk around all day and night like a freakin' war party. There would be armed guards every time the main group needed to stop for something. And the guards couldn't smoke, or they would give away their position and a sniper would blow their head off. If a chimp guard got distracted and started playing with himself for just a second, a lightning fast cheetah could come out of nowhere and take his face RIGHT OFF.

RIGHT OFF!!"~

Well, this isn't looking good. We can see that no primates could have lasted more than five seconds out in the grasslands beyond the safety of the jungle, much less have been evolving there, unless maybe they had spears. But even the most fundamental logistics of them trying to live with spears is a complete and total shit show.

It's really not looking good for the Savannah theory tonight, Mike. He's got a swollen black eye that he can't even see out of and he's missing half his teeth, but let's see if he's got any surprises left for us that he's been keeping up his sleeve, Mike!

That's right, Brent. What a night for science fighting! But you know what they say -- it ain't over 'til it's over.

So, what are the chances of them even having spears to begin with? Could they make some? Could they find any laying around?

Well, let's talk about making them. The only way a spear can actually be useful is if it has an extremely sharp end and if it is the right type of hard wood for the job. You can't just randomly make a spear out of anything. The wood you use has to be exceptional, otherwise you'll just poke and annoy your adversary instead of stabbing and impaling them.

I know that generations of people raised on only TV who've never gone out and touched stuff might not quite appreciate this. But selecting the right type of material or wood for the right type of job is a very big deal. Indigenous people from around the world go through an intensive process to get just the right wood for their spears.

THEN there's the point. The end has to be really sharp! They definitely weren't using sharp flints and flakes to cut things and sharpen spears. Are you kidding me? Go watch the way a chimp uses all his ability to think and focus just to smash a heavy rock down onto a nut, and then tell me that guy has the coordination necessary to wield the stone equivalent of a meat cleaver or a scalpel. Even if they found one laying around and then tried to use it, things wouldn't end well.

How else could they make a point? Well, they could sit there for DAYS and grind the end against a stone until they filed it down to a jungle shiv.

It's worth reiterating now that chimps have their own dark side. They occasionally do things like get into huge wars and start killing each other until one entire group is completely wiped out.

When one observes the ferocity and intensity of these wars, it leaves little doubt that if they had the knowledge and ability to make basic shivs, they would be speed shanking the hell out of each other, prison style, on a semi regular basis.

It would take far more patience and focus than any ape has to make a well sharpened stick point.

The deferral of gratification is incredibly important.

They can take little human kids and leave them in a room with a marshmallow and tell them that if they leave it alone and wait five minutes then they will get to have two marshmallows. Let's just call it what it is, the higher IQ children choose to wait because obviously two is better than one. But the lower IQ children can't handle that. They can't just sit there and watch that marshmallow. They gotta eat it right away. They do not have the higher deferral of gratification.

That's okay. We love those children too. I had a dog who was like Garfield the dog and loved to eat more than anything in the world and he had dogtism and no deferral of gratification. One time I left a big breakfast burrito on the table and I came back after about ten whole seconds and he was up on my chair plowing through my burrito. And I said "HEY" and when he realized I saw him he didn't run away. He just started eating faster. Then I ran over and grabbed him and said "That was my burrito, rascal!" And he started grumbling and making devil noises and other noises that aren't dog noises. It's okay though. Every child is perfect.

But all of those human children are bloody rocket scientists compared to the smartest chimps. You think a chimp has the deferral of gratification necessary to meticulously select the right type of wood in the right shape and then painstakingly file it for days to the point of being an effective weapon only to have to wait days more for the perfect opportunity to go get into an insane spear showdown and chase some lions off of their own kill?

There are too many moving parts. It takes a huge leap in intelligence to even reach that primitive point of weapon making. This is so far beyond finding the right twig to catch ants or the right rocks to crack nuts. Neanderthal got only to this basic spear making level and no further, albeit with stone spear heads, but that's where they topped off and they had a brain case on the order of ours!

The idea that *early* hominids were building effective spears is another mind-bending stretch. There is a lot more to it than just being a pointy stick, and even that is trickier than it sounds. It takes patient craftsmanship that is actually fairly advanced.

Apes don't think that far into the future. They don't plan to that extent. Some humans haven't even reached that level yet. Even some of US don't often plan far enough ahead for the simple and obvious consequences of our actions.

I once lived down the street from an illegal child daycare, or as I called it, the pregnant smokers club.

I myself have not been immune to the obliviousness of my own ill-conceived actions, like when I was a small child and there was this older bigger faster bully going across the monkey bars and there were these two orange traffic cones sitting there stacked on top of each other and the top one was sliced up. So, I peeled it back to expose the tip of the one beneath it and said "Hey, Percy! Is this your dick?"

Why do you think I know so much about getting persistence hunted?

21

XII -- SWAYING TO THE CHIMPANZEE OF DESTRUCTION

Do you remember your older friend Charles who may or may not have been on the Illuminati's hit list?

Remember that time you guys were on your way to the theater to see a movie and you swung by the museum to pick up Charles who worked there as a curator and researcher? And just as you guys were approaching the museum a huge explosion erupted and completely blew the entire front off the building and you were like "CHAAAAAAAAAAAAAAAAAARLES NOOOOOOOOOOOOOOOOOOOOO!!! CHAAAAAAAAAAAAAAAAAAAAAAaaahh... fuck it, let's go see a movie."

Sometimes we just need to let things go and move on. We can't always dwell on the past.

The idea of primitive apes making usable battle spears is just one of those bad ideas that has to be let go of, like that random broke guy you had sex with on that ship and then he froze to death holding onto your floating raft thing. Don't worry. He's just a poor person. There's like a billion of them.

Alright then. So, they couldn't make quality spears. Well, then what if they had some simple pole sticks that were kind of jagged and sharp on the end?

Well, there's a reason every successful spear maker including us used sharpened bone or extremely sharp shards of specific types of rock. It's not as easy to stab something as the movies make it look.

Without highly sharpened hardwood points on the ends of the spears to give them teeth, so to speak, the best you could do is superficially injure something. Most likely though, you would just annoy them and the lionesses would just be like "What are these ret***s doing?"

If they came at the lions with just a bunch of long sticks without serious edge weapon tips on them, then the big kitties would just get pissed off from getting poked and then swat their stupid sticks down and pounce straight through onto them and just start slashing them to ribbons. At this point they would all be screaming bloody murder and completely break ranks and then just get utterly destroyed by the lioness hunting party.

Maybe some apes DID try this at some point, with sticks that didn't have good points. But they aren't your ancestors or my ancestors. They aren't anybody's ancestors. They're all fuckin' dead.

Well, this isn't gonna work. We need another plan, guys, or we're NEVER gonna get to evolve on the Savannah.

Ah, yes. Then it's time to play ye ol' "tallness intimidation" card. They say that they would just scare the lions by making themselves look taller. Too bad lions don't give a fu**. They attack giraffe and huge water buffalo in big herds. Don't tell me a chimp waving his arms in the air and standing up really awkwardly like an old man with bow legged rickets trying to balance without his walker is going to scare away the king of the jungle.

Lions aren't that dumb. They know you are the slow snack. They're not fooled by you waving your arms in the air like a moron. The only time tallness intimidation works is when a CHILD *human* makes themselves look taller like an ADULT *human*. And this only works because these animals have already become well acquainted with danger human adults and their stabby death sticks.

The key though is that it is a child using height to emulate an adult.

It's not just tallness for tallness's sake, or you'd see wildebeest and things walking around on two legs putting their hands straight up in the air and walking all funny like your dog dancing and praying for treats every time a lion rushed them. Just a whole herd of wildebeest doing the "I want a snack" dance.

I know a thing or two about intimidation. I grew up in desert outback and we would play dijuridoos, which are long tubular instruments native to Australia and have a far reaching heavy resonant droning sound that often sounds like BahWah wahwah EEEWahwah WahEEwah...

We would often take them with us way out into the country or bush.

Sometimes we would see people in the distance and we would start playing the dijs. When they heard the dij drone they would freeze, and the one in their group trying the hardest to be the alpha would stop suddenly and put his arm out as if to signal to the others to stop. Then he'd say something priceless like "This could be trouble."

They would start to take off and get out of there, and we would see that and start giggling to ourselves.

Then deep behind us in the brush we would hear banjo music start to play and as the first group of people were hustling out of there we would come running out past them even faster.

"THIS COULD BE TROUBLE!"

Then twenty seconds later the banjo guys would come sprinting past us even faster.

"What the?! ... Dufuq's behind THEM?!"

So, these anthropology cunts say we used large numbers and putting our arms up in the air as psychological warfare to intimidate and scare these others animals. Hmmm. Alright. And this is supposedly why we stand upright? Is that correct?

Uuuuuh, okay. Why aren't our arms stuck up in the air all the time locked in the upright and scare position? That's why we're weirdly standing up big and tall, right? Why not have our arms up too like vertical legs sticking straight upward to take this fright game to the next level? We could just have ligaments keeping them stuck like that with our hands dangling like zombies, and just constantly intimidating all the other animals. We'd be the bullies of the Savannah.

Riddle me that.

That's funny too because the Africans that use intimidation tactics now on lions never stick their hands up like a bunch of posers. They just walk straight at them like a boss.

Let's put the hands-up idea to rest. Intimidation works because they've seen you kill some of their pride or pack before and they know you are dangerous. It has nothing to do with being tall, much less a little taller with your hands up. The theory that we developed upright walking to be tall and intimidating is nonsensical. And the mad assumption that this party trick could keep us fed and scaring super killers off their kills consistently for hundreds of thousands of years without them ever getting wise to it is a complete cartoon.

But let's have some fun with that. Let's draw another black box of mystery and just say tallness intimidation actually did work a little. Let's just make something up like maybe there was another big animal not seen in the fossil record that was super dangerous and by making ourselves look taller it made us look a bit like that other animal to the lions.

Alright. Can you imagine eons of time going by where the very survival of your species is dependent on this insane game of poker for every meal? Every time you want to eat you have to play a game of chicken and get into a car going high speed straight at a huge semi lorry truck going just as fast and hope to God that it swerves first and never calls your bluff and never figures out you're just a dumb slow snack trying to look big, holding your arms up. And this better consistently work nonstop for hundreds of thousands of years or your whole species gets flattened off the pages of history? Come OFF IT, man!

And you're going through all of this Madd Maxx insanity in the scorching wasteland Savannah when you could just mosey on over to the coastline, strummin' ukuleles on the beach and laughin' about the bad ol' days when every time your girlfriend asked you if you could grab her a wee snack you had to spray toxic chrome aerosol paint into your mouth and scream WITNESS MEEEEEEEEE... [|X

Primates living on the open Savannah without deadly weapons? Fail.

Basic hominids making spears? No way.

Using jagged wooden poles as spears? Not a chance.

Bluffing with tallness? Not a thing.

Hands up for extra tallness? Still not a thing.

Well, then what the HELL are we supposed to do now?! DAMNIT! I'm gonna start throwing textbooks across my office! There's GOTTA be another way. I spent my whole LIFE believing this horse's ass Savannah theory!

I KNOW!

Wait. No. That wouldn't...

Hey! What if theyyyyyyy... No, that wouldn't... FUCK... THAT wouldn't work either.

I never thought this day would come. There is *one* thing we still haven't tried. I know, I know, we were never supposed to use it, but do you have any better ideas?!

It's a long shot. But damnit, it might just be the miracle we need. Screw it. Maybe they got help from ancient aliens.

I KNOW! I KNOW! But just hear me out.

The only way they would NOT have had to undergo this constant harassment by predators is if they DID become effective killers using the spears and thus caused the predators to learn to be weary of them.

The predators are pretty intelligent animals. They have to be. They will learn if something is dangerous, just as they have now with humans. They avoid us unless they have some extraneous circumstance like an injury that has prevented them from catching their normal game and they have reached a point of desperation.

So, let's talk about that. Let's talk about the hypothetical pikeman apes that somehow had effective spears. How capable would they be at killing in the early hominid stages when equipped with some serious spears in hand?

Who knows. Maybe the lost Atlantean civilization left some spears laying around. Maybe there were gigantic cactuses with huge needles you could pull off. Maybe they found some kind of deposits of huge finger bones or wing bones from a dinosaur or something... LOTS of deposits.

Or maybe God was like "I'm gonna need to do something in evolution so nonsensical that they know they couldn't have evolved on their own. What should I do? Hmmm. I know! I'll give them spears when they're still simple chimps. This'll definitely shake things up. Can't wait to see the looks on the lion's faces when the slow snacks show up with pokie stab sticks... teehee."

Alright, we got our monkeys some serious impaling weapons. YES!! Okay. Now what?

Well, even with legit spears they wouldn't stand a chance. You've gotta be more than half way to human before that's even a viable option.

WHAT?! All that work to come up with stupid ideas for how they could get spears and the whole thing is still dead on arrival?!

Yep. It may be true that small children in Africa do in fact kill lions by themselves with only a single spear, all Sparta style, but that is something that requires human level intelligence and advanced technique and nerves of steel.

It is not so easy to effectively kill something with a spear and not also get killed yourself in the process. Even Neanderthal weren't so great at it and they were way past the intelligence of midway hominins.

Spoiler alert. *No chimps will ever be killing any lions* no matter what weapons you give them.

Train them how to use a belt fed Gatling gun from Terminator 2 if you like. When they finally pull the trigger they'll just lose control and spray ballistic lead across half the Savannah including their own family and parents until they finally regain executive function over their own screaming and panic and drop the gun, running away shrieking and scarred for life, which will only last another 17 seconds until the lion pummels them.

You can't just start giving weapons to primitive apes. God figured that out a long time ago... the hard way.

The reason primitive apes couldn't be effectively wielding spears, or any other type of weapon, is because they don't have the right type of muscle fibers for fine motor control like we do. Their muscle fibers are for raw strength. That's the tradeoff. Not to mention they still lack the temperament and necessary degree of intelligence to wield a spear.

Maybe if a bunch of apes all went out together standing shoulder to shoulder with long sticks advancing forward like a Roman legion. The sheer density of the spears together could be just the ticket.

But even more advanced apes simply didn't have the precision motor control and detail coordination to make this Roman legion strategy work. When holding a long pole, every small movement you make with your hand is amplified and much bigger out at the far end of it. They would just be thwackin' and crackin' those things all over the place.

Chimps can actually get running pretty darn fast. When trying to charge forward and sprint with their spears, though, half of them would just run the end of it straight into the ground and make a perfect castle siege bungee pike in the ground pointing the wrong way, or maybe they would just pole vault themselves straight into the lions.

Can you imagine the total cluster fuck of a horde of apes trying to get all planet of themselves and charge out together with their waddling squat little bodies all screaming and trying to stay packed together tight while running and holding their spears steady all at the same time? There's no way in hell they could master all that coordination. Even we humans have to train to be able to synchronize things like that.

Shoot. Well, what if they just came in all slow and sloppy as a team, just approaching the lions nice and easy, whimpering softly and nervous, but kept all their spears pointed forward so they wouldn't have to have good aim or skill or anything?

YES!!! Now we're getting somewhere.

Let's take a quick tally. We don't have super running speed. We couldn't use bows and arrows. We didn't have the coordination to accurately throw specialized projectile spears over a distance. This is still a level much less sophisticated than even Neanderthal who themselves didn't even throw spears accurately from a distance.

So, we didn't have nearly enough skill to actually kill things for ourselves yet. Well, then at this level all we can do is go scare some lions off of their kill using our spears.

Plains predators like lions are extremely powerful, so we have to use strength in numbers with all our spears like we're in some insane Napoleonic bayonet battle, and if we go nice and slow we just might do it, damnit.

Those things know they are the kings and queens of the jungle, by the way. They are *not* going to retreat. We are going to have to kill at least one of them, and ONLY THEN will they learn that we are dangerous. Only after that will we be able to try to psychologically intimidate them. And this will only work on the same specific groups of lions who have seen us kill one of them before.

This idea of scaring lions off of their kill is actually a lot like what some modern humans in Africa do, but these people do this very intelligently. They are not doing it all the time and constantly trying to rob super predators of their kills. They know not to overplay their hand. And they also only take a small portion for themselves. They don't take everything and then leave the lions hungry and pissed.

But our early spear apes trying to survive on the Savannah would not have had all the options of food acquisition that these modern humans do. They would be almost completely dependent on trying to steal the hard-earned catches of others.

So, our early spear apes would have to be constantly going into battle every few days, at least, just to eat. And you never know how it's going to turn out. Maybe you've been overdoing it and taking too many of their kills and now they're hungry and desperate and they just decide "You know what, spear monkey? Fuck you! I'm fighting you to the death for this one."

This state of existence would be a constant cortisol dump of never-ending stress hell. In real wars and ongoing conflicts they have to cycle the soldiers out in so many months or they will start to break. And yet these chimpanzees are alleged to be effectively fighting full scale battles for their entire life? Absolutely not. In order to eat, predators go after things that want to run away from them, not things that are happy to oblige them in a death match.

Species that actually have to deal with these apex killers exist in large numbers, and so specific individuals don't tend to have to experience that massive stress of dealing with them too many times. The run-for-your-life experience tends to get spread out amongst the herd. If the same individuals had to keep dealing with that heroic dose of angst and anxiety constantly it would just kill them. They'd just get sick and die from stress overload.

We need that fight or flight stress response for extra adrenaline strength and focus, but it's for emergency use only.

The amount of stress on the apes of having to fight a full scale battle every single week just to get food would be insane. It would be like the battle of Gettysburg in the American Civil War with sticks for muskets FOREVER!

The sheer post traumatic stress disorder and chronic battle fatigue from all of that would destroy their immune systems and they'd be dead from disease within months. The only ones that would survive would be the ones that drowned their sorrows in rotting fruit and threw the empty apple cores at their children and yelled that there were never any goddamn problems on the goddamn Earth before you god kids were damn born.

Then those children would grow up to be f***ed in the head and end up dating hot crazy chimp sluts that they had no underlying compatibility with because no one ever taught them anything about life and how to choose a mate.

Their dating life would just consist of putting on a circus act the whole time in a clown costume but only as long as it would take to figure out how to line up their bathroom holes for going number one. Once they got that squared away they could finally pull off their sweaty clown masks and stop holding their breath and just show who they really are.

"Yeah, this is who I really am, yeah, you like that? That feel good? I'm not even this clown. I'm kind'a psycho with all kinds of childhood issues. Mmmmm. You like that? That feel good? Mmmmm. Yeah, you love it, you filthy..."

Then after creating powerful biochemical bonds from having sex they realize they actually have completely different world views on things like whether dishes should be done more than once a month or whether it's okay to get to the movie theater so late that you don't even get to see the f****n' movie trailers for the other movies coming out. Then they realize they f****n' hate each other.

Then you start turning to the rotten fruit just like your father and you get to slurpin' and laughin' about the good old days with your brother until the worm turns and he says something that triggers you and you start talking shit until you go into a blind rage of belligerence and try to throw a swing, but your reflexes are shot and your little brother studied mix martial arts in school, so he puts you down easy, and you start crying and apologizing and saying how much you love him and then start throwing up on yourself.

Then you finally realize that you're destroying your life and you need to make some changes and you start making big progress in therapy and getting really hopeful and thinking revolutionary thoughts and start questioning everything you were conditioned to normalize as a baby chimp and realize no one has ever just tried to talk to the other animals on the Savannah before and negotiate with them and show them some respect for the beautiful and wonderful sentient beings that they are who are just as deserving of love as all the rest of us, and so you spend weeks working up the social courage to go talk to them and rehearse all the things you're going to say and the things you know will touch them deeply and emotionally and the other things you know you'll both laugh about and share a bonding moment over. So, with a pep in your step you embark on the first day of the rest of your life and take a trip down to the lion's den and they kill you.

So, as you can see, there are lots and lots and lots and lots of problems with this Savannah theory. Trying to make it work is like trying to sky-dive down the side of a steep mountain in one of those flying squirrel suits along the roadway trying to weave through the traffic going up the mountain and avoid turning someone's car into a convertible.

Now I'm going to say this very slowly for all the professors and the academicians and all the other top a***oles. There's no way for me to represent the slowness of my speech through the medium of grammar text, so you're just going to have to read this to yourself very slow and condescending.

You
Can't
Make
Tools
Without
A
Big
Brain.
So,
The
Nutrition
You
Need
To
Build
A
Big
Brain
Must

Come
From
Something
Very
SIMPLE
And
RELIABLE!

... unless you're into some kind of Schrödinger's cat, paradoxical, circular reasoning, Möbius strip, colossal tit bollocks.

AND ENOUGH with the spear thing. You need a brain to make the spear. You need the nutrition to make the brain. So, the nutrition has to come when you are still dumb and weak! The nutrition has to be something you can originally get *easily*, LIIIIIIIIIIIIIIIIIIIIIIIIIIIIIIIIIIIIIIIIIIIIIIIIIIIIIIIIIIIIIIIIIIIIIIIKE
defenseless mollusks and simple crabs on the beach containing lightning bolts of super nutrients.

Whatever they were doing for food, it had to be something very simple and reliable.

It's a total myth that there were apes trying to scurry between super predators for scraps or running around persistence hunting on the open grassland plains of the Savannah in front of professional killers that could run them down in an instant.

It strains credulity to the point of breaking to expect rational people to believe that low intelligence early hominids had to be constantly threatening apex predators with spears for their very survival, never mind how the hell they even got the spears.

Yep. This myth is busted.

Some people ask how I could have such an advanced knowledge of apes while having never studied them in the field. Well, it's called books... But I myself have been a bit of a Jane Goodall in my own right, observing primitive instantiations of primates over the course of my life.

There was this one def retarded kid in my elementary primary school. This kid used to bully and harass me. He saw how others were doing it, so he thought he'd join in the fun. He had a rather sophisticated technique for someone of his mental state. He would wait until it was class discussion time in which he felt left out because he couldn't hear shit, then he would point straight at me with both hands from several desks away and make his best attempt at wailing the words "HE'S PICKING ON MEEEEE!" which of course sounded bizarre on account of being def and retarded. It was something like "HEE PIGGEEN AHH MEHEH!" The teacher, of course, would actually believe him, because why wouldn't you. This naturally made him utterly appalled by me — to be host to a genuine hurter of special challenges students. So, I became target number one.

I had a feeling this wasn't Darren's first rodeo. This had all the hallmarks of being a tried and tested formula. And what could you even do? Go walk up to him and crack him across the face? Become the monster he wanted the world to believe you already were? He had won this battle of wits. It was checkmate. And he knew it.

Besides, he had retard strength. His brain's natural inhibitors for limiting muscle force were broken. So, he could unwittingly tap into some serious levels of chimp power. I saw the teacher try to physically restrain him once from picking up his desk and slamming it into the ground repeatedly. The teacher put all his entire body weight down on the desk and Darrin was still able to overpower him in a fit of rage. It was like the teacher was riding a sled down a hill way too fast and struck a patch of boulders.

Alas, simple Darren never made it in with the cool kids or the hot chicks as he had miscalculated he would. But he gave it one hell of a go. I'll give him that. I hope you are reading this someday, Darren, because it's entertaining for people like you and it's in your reading level and you remember me through all the clutter and mumbling of your mind and you realize that I am EVERYTHING you will NEVER be, you piece of handicapped shit! I was BORN better! You crippled def ass cunt!

22

❧

XIII -- MEET THE SPRINTSTONES,

... THEY'RE A MODERN STONE-AGE BULL**** FANTASY

While you're having a leak at the urinal I'm gonna come up behind you and put my hands on your shoulders, because we need to have a little talk and I need your undivided attention.

Woe! Woe! Easy, bronco. Relax. I'm just here to have a talk. Quit trying to shake me. Just relax. Here, let me do a little massage.

You know how you've been told humans came from monkeys running around on the sunny grassland plains of the Savannah? Yeah?

Yeah, It's total bull****.

A sunny beach is more like it.

It's okay! Relax. You're all tensed up on me. There's a major knot in your neck right here. Hey, HEY, quit struggling! Everything's fine.

But let's talk about what an ACTUAL plains ape would look like. What WOULD we really look like though, if we had adapted to the Savannah?

If we had actually adapted to the Savannah as a persistence hunter, we would be big hairy sasquatches covered in blonde hair. We would have that light colored hair like a lion's fur or like a desert Bedouin's robes to reflects the sunlight. This is a far superior way to stay cool in the sun.

Melanin in mammal skin helps us block dangerous UV, but it ironically also makes us darker, such as giraffe tongues, which are a vibrant purple to protect them up in the sun all day. They have melanin in their wet tongues because they can't have hair on them. Same thing with hippopotamusezezizesis, however you say it, that spend their days in the sun and water.

The counterintuitive reality of darker skin, however, is that it absorbs more visible spectrum light and converts it to heat. This is no way to go about running around in the sun all day. This only works if you're staying *totally wet* and able to shed your heat into the water. For an all-day fun run in the sun, you need light colored hair to protect you and then dark colored skin underneath, but you need that hair first.

They say, however, that we lost all of our hair so we could sweat and stay cool for persistence hunting. This is complete bull****. It would be exponentially better for us to be kept cooler by reflecting more sun with blonde hair. If we were actually meant to be on the Savannah plains all day chasing things to death, we would all look like albino chia pets.

But people who are darker skinned are not made for that. They are made just like a giraffe's tongue or a hippo -- to be sunny and wet. Same thing for everyone, maybe with a little less of the sun part.

Now, I don't like to get rumors started, rumors like that time you were looking for your friend's weird address and you overshot it WAY past his house, cuz maps weaken the mind, and you ended up in the hardcore ghetto part of town. Then you were just driving around lost, going down residential streets with lowriders and couches in front of people's houses in the middle of the night while people were trying to sleep. Then you suddenly realized you were still driving your off-road truck and blowing out the speakers to Sweet Home Alabama.

"OH, SHIT!"

Some poor lady in her bed... "Oh my gaud! They're coming!"

I wonder how many misconceptions originated from pure stupidity.

So, like I said, I don't like to get tales started out of school, but there is a point that needs to be made.

The whole idea of an open grasslands origin of human development supposes we came from these ultra runners on the plains. Again, we certainly do *not* have any mega runners capable of consistently doing what they claim early hominids started doing, but we do have amongst humans a group that is simply unmatched in running.

You know the super runner people I'm talking about. Yeah, I'm talkin' about you people. Yeah, YOU PEOPLE! -->

You know exactly what that's supposed to mean!

If you didn't catch it in the earlier chapter, this fact about being super runners and uber jumpers is just an established reality discovered by sports scientists. There's a much higher ratio of fast-twitch muscle fibers. Anyway!

There is one thing that I do want to be very clearly established on the record. The super running ability of SOME PEOPLE is a very new and advanced characteristic. This is by no means an older or more primitive trait. This feature is cutting edge.

The Savannah bogus theory of human development makes it seems as though super running is a more basic trait. It is not. Quite the contrary. It still has new car smell.

Each and every one of us has fancy random new things like extra running speed or lactose tolerance, but these are all very new exterior paint jobs on top of an incredibly ancient human form.

So, how do I know our original state as fully modern humans did not involve mega running? Well, I already shot that idea full of holes in an earlier chapter, but now let's take a flame thrower to what's left.

I know the first humans and what came before them were not a hardcore running species, because a serious running species would not look anything like us.

So, what WOULD a mega running plains ape look like? What body structure would the Sprintstones have?

Well, our hypothetical sunny endurance hunter would have had dog legs. He would have had legs like your dog has where their whole foot is just their toe area and their heel is up higher on the leg like a backwards knee. They don't walk on their heel. Their toe area, or balls of their feet, is just a little larger and is their whole foot.

We would look kind of like the pan pipe creature with the human torso on top and the goat legs on bottom. It's called digitigrade legs if you want to go see what it would look like. We basically would have legs like the popular portrayals of werewolves.

If running and sprinting and pacing were that massively important to us as a species we would absolutely have dog style legs, because those legs are like running on springs, like those "jumping stilts" toys people have been making. Whereas our legs are just straight up and down like tree trunks or like columns of a building and not very springy at all. They're like running on two wooden pirate legs.

It's a complex topic, but that's the basic idea. We have these legs like we do, because running has NOT been a huge priority for us during development. And we walk around on our oversized feet that go all the way up to our heels because our foot is a multipurpose platform for walking and, even more so, for use as a swimming flipper. And it ain't from growin' up 'round no nuk'yuh'ler radiation.

So, a bipedal Savannah ape would be like the goat legged pan creature mixed with a Himalayan sasquatch and have wicked body size combined with chimp strength and claws and teeth to defend itself against all the heavy hitters of the grasslands. And it would only be as intelligent as a lion, because you simply cannot get enough brain food on a purely land diet.

That is your Savannah ape.

I won't say that the idea was a bad one, or a waste of time, because we learn from our mistakes, and there is great value in that. But sometimes things have simply outlived their usefulness and just need to be put down.

It's basically the same as when your older brother borrows your expensive field camera to take on vacation and swears on his life he'll take good care of it and then you go over to his house a couple months later and he forgot that your camera is still laying on his table in a million pieces "being fixed."

"Uuuuuh. I fell when I was holding it. The river rocks were too wet."

Oh, the river rocks. Okay. And wetness. Makes perfect sense.

Well, time to make good on your promise then -- my camera for your life. Do you want the rope or just execution style on your knees?

23

XIV -- SET PHASERS TO DUMB

My little brother used to have this friend, Rob, and he would need rides places sometimes because his stupid grandma couldn't drive him.

I told my brother to tell his friend I would only drive him out of the kindness of my heart if he paid me twice the going rate of taxi cab services.

When we would get to Rob's house I would keep all my doors locked and only crack the window a thin slit, and when Rob would walk up to my window and start trying to talk I would just say through the crack, "MONEY FIRST!"

Sometimes there are tough realities that those of us in positions of knowledge with also our own cars have to bestow down upon the rest of you out of our own goodness and our own betterness than you.

So, let's just relax and take ten deep breaths and go to our happy place. Just remember back to some adult that was authentically kind to you as a child like that teacher's assistant that noticed you didn't have any markers and smiled and gave you some, or that one odd gentleman in the top hat that let you and those other children attempt a series of questionable psychological challenges to see which one of you would get to inherit his chocolate factory.

Where are we going with all this?

Well, one real tough reality for all of us happens to be that we as modern humans are remarkably far behind developmentally. We are severely retarded in our progress from where we already should be both technologically and societally, hell, even spiritually.

We went from wagon trains to moon walks in less than a century when we eventually got our shit together just barely a little bit finally. And yet we've been modern intelligent humans for AT LEAST one hundred thousand years. Something has been artificially crushing our potential. What the holy hell is going on here?

It turns out that the problem is not actually our brain power but rather our personalities. We've got a serious attitude problem.

Delusional people outperform honest people. Yep. People that just believe whatever the f*** makes them feel better have been psyching themselves out and pumping up their personal morale and going into the big games strong for a very long time. And they've also been sleeping better at night... literally.

Despite our species selecting from within itself for the fruits and the products of higher and higher intelligence in the form of higher quality providers, nature has also been selecting us to be more and more fantastically magical thinkers.

We've had this crazy push and pull, tug of war inside our own species between the poles of raw calculating intelligence and pure baseless delusion.

We have had this feature bred into us for being able to lie to ourselves in order to self comfort. To be clear, though, it's not all that bad. This feature can help us fuel our imaginations and ignite our dreams and can help us find hope when times are dark and when it feels like the storm will never pass.

The key, however, is that we need to be *consciously aware* of this aspect in ourselves so we can use it to our benefit instead of being blindly used *by* it. It's essential to remain capable of challenging our own beliefs and not turning something into an irrational religion just because it "feels right."

Sometimes our feelings have critical guidance for us, and other times they are just straight up bu****** because we aren't even interpreting them anywhere close to correctly.

Humans could have already been building smart devices and rocket ships and secret schools for wizarding families eight thousand years ago around the beginning of the Holocene period if not even twenty times farther back in history if we weren't constantly killing off the smartest monkeys amongst us every time one of them goes and questions the gods.

Every time some forward thinker starts trying to question the sacred joojoo of the tribe, all these other sons of bitches start throwin' around accusations of witchcraft, and hootin' and hollerin', and squealin' and coughin', and shakin' on the floor, while claiming they are being tortured via some heretofore unexplained mechanism of psychic attack... ... And then someone in the back yells, "CUT HIS F####NG HEAD OFF!!!"

And all this because I just questioned how Muhjubjub made the world on the back of a turtle and an anus fish. I'm just curious about some of the mechanics involved is all. I'm not saying he didn't do it.

Curiosity is absolutely essential for our progress, but so often it ends in disaster. I really don't know what it is about us big thinkers and curious people that is always so triggering for the rest of you.

I was just at an automobile fueling station a few weeks ago and there was this gentleman filling up next to me. He was certainly quite bright eyed and bushy tailed like things were finally starting to go his way. He looked like his name could be Stewart and he was just doin' the best he could with the physical body the good lord gave'm.

He had one of those cars that was lower middle class fancy but you could tell he was really proud of it, like when he got home and was coming in from the garage he would stop for a moment and turn around for one last look to whisper "At'a girl."

He was there at the pump next to me filling up and getting a bit of a pump of his own on. He was cleaning his windshield with one of those squeegee brush things and he was really goin' at it with a serious grip on the handle using both hands like he was stirring a big pot of spaghetti. He was just leaning out over his windshield facing towards my car, with his tongue out to one sided for enhanced focus, just grippin' and rippin' that stick with both fists and pumping away like a madman.

Now, the scientist in me wanted to walk up to the front of my car and stand directly facing toward him over on the other side of his car, with him still yankin' and crankin' on that pole, and then lean myself back onto the hood of my car with my legs spread apart and head tilted back like I'm in pure ecstasy and start gyrating my hips to his rhythm.

Eventually, probably Stewart would start to look up, still quite pleased with the work he was doing, givin' that squeegee a good jackin'.

At this point, just as he was raising his eyes right into the path of oncoming visions of what was happening on the front of my car, his mind would be confronted with computational challenges it had never even imagined it would ever have to face.

But these are the questions we need to be asking.

They say that when the first sailing ships of explorers arrived on the shores of the Americas, it was such an alien phenomenon to the native peoples that they simply could not see the ships.

Would Stewart even be able to see me? Would I be completely invisible to him?

Or would he still retain the power to behold the site which lay before him? Would he think I was a god?

One thing is certain. We will never unlock the mysteries of the universe and continue to plunge deeper into the heart of discovery if humanity cannot one day quit calling the cops all the time and COMPLETELY lying about what was even happening.

24

XV -- MORE DOCTORS
TRUST THEIR FAMILY'S
HEALTH...

... TO LANDERS BRAND CIGARETTES

Remember that time you tried to bribe the guys at the National History Museum to let you saw off one of the legs from the frozen wooly mammoth specimen so you could grill it up for the Super Bowl?

But you told them it was for science at another museum but that you forgot your science clothes and name tag, and they said it was okay because they could just look you up in the system, and you said you were too new and you wouldn't be in the system yet, and they said it was okay and they could just check anyway, and you said it wouldn't work because all the computers at your museum all got hacked and if they could just help you out you'd make it worth their time, but you didn't realize there was that one museum director lady standing right behind you the whole time and she saw you trying to give them a weird handshake with obvious cash in your hand and she asked what you were doing, and you realized she saw everything and you started acting like you were struggling with emotions and you made up that whole story about how handshakes were very important to you because your family was never allowed to give them growing up as a small child in eastern Europe under the iron grip of the Mongols, and she said "Yeah, but why are you trying to give money to my staff?" And you were like "I don't even own money!" And you made up that whole story about how no one in your family ever carries money because your great great grandmother told stories about how she would be beaten by the soldiers if they found her with money growing up as a small child in Holland under the iron grip of the Mongolian Empire, and the director lady asked why it looked like you were reading stuff off the historical plaques behind her and trying to sound out words and you got all mad and made some vague reference about American rights and implied you knew a senator? Remember that time?

That was crazy.

Then I started to run and it triggered your instinct to run and then you tried to Michael Jordan the security camera off the wall on the way out but still came a whole arm's length short. So, we just laid low for the next couple months, until you got tired of layin' low and said you didn't care anymore cuz if it's God's plan for you to go to prison then who were you to question God.

Do you remember that time? That one time?

That was pretty paleo. I wonder what would've happened if we actually ate that mammoth. I bet we'd get crazy strong.

Maybe not everyone even knows what the paleo diet is though. What is the paleo diet anyway? What exactly is the pre-Holocene late Pleistocene Paleolithic Epoch diet? The paleo diet as currently popularized has attempted to approximate what humans were eating during the Paleolithic era prior to agriculture and a heavy reliance on grains and simple carbohydrates, or as most people understand it, "CARBS BAD!"

I know what you're thinking, Smuggy McPaleo, with your morally superior cave diet and your several sticks of butter blended into your thermos full of steak sauce. You're already on it, doggonnit, and I'm proud of you.

There's just one little problem.

Remember how we talked about how easy it is for us as humans to get things completely wrong while simultaneously feeling like we've gotten it all totally right?

Well, we need to have a little heart to heart, my pedigree paleo chums. IT'S OKAY! BREATH! BREATH! DEEP BREATHS! Yes! Carbs bad! Yeeeees, carbs baaaaaaad. Okay. Are we calm? Everybody calm? Remember what we talked about, about how if we're good in the book today we'll get to go to the CrossFit tomorrow? Yeah? There he is. Still cave buddies?

Okay. I'm back. And don't worry, paleos. I'm not trying to sneak a vegan shaped baseball bat handle around your backside when you're not looking, like I'm some kind of maniac that sees you shuffling down the aisle in the bleachers trying to squeeze by people, coming my way and I'm just sitting there thinking to myself I wonder what would happen if I made my hand into a karate chop shape as he's going by me and just shoved it up...

If you actually are vegan, though, or vegetarian it's all the more critical that you understand what our real ancestral diet was so that you know exactly what template of nutrients you need to aim for in designing your own plant based or lacto-ovo diet. I know that technically you can actually be a fully healthy vegan your whole life, but that ONLY works when you are still getting enough DHA & EPA from the algae that make it. So, you are still dependent on sea nutrition. I just have reluctance, though, about the idea of people being completely 100% vegan, because the human animal was built by the pescatarian lifestyle.

But we got to be big boys and girls now and acknowledge some realities.

The sky is blue, water is wet, and *land meats and many plant foods are simply NOT native human foods.*

Oh, we can definitely eat land animals, alright, and even derive an okay amount of nutrition from them. But as we discussed earlier, land animals alone simply do not have everything we need to be healthy and have a *fully* functional body and brain.

We went from a clever jungle ape all the way up to a fully modern intelligent human by living on the beaches and seasides and primarily eating seafood and eggs and nuts and fruit.

It wasn't until we became these super-brained dolphinized swim-chimps that we started making advanced tools like spears and then expanding out from the coasts and began hunting land animals and eating different varieties of land plants.

I have to emphasize this point, because people seem to have a lot of trouble with it. We got our brains on the beach, and THEN we were smart enough to make the tools, and only THEN were we able to hunt these land animals and start migrating inland to eat these new plants and animal types.

Our sea meat and jungle food has been with us for a vastly longer amount of time than anything else. These original ancestral foods from the cradle of our species have been time tested for performance and healthy aging!

If you look at all the places on Earth with the longest lifespans and lowest disease rates, like "blue zones," you will find that they are places on the *coasts* and that they eat nearly the exact same diet as I have mentioned. They eat fruit and nuts and eggs and fish (sound familiar?) and then beans and greens and some dairy and things. And they only eat a very small occasional amount of land meat or simply none at all.

That right there should be blowing your mind right now. Or as my former eastern European female roommate would say – "This is blowing me right now." That's the same eastern European roommate that met my super tall little brother once and was like "Woooow, you are so looong!" He was like "Thanks. How did you find out?"

But it still blows me too that there is such a striking correlation between the jungle beach diet and the diets that are clearly shown to be the best for our human health and aging.

I also have to stop for a second and just mention something before any vegans reading this go and blow a gasket. Yes, the blue zone diets lean heavily toward plant-based foods, BUT you still have to meet me half way and acknowledge that their go-to staple source of meat is *fish*.

Some folks that were definitely not vegan, however, were the Neanderthal. Apparently, there were some small pockets of Neanderthal that actually did eat a vegan diet, but GENERALLY that is not the case. There were also some Neanderthal populations still living in the most ancient traditional way on the beaches around the northern Mediterranean and thriving on seafoods, but these populations too were very much in the minority.

The main bulk of the Neanderthal population lived inland within Europe and some into Asia. The average Neanderthal ate a diet that was heavily land meat, as they have become most well known for.

Even though our old neander friends are all gone now, they actually existed as a species for FAR longer than we have – around a quarter of a million years longer. But we are the ones doing math and science and exploring space and building a world-wide digital internet that we are mostly using for productive and responsible purposes, and they are the ones that never got past making big heavy spears, albeit with glue and string and flake.

Why was it that their species existed for nearly a half of a million years but was just frozen and stagnant in its development the entire time? Why did their intelligence never increase to the level that ours did?

Well, the answers are likely multifaceted, but the vast majority of their populations lived inland and functioned like wolf packs and basically lived on land animal meat. As we discussed before, the pack hunting strategy is very inclusive and does not allow for slightly smarter individuals to stand out and be more favorably selected, as is the case in omnivorous forager populations. And since Neanderthal subsisted on land foods, they were largely deprived of the critical ocean brain foods. There is even evidence that they too had their own cannibal hillbilly moments.

Now, a lot of people these days have come to idolize the Neanderthal lifestyle and worshipfully hold them aloft in esteemed reverence as a standard and model to live by, but that's a strange fetish indeed since Neanderthal were a lot dumber than us. They would just run straight up to woolly mammoths and shove their spears into them. Then the mammoths would freak out and start smashing them. After getting a can wool ass opened up on them they would be left with massive fractures and injuries that would last them a lifetime if not just kill them immediately.

These guys were admirable in their own right, and we should honor their memory, but this is no model to live by.

Their land based lifestyle was a dead end. It's no moral judgement of them or anything, but their diet was a losing strategy. Extreme old age for them was 40 to 55 when they were developing degenerative diseases. That's a far cry from the double or triple long lives of people on blue zone diets.

As it is, however, they lost... and we won.

We did much better for many different reasons including that we brought our *spear fishing* technique onto land. It was a sleeker designed spear or "dart" and an atlatl throwing handle for more range and power, which kept us well away from the mammoth and other megafauna.

It was a FISHING technique. Tribes are still using it to this day, just as they have been since beyond the horizons of knowable time.

Our species trekked the entire globe and mainly traveled by coasts and rivers, thus retaining access to critical marine sources of food. Our superfood is in the water. This is where we came from. And the blue zone populations from all around the world are an echo of this past, still reverberating from old times and bidding us to remember what our real early paleolithic history truly was. We are not Neanderthal. We are human.

In modern times, many paleolithic diet enthusiasts are people who have had some serious health problems that they have had to battle with, but they said "To health with that!" and they went out and started figuring things out for themselves. And these people deserve incredible respect and even some sympathy too for not just letting some professional shove a bunch of symptom treatment down their necks.

Same thing on the other side. Most plant-based dieters are trying to not be a burden on the other animal life and they're doing it out of immense compassion for the animals. How can you fault them for that?!

One of my friends had this girlfriend for a while and all her toes came out to the same length, so the end of her foot just looked flat, like the end of a board, and she was really embarrassed about it. And so, she never wore sandals or anything that would expose her feet. My friend who was dating her, of course, was always very respectful of her condition, and as such he made sure to never forget to call her "two-by-four" in front of everyone else.

No matter where any of us is coming from and no matter what we may disagree on, we can all try to see some common humanity in one other and just be grateful that none of us is a two-by-four feet girl.

We can do stunning and awesome things when we put our minds to it. Our mnids hvae a lot of srupirsing aibilites and ptatren rcegoiniton wihch can be uesd for eonromus ceraitvtiy.

Did you know they moved the entire London Bridge all the way around the world?

Yeah, the actual London Bridge with the original bullet holes and everything from when London was strafed and bombed in the war. Pretty wild.

They used it to connect the mainland to an island on a lake in this desert in the middle of nowhere out in the back of the U.S.. It's a huge lake that didn't even exist until we made it. And now there's a thriving economy around it where there was absolutely nothing before.

I know people out in the bush or country who have made their own ponds and lakes so they can have fresh fish.

Has your family been ranching for generations? Well, why don't you look into using some of that land for artificial lakes and learn how to ride a seahorse?

People have made large inland pools to hold water and grow algae. We can make these pools to aesthetically resemble natural lakes and be a pleasant home for fresh water and salt water marine life.

I know people who have made their own ecosystems by adding basic mineral fertilizers with trace ocean elements to feed algae which itself is a superfood but also feeds fish and shrimp, which of course are a superfood themselves. And the waste they produce then goes on to feed plants which grow things like fruit with vital vitamin C. They have everything they need right there by just adding some sea minerals and fresh air and sunlight.

If you live far inland you will have to import at least the basic essential trace sea minerals no matter what, or just suffer the health consequences. If you're worried about self-reliance then, I don't know, buy A LOT of it.

When we are living far inland away from the ocean we are like astronauts going into an inhospitable planet. We have to take life support.

We can do agricultural style production with all sorts of setups like industrial greenhouse style operations and man-made inland seas where we can control the quality.

We need to always consider quality, especially now days, and we should do simple things for ourselves like maximize our body's own natural self-cleaning mechanisms with foods high in things like sulfur. Just make your own broccoli sprouts or simmer whole broccoli heads for five minutes from a boil and then give it a light fry in butter or coconut oil and salt. It'll be delicious. Or put it in an omelette and quich'yuh belly achin'.

And briefly, if you REALLY want to take the self cleaning mechanisms of your body to the next level, pick one day a week where you can fast and not eat any protein or carbs – only fat, like tasty avocado or coconut cream powder. Your body will clean all of the junk and debris out of its cells, which is one of the main causes of aging. This natural process is called autophagy, if you want to go look it up. And you can do this for free. So, you're welcome, sonny Jim.

But, one incredible example of ingenuity was a couple of enterprising and astute men in the American south who are brothers and figured out how to grow rice and Cajun crawfish in the same field at the same time. Well, I'm not gonna say their names in here, but these two guys are both Cajun and exceptionally short. So, just know that I am using the term "men" very lightly.

In South Africa they section off huge areas of ocean with nets, and little boys from the SASLS, South African Surf Life Savers, go out and swim in the water with the biggest of sharks. And they get cheered on from other little boys, one of whom may or may not have been named Simon, in the SASLSSPSLHA, South African Surf Life Savers Simulated Participation for Slow Lad Hopefuls Association.

We need to turn our focus back to the ocean marine life and consider that just because our most recent human ancestors have been surviving off of inland foods for a veeeeeeery long time, doesn't mean that those foods are either optimal or ideal for our human body. It was within the living memory of people alive today that licensed medical doctors were letting us know which brand of cigarette they preferred, and we as a society were oh so sure of how very good smoking was for our health.

I am not worried about what humans can do when we roll up our sleeves and get to work and when we don't have two-by-four feet. We're a creative species. That's kind of our thing.

For the last two thousand years we have been in the astrological age of Pisces, the fish. Now this age is drawing to a close and we are about to enter into the age of Aquarius, the water bearer. Will we finally come to remember the true nature of our fishy past? Will we eventually come to find our place as the water bearers and begin to move the waters of Earth wisely and intelligently for the betterment of ourselves and the natural world?

Usually, we are more limited by our ability to dream rather than by our actual means of achieving the next level heights within our reach.

That's "ability to dream" as in aspirations. Not "dream" as in involuntary experiences of consciousness at night where you're on a plane and your friend from Mrs. Meehag's chemistry class is there and still a teenager and wearing a full body suit made out of meat and laying on the ground with his legs spread apart and your mailman is between his legs doing God knows what, because you can't see what's going on, but your friend is just staring off into space with a melancholy disassociated look in his eyes, and you're like "Ugh! What the fff... BRIAN! ARE YOU OKAY?"

Then Brian turns to you slowly with a look of satisfaction and whispers, "This is what I like now" and you're like "UGH! What?!" And then you start to grieve the loss of normal Brian and a voice from behind you says "I am NEVER letting my wife marry this man. He is WEAK!" And it's that severe hillbilly couple that lives a few places down from you and the guy wants you to marry his wife inexplicably, but he has to live with you together in a nasty little mobile trailer home house thing for a year first to make sure that you are an honorable man, but the beds in the house are just these huge stacks of mattresses piled all the way nearly to the roof. So, you can barely fit into the bed between the mattress and the ceiling and all that gross ceiling texture paint is always flaking onto you, and the whole house is filled with old blankets and newspaper and you have to crawl everywhere.

Then your wedding day finally arrives and the ceremony is on top of a huge skyscraper that constantly sways in the wind, and after the reception part of the ceremony you and your hillbilly neighbor's wife that you just officially married now have to swim way out into the ocean to get to your honeymoon sailboat, because it's a timeless tradition that you must honor. But she can't swim for shit and she starts drowning halfway out and a part of you doesn't care because of what she looks like, but another part knows that you *must* save her because it took you an entire year to prove that you are an honorable man.

Not like those types of dreams.

25

XVI -- ALL THAT AND A BAG OF DICKS

Number one, the big swinging dick is the aquatic dick.

The aquatic vagina has to be deeper, so the penis has to grow in length to accommodate it.

If you want to see it in action go watch the film The Blue Lagoon from 1980.
And I mean go see it "in action" as in go see it in the water -- not go see it... uh... in the uuuuuh... ... whatever.

The thing you already spend half your day thinking about.

Anyway, the big swingin' tits are the aquatic tits. They need to be inflated all year long so they will float and make it very easy to breast feed chubby little babies which also conveniently float.

Big awesome lady butts are also great for small children to be rested on and to stand on in the back while in water and go for rides with mom. Mom's head hair is also pretty tough for little monkeys to hang onto.

You really think we were running around all day on the Savannah with our big butts and tits and dicks prior to sports bras and underwear, just floppin' and ploppin' and wackin' and smackin' all day?!

SERIOUSLY?! That ALONE tells what a joke the Savannah theory is! Dudes racking themselves constantly. Yeah right.

And, no, dicks were not selected fetishistically to be bigger. After a certain length the thing just becomes a ridiculous burden. Despite modern running jokes and memes, Johnsons only need to be long enough to accommodate a vagina.

This chapter is not for the faint of heart. If you are pregnant or nursing or just bluffing like you're tough when women are around, you may want to sit this one out.

NO... HEY... NO. Not you, gym bro! You stay! You're not going anywhere. I know you're still mad about last chapter when I implied that land meats are not paleo as fuuu...

This chapter is for YOU!

This ends TONIGHT!

Or, if you wanna fight about it, then you can come to my house in real life and we'll take it out back.

But come alone. NO cops!

Okay, maybe one. But then I get one!

Yeah... that's what I thought.

Anyway. One of the first things that happens when you imply that land meat isn't actually so great for your long term health and performance is a lot of people automatically assume you're part of some Illuminati agenda to keep the peasants from getting the *good stuff* and rising up! Afterall, many of those pompous elite billionaires loooooove them some steak dinners, don't they?

Well, you know what, conspiracy theorist?! I don't blame you. Yep. Those Illuminati are definitely some sneaky bitches. But most of those animal mask wearing, sex orgy having, secret reptiles are under the same exact misconceptions that you are. They also loooooooove them some lobster dinners and snow crab and shrimp cocktails and tuna steak and ice cream.

If there really was a secret agenda to dumb you down, then it would actually be better to redirect your focus AWAY from seafood and over towards alternatives like red meat and bird meat, which lack the critical sea nutrients that are absolutely essential for higher cognitive ability. The real agenda would be to make seafood look all wimpy and lame like something only delicate British fancy lads do, while using all of their power in the media and Hollywood to make television commercials and movies that idolize beer parties and sports games and grillin' up SOME CAVE STEAKS! AAAAARRRR! I'm a CAVEMAN!

I wonder if anything like that has ever happened or been exactly what has been going on constantly over the last century. Hmmmm? I'll have to think about that one.

But this of course isn't to say that our reptilian overlords don't also want to deprive us of land meat too now. Anything containing good protein will make you strong, and strong peasants are uppity peasants. But that is a feature of protein – not just uniquely land meat. You can get quality complete proteins from all kinds of stuff.

So, could there be an agenda in the future to deprive you specifically of land meat? Sure. But that would be an agenda to make you physically weaker – not mentally dumber. They've already completed *that* stage of... The Plan.

But what if the Illuminati knows something that we don't know about land meat? What if there is some mystical magic cave joojoo in there like rhino horns and tiger penises and albino human flesh?

Well, to make this debate REALLY short I can just say this. Do an experiment. That's right. Put on your lab coat and your chemistry goggles and go watch some cool movies where the main characters are scientist /slash/ secret super heroes to get yourself all pumped up, AND LET'S GO DO SOME FUUU#### SCIENCE!
Just switch out your land meats with an equivalent amount of protein from good quality seafoods for a few months, and then tell me if you feel like a weaker and dumber more manageable peasant, OR if you feel noticeably better with sharper thinking.

I haven't even personally eaten land meat now for two freakin' decades (I started at an early age). The only time I've had it was to do the inverse of the experiment I'm telling you to do. I did the complementary trial and went several months eating the highest quality grass fed beef organs, and I did this same trial three different times over the course of five years. During those three month periods I had replaced 100 grams of protein from some of my normal seafood and eggs with 100 grams of protein from grass fed liver and ground beef and some other cow guts. When I did that I suddenly got super smart and super strong and super good at sex and super...

NOT. I never perceived any advantage. I just felt more bogged down. It takes way more time to mechanically chew and grind through the much denser land animal tissue, and then it sits in your stomach a lot heavier from the increased work needed to break down the coarser land meat. I didn't sprout any wings or get lasers for eyes or suddenly get much better muscle gains in the gym. The word I use to describe it is I just felt more bogged down the whole time.

I'm personally not going to eat any land animals again until the apocalypse happens and I'm forced to eat whatever I can get my dirty survivor hands on, but until then I'm not risking it. And even when that happens, I'm still going to assemble a band of lovable and quirky misfits to go on an adventure to some place that actually has the food I'm looking for.

I'm still not sure though how we'll be able to tell the difference between the normal people that survived the apocalypse and the ones that were already homeless people beforehand.

I guess the giveaway will be the guy with the one and a half shoes doing the marching band strut down the side of the road with a woman's silk sashay secured tightly around his head like a sassy bandana fluttering in the wind, long and free and far out, well into the road having a cheeky flirt with getting wrapped around someone's sideview mirror flying by.

"A Florida man..."

But anyone can do these experiments for themselves. You don't have to trust ANYONE else. If you think I'm part of an agenda to get you to voluntarily discard your super secret cave sauce, then you can just tinker with these diets yourself and see which ones really make you feel and think and perform better.

The last point on this is that some people think that certain foods could contain some critical x-factor that our science still hasn't been able to pin down. Well, that concern is actually perfectly legitimate. We discover new things every day, like some new bioactive property of a particular protein complex or some new sneaky organic compound we hadn't been paying much attention to. You never know.

Here's the thing though. We can still look at population examples of people who consume primarily seafood diets like Mediterranean and blue zone and coastal diets, and we can compare those populations to other ones that consume heavily land animal diets. And we can see from all of this data that the land protein diet is the under performer with generally worse disease rates and shorter lifespans. Some people try to explain these statistics away and claim there's some OTHER reason for these correlations, but I believe that there really *are* still unknown x-factors in the land meats -- I just think that they are *negative* deleterious x-factors – not positive beneficial ones.

We've already seen things like red meat implicated in chronic inflammation and various cancers. And every time there is a forest fire it disperses all kinds of lead and mercury and radioactive elements into the atmosphere which then rain down on the soil and crops that these animals eat and those toxins get concentrated inside their bodies at *exponentially* high levels because of the large size of the animals. Even quality land food is often not nearly as pure and pristine as many people have believed it to be. Does that mean it's the worst thing in the world for you? No. But it's definitely not the best either.

Some people say it is the best, though. They say eating all the inner guts of the cow makes it a super food. They get really into it too. They eat the whole-animal from feet to face to anus, and many of them like to point out that there are more vitamins and minerals in the liver. But then you go and look it up for yourself and see that a big block of those purported vitamins and minerals are actually only in there in super small amounts like one percent of your daily value. I just gotta say, we don't get to claim a vitamin or mineral is in something if it's only in there at one flippin' percent. "Sweet! I only gotta eat thirty three cows today to get my daily requirement of omega-3! SUPER FOOD!"

I mean, I do have to commend people for that whole-animal diet. At least you're not letting anything go to waste, and you're actually eating parts of the animal that do have some marginal nutritional content, unlike the muscle meat steaks that are barely anything more than just basic protein. I SALUTE you, whole-animal eater!

Here's to you, whole-animal eater! We honor you for trying to squeeze as much usage out of land meat as the citizens of Texas try to squeeze out of the word "deal," like it's some kind of Swiss army deal.

"Dang! Y'all seen that one deal where this guy made a deal to get this deal from this one dealer cuz he couldn't deal with that other deal no moe'?"

"I was dealin' over to see my cousin Deal down in Dealton in my auto-modeal and I saw a crocodeal up on the road. So, I dealed 9-1-1 without dealay and I swear I told'm there in that there dealspatch that there's a WHEELIN DEALIN DEALLIGATOR UP ON THE DEALWAY!"

But if you're new to the whole, whole-animal eating game and you've been listening to some charismatic guy with a great physique of questionable origin who starts to go into a religious trance when he talks about eating the whole buffalo, I'm just doing my neighborly duty to let you know that you don't actually have to do all of that. I mean, if you in any way felt uncomfortable or like your personal boundaries were being crossed by having to chew on cow buttholes or shove dicks in your mouth, I just want you to know that it's not really necessary to do all of that and we are all here to support you. We can get through this together - one day at a time. Let me hear you say it, "It wasn't my fault. It wasn't my fault."

Att'a boy. Come on. Bring it on in. There we go. It's okay. It's okay. The things that went into your mouth don't define you as a person. You got that? You *got* that?!

There are also some serious safety issues with these whole-animal diets that you need to look out for. If you eat the brain matter, it can have pathogenic proteins in it which cannot be cooked out and end up giving you the human form of mad cow disease, known as spongiform encephalopathy, which eats your brain. And as it's eating your brain alive you start to get the shakes and can't stop laughing uncontrollably. Sounds funny – isn't actually funny. Well, for us it's funny, but for you you're actually screaming inside while outwardly laughing. Yikes. It's a dangerous game of Russian roulette.

Some real over achievers have even taken to eating land animal meat straight raw because they think this is an even more extra awesome paleo way to live. Uuuuuh... I don't even know what to say about this. First of all, we humans simply do not have the extra intense level of stomach acid necessary to eat raw land meat like a true land predator. And secondly, this is a really great way to give yourself trichinosis parasites.

Plus, they always claim that they are really enjoying eating the raw meat, but the look on their face while they're doing it always betrays them. They have this look like "What in the actual FUCK am I doing?!"

Also, when you're eating the internal glands and organs of mammals like us, you're getting a lot of potent and powerful biological signaling molecules from their body into yours, and it's a lot like randomly flipping levers and mashing buttons in the control systems of your own body. Your body makes its own signaling and control molecules, and it can be really dangerous to start disturbing those balances. One of the things that concerns me most about that idea too is if you were ingesting some hormone or something and then your body got lazy at making the hormone itself and then you just developed a dependency on that external input. I don't want to be critically dependent on eating a cow's thyroid every 48 hours just to prevent my body from going into seizures or something. It's a valid concern. We don't have the same intensity of stomach acid as these land predators to break that stuff down.

If you really are THAT into the idea of eating raw meat, then just get yourself some raw shellfish. That's something we actually CAN digest raw, as people around the world have been doing since antiquity.

I'm not trying to spread fear porn about land meat. I'm just saying these are all actual realities that people need to be able to consider, and it gets a little old for me personally to see people trying to romanticize the Neanderthal diet like it's this super duper healthy way to live.

Don't get me wrong, I don't blame people for thinking this way. We haven't known any better up until now. But it's time to start waking up. It's verifiably not the most ideal way for us to live.

I don't care how many tribal groups around the world are eating all the parts of the land animals. With all due respect to these groups, I'm not modeling my life after people that trundle around lethargically all day and are so deprived of brain food that they all look like they're stoned out of their gourds constantly. I want to be like a fit Italian grandpa that dances with his great, great, great granddaughter at her wedding. Except I don't want to turn into an Italian.

I just want us all to live long and strong and take care of our world and have more time for the things that matter most – our kids and our critters and... Oh shit... Oh, son of a bitch... It's all wrong. Everything in this chapter is wrong. None of it is true. Damnit! I'm sorry. Please, just memory dump everything we just talked about. Yeah, forget it all. Just erase it all from your mind for the context of dogs. None of it applies to dogs. It's all accurate for humans, but not for dogs. Sorry about that. It's still good for humans though! It still applies to humans – just not for dogs. If you were successful in memory dumping all of that just now as it applies to humans, then I apologize. Please go back again and re-read the entire chapter from the start. But keep telling yourself "This does not apply to dogs. This only applies to humans."

Land meat actually IS a superfood for dogs, the literal descendants of wolves. There's no mystery or ambiguity there. Dogs were made to be land predators. They can also live on seafood too. There are wolves that live entirely on salmon and beach food and river fish.

I usually feed my dogs seafood like I eat, but occasionally I'll give them some land meat since that actually is their ancestral diet — of course ethically sourced, humane, happy, healthy, clean, green, grass fed, cage free, no nonsense, all natural, and so on and so forth and whatnot.

I just cook their meat on low, barely to the point of being cooked and mix in a variety mix of sprouts I make myself with added concentrated trace minerals. It's super cheap to make yourself and I mash the sprouts up and add it to their meat to approximate the stomach contents of the prey they would be naturally eating in the wild.

Some people say I take things too far with my dogs. Well, I don't know about that. Do I tether them with body harnesses and static leashes so they can hang their whole upper bodies out the windows while I'm driving?

Yeah. So what?

Do I put expensive panoramic visor goggles on them so they can enjoy a face full of unnaturally high wind velocities?

Sure. What's wrong with that?

Do I take them down popular jogging streets so my dogs can remind all of the exercisers there that they have received no such permission from my dogs to use this road?

You bet your sweet ass I do.

My dogs find their lack of faith disturbing. They must be shown the power of the bark side.

But to answer your question, no, I'm not taking things too far with my Pomeranians. These are common courtesies that should be extended to all dogs.

Yeah, I just said I have Pomeranians. You gotta problem with that?! And, no, I'm not gay! But even if I was,
YOU GOT SOME KIND'UH PROBLEM?!
YOU GOT SOME KIND'UH PROBLEM?!
YOU GOT SOME KIND'UH PROBLEM?!
MAKE YOUR FIRST MOVE!!

The one thing I don't feed them though is poop. Yep, I said poop. Wolves eat shit. They found a five thousand year old preserved human man in the snow in the mountains of Turkey - Ötzi the Iceman - but some wolves at some point had actually found him first and ate his ass out. One of the first things they do is go right up the butt.

That's also one of the main ways I know YOU are not a legit adapted land predator. The stench of poop does not smell appetizing to you.

A persistence hunting ape would have desperately relied on everything he could shove in his mouth from his prey, just like the land predators. And this hypothetical ape with super runner legs would be able to run down his prey long long before he had a big human brain. So, he would have no qualms whatsoever about eating that ass.

A swimming hominid on the other hand wouldn't graduate to taking down big prey like sharks until it was intelligent enough to create effective spears, and by then he would have no history of eating big game crap, so the concept would be alien and unappealing to him.

A land predator ape is a butt hole eating, poop ape. Go watch predators hunt. Watch the full thing, not the one edited for little kids.

This is another dead giveaway we didn't come up as a land predator. You don't see families of country folk with the youngens in their Sunday's best heading down to a restaurant with crazy crap on the walls to order'm all up a bunch of plates of steaming pig shit.

When I see hot chicks with their little dogs, I see them trying to *pre-*

vent their dogs from touching poop on the ground, not trying to get up in there *with* them.

And when I see gym bros exceeding the carry limits on their grocery store hand baskets, sure, the poor guys are usually all loaded up with a bunch of bro-science wastes of space. I get that. But rarely are they buying ACTUAL bodily waste.

Rarely.

Watch, I make this point about feces to help snap people back to reality, but our world is so brilliant now, people will take it the other way. You'll see gym bros munchin' on protein bars infused with elk shit.

"You think you're paleo?! Well, you ain't shit until you try our new line of paleo literal shit! My fellow gym bros and I had a vision to get back to our ROOTS! That's why we started Turd Gobbler and put together a complete line of shit sandwiches..."

We're not ass eaters. No part of that is appealing to us. "But some people..." YEAH, yeah, yeah, don't try to twist what I said. We are an adapted free-dive predator and beach scavenger - not a feces connoisseur land carnivore.

Real human paleo isn't bringing home a buffalo, it's bringing home the shark.

We are *not* a blonde pubes covered sasquatch with dog legs running around with tiny dicks. At least I hope not.

It doesn't matter if you're the biggest Samoan or the smallest pygmy or the bounciest ninja or the most lumbering Viking. You are Man-ApePig, the nerd fish monkey with a big aquatic primate dick and a torpedo head and flat face and a diving nose and a fin chin, walking around pig naked on your flipper feet and descendent from thousands of generations of beta males paying for sex with seafood.

And if that's what it took to make us, then I'm glad it happened.

Not a lot of ice wedding rings in our early ancestry drink. It was mostly horchata.

Just how I like it.

Or maybe you want to be like my dad and know all about these things and still keep eating land meat. But do you really want to be like the guy who introduces himself every single time he texts like he's writing an old letter and gives a fond farewell and salutation after almost every last one, like eight times a day?

Is that the kind of company you want to keep?

I should call him out for the few times he doesn't put a whole sincerely farewell ending on a text and ask him why he's conducting himself in such an informal and illiterate manner.

Wow! Just as I was writing this he texted me. Woe. This is a sign from the universe... not that he texted... he does that every five minutes. But as I speak to you from this final month of our Gregorian year 2021, my father has finally discovered how to do emojis at this very moment I write of him.

The universe has sent us a sign. SEE?! I told you, ATHEEZ! This proves everything! You'll just have to take my word for it.

And congratulations, Pop, for figuring out emojis, since I know you'll be reading this later.

Alright. Enough of that guy.

Anyway, you can still get yoked and swole on any diet. You just need lots of complete proteins is all. There are even tons of vegan body builders that are bigger than I am. I had an old-school trainer once that certainly wasn't a vegetarian by any stretch of the imagination, but he just used tons of beans and rice mainly to get hulked up (What's up, Julius!). Some good examples of complete proteins are incredible edible eggs, all kinds of fish, whey extract, cheese, beans & rice, pea protein, tons of different plant combos. There's all kinds of stuff. You can even eat insects if you want. They have tons of protein and they are the most paleo food you can possibly eat. I myself will not be doing any of that... but YOU can. :D

I like how we always talk about what's hardcore and paleo and cave, but if most of us had to spend the night in an actual cave we'd be like "Uuuugh, it's cold and scary in here! EWWW! Something just touched me!"

Oh, and why don't beach caves get any credit? We find all kinds of super ancient remains of clam shells and fish bones and camp fires in coastal caves.

Why is it all about saber tooth tigers and leopard togas? How about climbing into the water with some great white sharks like traditional women free-divers do? Go slip into the deep blue and have a wee merry dance with some hunting reef sharks and THEN I will fully recognize your all-access brag stage passes.

I'll even be your hype man when you tell the story to women at the bar and I'll run around behind you hootin' and hollerin' like I can barely handle the emotional awesomeness of what you're saying, and then I'll fall down on the ground shaking violently next to your feet, holding onto your leg with a death grip and foam coming out of my mouth from how insanely wicked sick your story is, and then one of the girls who's a nurse will ask if they should put a stick in my mouth and you'll say "No, he's just excited. So, anyway..."

Then I'll just stay like that convulsing on the floor with my eyes rolled back murmuring something unintelligible until medical workers come pick me up and take me away, and you'll be like "Dang, I guess he really was having a problem" and you'll keep telling them the story until you start to get to the best part, and then I'll come running back into the building yelling GOOOAAAAL with my shirt medically cut off and IV hoses all in me and jump straight up onto your back like a horse, as you bring it in for a strong finish and as I attempt to climb up even higher on you with my shoes on, singing the La Macarena.

You'll be a hero.

26

XVII -- TO INFINITY AND YOUR MOM

Whenever I see one of those student driver cars in traffic with a teenager behind the wheel and an instructor in the passenger seat, I always want to pull up next to them as we're driving and then slowly drift into their lane until our cars crunch together so that the teenager girl thinks she did something wrong.

Right about the time you're bouncing off the side of their car you'll be able to see the instructor guy's lips through the window yelling something like "HEEEEEY!! Ooooooh myyyy GOD!!"

Then you gotta slow down and line up the side of your car towards the front with the side of their car towards the rear and then crunch back into them again and hit the accelerator so that you slowly start grinding up the side of their car with that metallic shrieking noise. You should be able to see the lips of the instructor mouthing the words "OOOH MY GOOOOOD! Just keep driving straight, Stacy! You're doing great! THIS guy is WAY OUT'A LINE!!!"

Keep scraping the metal of your car up the side of theirs until your driver's window is aligned with the instructor's passenger window and look over and motion for him to roll down his window. As soon as he does and is screaming something about you being out of your mind, reach out across with your free hand and smack him in the face. With all of his adrenaline pumping, this is surely to ignite an all out slap fight across the cars still hurling down the road.

Now, just go at it with both hands until you've taken plenty of hits. Then simply retreat back into the window of your car and pull back over into your lane, and look back over at him with your hair all messy and the side of your car all grated up and a bunch of red hand prints across your face and then start quivering your chin and letting your eyes water while staring at him with a heart broken look in your eyes like he is the bad man and you are a victim. Then just let yourself start to cry while still looking him in the eyes as you slowly drift away and back out of sight into traffic.

It's okay to cry sometimes. It's not your fault what he did to you.

Some of my friends say that I'm crazy. I beg to differ.

It's actually most humans that are really crazy. They will follow along with the absolute most obvious nonsensical bullsh** simply because others are doing it or because an authority figure said to. At least I can come up with MY OWN fun ideas of things to do in traffic.

I'm talking about these phenomena like the experiments where a bunch of actors will stand in an elevator facing the wrong way and then an unwitting stooge comes along and gets confused and just does what everyone else is doing. Very few people, it seems, actually think for themselves.

Why are there so many people like this? How on earth is it possible that so many people could accept an obvious unreality as real? Well, most of the story is pretty simple.

Throughout our history, one of the most critical factors for our survival was gaining social approval from the rest of the group, lest you be banished from the tribe.

So, if the social leaders of the group all started believing in or even just saying that everyone had the power to jump off of a cliff and fly, then you better damn well start believing it. If there isn't at least one part of your brain that can lie to itself and convince your own mind that what they're saying is real, then there will be little giveaways and tells that you are only pretending to believe, and people will know you're not on board with the tribal cult. Then it's bye bye for you, heretic. Not to mention, you'll be under constant psychological stress with the knowing that everyone around you is half bat-shit crazy, and that stress will disadvantage you and your odds of survival - unless you have some other exceptional quality to compensate.

Briefly, it is worth quickly mentioning the basic pedigree for this cult delusion and bizarre ability of ours to just live in total denial and even feel all warm and fuzzy inside about something that our subconscious still knows is bullshit.

It all begins with things like early alpha leaders that were hyper aggressive and then started killing off rebellious contenders to the title. This began the deselection process for nonconformists. Then, overtime, with an ever increasing concentration of genes for conformity, the tribal group itself begins to self-enforce this uniformity by killing off and ostracizing all of the deviants that are openly thinking for themselves too loudly and who are able to march to the beat of their own drum.

This then became yet more reinforced and selected for through the phenomena by which conformist tribes proved to be highly effective and functional teams, like a military unit where people aren't always arguing with each other and off doing their own thing - but instead just shutting up and following orders. There is a reason why militaries function in this manner - it works.

This is a bit of a simplified generalization of the development of tribal culture which attacks novelty. But it serves to illuminate the basic idea. We have been HEAVILY selected for attributes of cultic tribal conformity. We are born and bred to obey, even if that obedience means believing something completely insane. After all, you can't spell "culture" without "cult."

I just made that up.

Sounds cool though.

And this vulnerability to herd mentality is also a big part of the reason why most sober and sane people are completely terrified by the prospect of doing public speaking.

If you think about it, public speaking is a strange thing indeed to be afraid of. All you're doing is standing up in front of a bunch of your fellow upside-down mop looking beach apes and waving your arms around and making a bunch of face sounds. It's not like you have to go do something like crawl through a dark cave and then come face to face with a pissed off rattle snake because your little brother would rather take the cave route than just climb up the waterfall which de-spite its difficulty still isn't a waterfall WITH RATTLE SNAKES IN

IT!

And yet most people, it turns out, would rather just take their chances in a Mexican standoff with a deadly snake than to simply just get up in front of a small audience of their fellow Rapunzel monkeys and start making mouth noises.

This bizarre phobia still conspicuously persists within human beings though. People by themselves tend to be quite reasonable, but when in large numbers, they suddenly become readily capable of stampeding over anyone daring to challenge the latest trendy dogma. Quietly questioning the gods in a private conversation with your friends can get you hushed, but projecting that same idea out loud in front of your entire tribe can get you killed.

The wrong type of thinking for yourself can quickly become a hazard to your health. And so, we are literally hardwired to be capable of living in a complete and total fairytale fantasy and happily believing whatever magnificent horse sh** is currently being dictated to us. The trance of conformity has its own particular advantages, but it also has its own degenerating disadvantages, and this is a large part of why we must always be double-checking ourselves, *especially* in the moments when we are the *most* confident.

To put a capstone on this idea, there are of course some genetics that trickle around in the gene pool that will occasionally make certain individuals a little more capable of seeing through the spells. These are individuals who, throughout history, end up in the priest classes and the advisor echelons and become the power behind the throne. They are the ones that understood "the game," and they instinctively considered this to mean that they were of "noble blood."

Well, I think that it's time that we *all* got some noble blood and started learning how to trust our own personal ability to patiently problem solve and our own natural heritage to truly think for ourselves.

To be clear though, our subconscious mind deep down is a no-nonsense subconscious and it will still know that our upper level conscious mind is up to some nonsense. That split between parts of our mind will put us at war with ourselves. That's why cult people seem like they have had a lobotomy and why their weird music always sound really awkward and unsettling. It's because they've disinvited an entire portion of their brain to the party.

This cult brain phenomenon shouldn't be confused though with ignorance confidence. That's when especially teenagers think they've got it all figured out, even though they only have enough of the pieces to put together one corner of the puzzle's picture.

"Oh muuuuh gaud! It's all so clear to me now! The world is half a FLOWER! Hahaha! Can't you see?! Why doesn't everyone see this?!"

People like this can still make cool art and music, though, despite being ignorant. This is because their instinctive genius can still leak out through their subconscious mind. That's because they haven't shut it all down with cult delusion.

So, we covered the two distinct concepts of cult delusion and ignorance confidence. You yourself may even be potentially suffering from one of these serious conditions as we speak if you've ever been diagnosed with "trippin' right now," "man, you ignorant as hell," or "Ahhh HELL nah."

We *do* have to mention one partial exception to the cult mentality though. For nerds, they are inherently forced to think outside the box and view the world more independently just for a basic chance at genetic survival. That's just nature. But the general rule still applies to everyone, including them. Our only major defense against the dark arts of demonstrably irrational group think is our powers of objective and careful consideration.

And, no, I am not the one that is crazy. You can't say that I am the crazy one just because I've planned to do things like become a stalker-stalker and crawl through the bushes with a big telescopic camera and take pictures of a stalker while he's taking pictures of someone else and then send him pictures in the mail of him sending his stalker pictures to the other people in the mail and then take a picture of the exact moment when he opens my letter in his house and sees my pictures of him and then send that picture to him too.

Then I'll send him a bunch of dark blurry pictures that get weirder and weirder, like a picture of hundreds of photos of him scattered all over the ground in my basement; then one of me laying in those photos naked; then me all tensed up backwards and shaking like I'm having a grand mal seizure on his pictures.

The note inside will just say, "I caN mAkE yOU into SOMEthing beauTiFul your sKin will seT us Free

Fhi'fhi'FHI'FHI'FHI (O),..,(o)"

To answer your question though, no, I do not do drugs. And I find it highly offensive that you would even assume that of me. I don't need drugs. I'm already strung out on superfood and dubstep. I don't need an artificial virtual reality for my brain. I don't smoke or toke or drink or tweak or shoot up or huff paint or see who can hyperventilate the longest with a friend before someone passes out and hits their head on something on the way down and wakes up with no memory of the event.

Heck, I don't even hardly drink caffeine, even though my niece often invites me over to come have air-tea. I just lift the cup up to my lips and make sipping noises like I'm having some while muttering sounds of tacit approval of the flavor. But I never actually inhale.

Quit distracting me. Alright. So, why are we talking about delusions and ignorance and student drivers? What does any of that have to do with anything? Well, the reason for needing to include it is that it is the final piece of a simple but grand puzzle. This book was not only made to show us who we are and where we come from, but it was even more so made to catalogue *the recipe for higher intelligence*.

The two basic components an animal needs in order to continue to grow in intelligence are first to give their brain the proper building blocks to grow, and second to select for intelligence in the form of effective providers.

For the first part, the building blocks necessary to grow and maintain our brains are mainly sea nutrients. For the second part, the way we maximize the selection of intelligence amongst us now is to maintain free societies. The more that human beings are able to freely associate with each other in a state of being that is free of coercion, the more that genes for *talented and intelligent providers* will organically increase within in the gene pool. In other words, the more you leave people alone, the more they will simply naturally select for intelligence and the more intelligence will multiply. This will be to the benefit of all because the genes for higher intelligence will spill out into the rest of society, and the intelligent will invent and create new things, and finally, the only way for the intelligent people to get ahead in a world of free association is for them to create value for others. And so, everyone wins.

To conclude, for us to continue advancing from our level here we need plenty of brain food and plenty of freedom.

So, what happens after you have those first two basic principles covered and established? Well, there are three critical things for a species to do once it has reached the level of having the power of speech - stop killing people who question the gods; start recognizing your own potential for ignorance confidence; and don't forget that you are selected for delusional cult thinking.

In the same way that there are occasionally popcorn that jump extra high off the skillet, there are occasionally children born that get an extra dose of intelligence. These children will be naturally inquisitive and question things like societal norms and the gods. When this happens, DON'T KILL THEM! We've been doing this shit since the dawn of time and thus destroying new genes for intelligence and constantly f**king ourselves. No kill-kill. Debate them? Sure. Tell them to get lost when they're being annoying? Go for it. Tell them you're gonna kill them if they don't shut up? Uuuuuuh... I guess you could say that... but don't actually do it.

So, to summarize the three most fundamental things after acquiring the power of speech - don't eliminate the annoying smarty pantses from the gene pool for questioning the gods; don't forget we get ignorance confidence and there might always be more to a situation than what we currently see; and finally try not to succumb to cult delusion and end up valuing charisma and groupthink over actual facts and reason.

Now, in total we have the two basic principles for increasing animal intelligence and then the three fundamental principles for accelerating intelligence in animals with the power of speech. So, what comes next?

The final three principles are the most enlightened. These three enlightened principles don't necessarily come last though. They should be adopted as soon as they can be understood. These three principles are as follows. Protect the children; follow a compassionate morality; and identify with something greater than yourself.

Let's talk about protecting children. When children are traumatized growing up, it hardwires their brains more for living in a fight-or-flight mode, and this causes a significant functional IQ reduction. The more patient and peaceful children are raised, the higher their operating IQ and compassion will be. I'm not saying turn them into pansies. They need grit and appreciation for all of the simple things. But instilling this bearing in them cannot come in the form of abuse.

One very overlooked form of abuse is neglect. Some people don't take it serious because it's not an active form of abuse, but if a child does not get a lot of attention from their parents it trains their subconscious mind to feel like they are alone in the world and that there is no safety net, and so they get stressed out much more easily because they feel like everything is do-or-die. They don't have the natural sense of security and confidence of a decently cared for child. This engrained low-grade panic and pseudo paranoia are terrible for their cognitive performance.

Unhealthy trauma caused in children has been going on since the dawn of time and it needs to stop because it has many disastrous and far reaching ramifications including major reductions in intellectual performance. It's not normal; it's not okay; it doesn't make children "stronger." If these things were normal then you would see television shows depicting physical and verbal abuse of children and audiences would be perfectly fine with it. But that's not what we see. If there were shows that tried to normalize that kind of treatment of children then viewers would all be outraged. It's not normal.

Besides, if I can potty train and raise dogs without so much more than a very rare need to use a firm voice, then you can raise your big brained human children with peace and compassion too. I had one Pomeranian boy who was a wild child and he just refused to stop being naughty. Whenever it was snack time I would gently say "Oh, I can't give you any. You've been a baddy, not a little laddy." Then I would take him to where he had been naughty and remind him, and he would understand that's why he wasn't getting a treat. He did not like that one bit. He would bark at me like I was some kind of monster and I would have to hold a straight face and stay serious. But after a few months he finally just started behaving.

The second of the three enlightened principles is to have a system of compassionate morality. Societies which function like this will always outperform an analogous counterpart. If you take a homogenous population of people and split them into two groups and have them live separately under identical conditions but one group has an empathic moral code and the other does not, the group with the moral framework will vastly outperform the other. There will be more trust and less fear, which will allow for many incredible things to happen including enabling the citizens to intellectually perform better. Afterall, fear is the mind killer.

Some might argue that you can get a lot done in a brutal empire with a bunch of slaves, but in truth you can get infinitely more done in an empire of the same size population with freedom and morals. Rome knew all about steam engines and harnessing energy, and they could have begun the industrial revolution two thousand years ago, but they didn't because they had plenty of slave power, and we all know how that turned out for them.

To conclude the point of morality, the easiest and most transcendent way to structure any code of ethics is to base it on the ol' Golden Rule. Simply expect others to treat you right, and in turn, treat them with the same consideration that you yourself deserve.

The third and final enlightened principle is the hardest to discuss. The basic idea is to identify with something greater than yourself. We'll get into what this means in a moment, but the reason that this principle is important is because it is the best glue for holding every other principle in place and it gives everyone their greatest sense of peace and purpose, which allows them to soar. This is by far the most psychologically healthy state for us to be in, and it comes as a benefit to all of the different areas of performance and life, including intellectual prowess. This is the way we were meant to be.

So, what does it look like to identify as something greater than yourself?

Well, the simplest and most obvious version of this is your genetics. You can see that your family and close relatives carry the same genetic code as you, and as such they are extensions of you. They are your same organic essence. Their biological success is your biological success.

If your own personal biology is the most important thing to you, then consider that your children and grandchildren and great grandchildren and so on will need happy, healthy, well adjusted people to create families with in order to keep perpetuating your genes. Now consider that the people living around you today will one day be apart of your family. Their descendants will merge with yours someday, and even if some of them won't, they will still live in the same world as your grandchildren and help benefit them by being compassionate and productive themselves. Everything you do in the world trickles through it and affects a vast number of people that you will never see – everything good and everything bad. When you do something good, you are helping the fellow ancestors of your own descendants, and likewise when you do things which are gratuitously harmful to others. These people are all your family and neighbors – you just have to go down through time through your own family and then make a u-turn and come back up through their family in order to see the connection.

Family and genetics tend to be the most important things for a lot of the true-blue die-hard atheists, but what you might not have considered is time travel. Yes, I said time travel. Our progeny could very well uncover time travel in the future and have the ability to come back and both clandestinely monitor us and even rescue our consciousness before death and then have the minds and DNA of everyone who has ever lived stored on thumb drives in the pockets of future people. This may have already happened... uuuuh... or it will have already have happened in the future... or.... uuuuh... if it's the future and you went back to the past and your mom is there flirting with you at the school dance and... uuuuuh... wait... is it still incest if you haven't been born yet? Uugh, nasty. You know what?! Just *don't* be a DICK, ALRIGHT?! You never know what could happen.

Alright, let's rewind this a little bit. Let's talk about even more expansive ways to be bigger than just your immediate self.

First consider the things in the world which to you are the *most* sacred and wonderful. These are things that emanate immense beauty and transcendent qualities to you – things that you want to continue existing in the world and thriving for all time. For most people that would be things like puppies playing with happy little babies or other things like entire towns that have been built just for little midget dwarf people and all the little houses and cars are perfect size for them and they live there the whole time doing their little things and you can go there and visit and they don't mind if you go around hugging all of them and sneaking up from behind and tickling them while they're doing their serious little jobs. :D

Let's pick a different thing to talk about actually, because some people might be getting too sad to think about the fact that Shortsville doesn't exist yet. Let's say one of the most precious things to you in the whole wide world is when someone goes into an animal shelter and rescues a dog and brings them into a wonderful home. You can now think about these other people out there in the world and how they carry a bright little spark of the same essence as you. They are, in effect, carrying a part of you with them because they love so dearly the very things in the world that you yourself love with all your passion. There are pieces of you scattered throughout the world. You are bigger than yourself. Your essence is everywhere.

We'll finish it all with this. The final way to become something greater than yourself is through spirituality. Not everyone believes in an afterlife, and not everyone who does believe in it is believing in it for more rational reasons. But if you want to approach this topic with humility and an open skeptical mind, there is a wealth of resources you likely have never even heard of. If you are seeking, you can find.

I personally have experienced things that have long since made it quite impossible for me to deny that there is something more to the cosmos. I can't go back to a state of not being very certain. I'll share a brief anecdote and you can tell me if I'm the crazy one.

When I was a teenager I was living very far from where my grandmother lived. I was over the horizon about one hundred times over. My gran back home was still healthy as an ox and she walked very long distances every day to make pointless geography observations and useless field studies. She was so healthy she would outwalk her dogs and have to carry her fat little brown Chihuahua dog on the way back home and we would yell "Score a goal, grandma!" There was nothing wrong with her health and absolutely no reason to consciously or subconsciously suspect any issues.

One day I was at home getting ready for my afternoon job and I was suddenly overtaken with drowsiness in the middle of the day, so I had to lay down. No sooner than I laid down I was "transported" away. That's the best word I can use to describe it. My consciousness was up in the air hearing and seeing the world in far more clarity and vibrance than waking life and watching a vehicle exactly like my grandmothers with a little woman behind the wheel looking and dressing exactly like my grandmother. The only difference was that the woman I saw was my grandmother in the prime of her youth with her same little pearls and flower dress and curled hair, a way I was not accustomed to seeing her.

It felt like I was only there for about five seconds when the scene was violently interrupted by an explosive light and deafening roar that felt like having a nuclear bomb going off in my face. I really can't over emphasize how insanely loud and bright it was. I was shocked awake and I ran around my house asking my roommates if they had heard "the bomb" going off. I went and looked out of every window because I was sure there would be a giant mushroom cloud within the vicinity. But there was nothing at all and my roommates all raised an eyebrow at me. The very next day I got the call that my grandmother had been excitedly driving to go see my new nephew being born when a large truck had run through a stop light and completely destroyed her car. Her passing was instant. This is the event I had seen.

I found many years later that these occurrences happen to enormous numbers of people and they are extremely common. Every type of people has experienced these – doctors, neurosurgeons, atheists, you name it. The term for them is "shared death vision." There was absolutely no way for me to see this event coming; I have my roommates as witnesses that I was in fact running through the house like a nut talking about an explosion sound prior to any knowledge of the vehicle accident; and the event itself was more vivid than anything I have ever experienced.

I've since had other extraordinary experiences and done years of research on this general topic. I could write an entire book, and it would be way sweeter than all of those other books. I've researched both sides of the debate and seen arguments like ones claiming it's just people trying to get attention. There are definitely the unfortunate charlatans like that who are trying to scam people and there are the others who sadly are just desperate for attention, but we know all of their tricks. And the vast overwhelming majority of witnesses to these things are just people like the ones you see on the local news being interviewed when something sensational happens in a low income part of town and we tell the story really awkward and just keep repeating the same points.

"Yeah, I seen the whole thing... We were tryn'a get out'a der and we were ruuuuunnin. Yeah, I was here duh whole time and seen duh whole thing with my own eyes. Soon as dat fire started we started ruuuuuuuunnin. Oh muh gaud! I seen the whole entire thing... I ain't nevuh run like dat without no pants on."

But suffice to say that I am not even beginning to scratch the surface of this topic. Like I said, if you seek, you will find. Either way, it helps immensely to have a prudent and measured spirituality that places love above all things. Granted, we need to learn the value of tough love and not being suicidally altruistic. But the motivating driver behind everything should ultimately be compassion and concern. When people see across the veil and ask them how we should live, the instruction is always very simple – "love." Treat others how you would want to be treated if you were on their side and they were where you are.

Oh, and we also get told that we often over complicate things. And also sometimes people ask them what they think of us down here and they say we're "funny."

Funny?! I don't know what they think is so funny. I'd like to see them come say that to our face.

Anyway. If you're such a staunch atheist that you don't need to go the spiritual route and you would just chalk my experiences up to miraculous coincidences, that's fine, but you still can't fault my assessment. Every time we turn around we are finding something we still don't understand, like apparent red shift and dark energy effects. Is it really so absurd to suspect the existence of unseen advanced consciousness giving us our space to develop and have experiences? Alright, that's enough of a dog bath of serious talk for me. I need to go roll in the carpet and pump my legs.

So, now we can finally catalogue the eight major steps to create an intelligent and ever advancing species.

There are the two basic components for building intelligence, the three cautions for species with intelligent speech, and the three principles for enlightened societies.

First - Get your brain food, which is primarily sea nutrients; then foster free societies so they can naturally select for intelligent and talented providers.

Second – Quit burning people at the stake for questioning the gods;

don't always be so sure you don't have areas of ignorance and pieces of knowledge you are missing; and keep in mind that you need to compensate for an alter ego inside of you which has been selected to be a straight up kool-aide drinking cultist.

Third – treat children with patience and kindness; maintain a system of compassionate morality around the Golden Rule; and never forget that you are immensely expansive and something far greater than just yourself in many more ways than one.

Do these things, and from the jungle beaches of Eden your children will one day reach the stars and beyond.

And it all starts with rediscovering what we really are. We are not the jock animal. We are fish and chimps.

We don't need to get way too hung up on a lot of the crazy stuff our ancestors did between becoming fully human and then traversing the globe. We want to focus on the things that matter most. Our ancient ancestors had ritual sex with horses and men sucked each other's nipples to show fealty. Uuuuuuh... What?

That's why we know so little about history. People kept erasing records out of embarrassment. "Grandpa did *what*?!?!"

It's only the important things in the world that matter - like how famous actors in movies really matter but all of the expendable extras don't. It's like a sad scene in a movie that makes us feel bad for the heartbroken characters because of our natural human empathy as the camera pans over the faces of the celebrity actors and then it pans over the face of some grieving extra we don't give a shit about and then back to another weeping celebrity that tugs at our heart strings.

Don't obsess over that extra. What? He's a real human being? I don't give a shit. His name is Perry and he has an adult daughter and a place on Reseda and he struggles to pay rent? HAAA! He struggles. That's funny. The only people that matter are the famous ones!

We gotta focus on what matters, and what matters is the admirable yet goofy thing that we really are – not the romanticized fanciful things we enjoy making up. We can't go on thinking we're something we're not. We can't be like that guy who thinks he's hard and walks into rush hour traffic right down the street from a perfectly good cross walk and starts strutting straight towards the lane of vehicle travel as if to say "You better stop when I'm walkin'." Okay, guy. What are you gonna do? Shatter your body against my car? Yeah, that'll teach me.

I noticed they go from thug life to bug eyed REAL quick when they realize I'm not even slowing down. Yeah, this where cars drive, dude. Don't test me, street walker.

Besides, we're pretty damn thug life as we already are. Have you seen people that learn how to do the Aqua Man swim?! Don't tell me that's not cool. We need to combine professional sports with Sea World, like how Rome used to flood the Colosseum.

We're just wasting our precious time when we live in a fantasy. We can't be like one of those motorcycle guys who rides a cruiser and revs the engine all loud like he's some kind'a badass but then has a bumper sticker on his truck that says "Look out for motorcycles! We're just some widdle bitty guys." As if we're not already looking out for your delicate little motorcycle that we could accidentally obliterate with a flick of our wrist. And like we're not already making room for you like at a concert when someone passes out and we all put our arms out to make a hole.

"Everyone get back! Everyone get back! Motorcycle! Give'm room! Him just a widdle guy. Aww, look. He's being a little tough guy. Yes, you are. Yes, you are. You're a big tough moty cycew. ...WOE! Careful, guys! Not too close. One little tap could annihilate him."

I never had a bumper sticker like that when I used to ride. I was like, yeah I might get fuckin' slaughtered on the road. I accept the risk.

Well, actually, I take that back if you're the dude with the full-size American camper size motorcycle with the hot tub in back. I didn't know they could put that much infrastructure on two wheels. Freakin' thing has autopilot so he can get up to go use the restroom. Yeah, look out for that one... for your safety, not his.

But at least with motorcycles you're actually doing something. You're out on the open road, going places, doing stuff, seeing awesome things, as free as a bird – a bird that likes to go to biker bars at night and hang out with your bikie friends until that one night when there was a song playing about "guitars and Cadillacs and hillbilly music" and that big naked dude showed up in the bar but he was really an exterminator robot from the future inside of a skin suit, and you just assumed it was a big naked body builder guy because you didn't understand the technology of the future.

Then the naked robot guy said "I need your motorcycle and your clothes and your boots" and you started laughing really hard and looked over at your friends to make sure they could see how tough you were to laugh in this guy's face. Then you really wanted to make extra sure your friends didn't think you were scared of this guy, so you reached over and put your cigar out on his bare chest. Then he reached over with his machine strength and crushed your hand, and you started screaming, and he picked you up and threw you completely across the bar back through the order window into the kitchen

575 - BARTROLOMEW MC INNCEL

area.

The concussive blow to your head and body from being tossed like a rag doll put you out of it for a second, but as soon as you came to again you realized you had landed on a burning hot kitchen grill and you started doing that hot-potato dance on your hands and knees while your burning flesh was hissing and crackling and you finally fell off onto the floor.

Then the naked machine came back into the kitchen and you tried to pull a gun on him with your broken fried bacon hands but he just snatched it away from you and your voice started to crack and got all high pitched like a sissy and just you gave up.

Then he used your motorcycle to chase me down on my dirt bike, and I thought he was trying to get me but he was really just trying to protect me. Later on when we were having a bonding moment at a junk yard in the desert, my mom was watching us and it suddenly became so clear to her that this machine would never stop – it would never leave me – it would never hurt me or shout at me... or get drunk and hit me or say it was too busy to spend time with me. It would always be there... and it would die to protect me. Of all the would-be fathers that came and went over the years, this machine was the only one who measured up. In an insane world, it was the sanest choice.

dvv'dvv dvvv dvv'dvv

dvv'dvv dvvv dvv'dvv

dvv'dvv dvvv dvv'dvv

dvv'dvv dvvv dvv'dvv

27

XVIII -- LET'S END THIS TRAIN REKT

YOOOEEE, DEEEEEEE JAAAAAAAAY!!

Drop that next level beat!

And gim'me some filthy funk...

Alright, now LE'MME AT'M!!

I didn't come from no freakin' monkey.
It was a monkey AND A FISH.
Whatever made me in its image
also got ridiculous!
I ain't from no runnin' monkey.
That shit blew out my hip & my left knee.

If you think we were persistence hunting,
you might already have spongiform encephalopathy.

I just hope that the Earth gets flooded
so we can have a water world in the future,
and I can buy half a dozen women
for just a couple dry sheets of pyepper.
I don't want some Madd Maxx inferno
where just one dude gets all of the chicks,
and he's some NASCAR hillbilly warlord
that's harvesting milk from fat ladies' tits.

I'd rather be GONE out deep sea fishin'
and gettin' these riches on those high seas.
I'm makin' it BRAIN on deez boat women,
struttin' those semi-aquatic bodies.
I ain't attracted to no chimpanzee,
unless it's aquatically adapted.
But also it would need like our intelligence,
or it's taking advantage of a mentally challenged.

I'm stackin' up books here like a mothuh...
but the cosmos is WAY BIGGER!
The Galactic Federation probably has monthly potlucks,
and we're not invited cuz WE'RE THE BULLSHITTERS!
Who even wants some dope at their party
who be ackin' like somethin' he isn't,
and who really thinks naked running is feasible,
which implies he has a micro penis!

PEACE!!

(Open parenthesis: Thugged out gangsta music fades out and slow-mo shots of us bobbin' our heads. Editor note: just use one of those singing holiday card music things with the little tin can speaker for the gangsta music to start playing when they open this weirdly thick page of the book. And for the head bobbing, use one of those things where the image changes as you move your head from side to side giving it the appearance of movement but it's only two frames, so the novelty of it wears off instantly and then they're just annoyed that you would waste their time and the world's resources to make that cheap bullshit and think it would entertain them for more than one microsecond, and the added guilt for them of having to throw it away because of the obvious thickness and materials that were completely wasted on making it and they're not even sure if they can recycle it or which type of recycling it would even go in because it's like paper and plastic at the same time, and now they basically have a mild degree of visceral hatred for you. Just put one of those on the back cover of the book so they can't rip it out and recycle it. Obviously delete this editor's note when you've done all that stuff, Dave, and don't fuck this up or I swear to fuck I will fire you. I swear I will go to your manager and talk to him and tell him you like tried to do something sexual or pinned me between your arms with my back against the wall in the hallway and put your face way too close and nasty to me. Don't FUCK with me, Dave. You mess with the BULL and you get the REGRET of doing it! ... ... close *paRENTHEsis!!!*)

28

XIX -- DREAM A LITTLE DREAM

Have you ever had dreams you were flying?

Did you rise up off the ground with your feet?

Were you gliding all around your own city? Could you even hover high above your local scenery?

Were you often just still and gently floating? Or did you shoot like a meteor across the sky? Were there times you'd get stuck in slow motion, or bogged down in a race against time?

Were there moments of deflating and sinking? Did you panic while drifting down from so far up? Was there something just too difficult to get away from? Or were your limbs ever dragging through muck?

The breadth of our scope here in this place is not to discuss the psychology or the labyrinth of passages and corridors through the meta-physical and unknowable topology.

One thing that we certainly do know, is that our *conscious mind* is actually quite small, like a little red and speckled black lady bug, atop a planet with enormously vast sprawl.

With its forest greens and aquamarines, and its ivory tan sands that fade to beige, our *subconscious* mind is like this large planet, with the broad oceans of majestic history it contains.

We know that our core dreams do swirl as they ascend spiraling up from below. They are born from deep within our subconscious, built on ages of primal instincts for survival.

There are the obvious surface layers of meaning and symbolism to each of these dreams, which I have catalogued here earlier just briefly, these all too familiar life themes.

Sometimes we feel free and unrestricted, or caught slogging and blocked off from escape. But how curious that our wise old subconscious readily chooses from these reoccurring templates?

Why indeed would we dream we are flying, or floating up and around our own home? Why not running like the wind across plains land, or out bounding where we can endlessly roam?

Why have dreams about fleeing a pursuer while pushing hard through dense sludge like thick mud? Why the depictions of impotence in this strange manner. Why not just tangled up or simply bush trapped and stuck?

It is clear this all happens for a reason. Our sleeping mind toys with things because they matter. It communicates in *the* old forgotten language and still uses these long lost words for our new chatter.

Our dreams
always speak
in these archetypes,
like burdened movement
or soaring
on wings,

yet never
before
in our history
have we known these
but for flying...
through... sea.

29

19 ¾ 1 -- THE POWER OF THE BARK SIDE

The character of batman was based on me originally.

I mean, not the comic book character, but like when they started making these cool new movies - they got the idea from me I think.

It was my idea first to go out dressed like batman in real life. I was gonna use the symbol of the dark knight to fight the darkness in my city and just cut out the middle man. It was a pretty dope idea too when I rolled out in the batmobile on the first night. I popped the top up too so I could wave to the people and bump some music while I was driving around. Dude, it was freakin' awesome. Until some fucked up little kid shot me in the mouth with a paintball gun.

That mother fucker.

Those things fuckin hurt.

Especially when they shatter into your mouth.

That bitch ass mother fucker.

That's one fatal flaw of the batsuit though. I never thought about it before. The whole mouth part. It's totally open. You need like... ... a shield... or something. Like, if you were fighting superman and you had all this armor on, he could just punch you in the mouth.

But it made me realize, fighting crime is pretty dangerous. Maybe most people don't realize that. You could get hurt fighting someone.

So, it's just better to find other ways to fight for justice. Like using your words and arguing with your friends and stuff.

I'm not retired from dark knighting though. I could have my car up in no time. So, if anyone knows about a kid who was ever bragging about something like that, or like about shooting a guy in the face, then just get in touch with me and let me know. Let me know what's goin on with this little fool, and like where he rides his bike and stuff so I can run his ass over.

Things don't always turn out the way we think they should. One minute you're just ridin' around, rockin' and rollin', cruisin' down town, lookin' for some criminal's ass to kick, and then the next minute you're trying to find a private investigator who's willing to hunt down a little kid for you.

You never know what could happen. One minute you're trying to make a brief mention of a coastal civilization in an earlier chapter, and then the next minute you realize that people just aren't quite getting it and now you have to write a whole entire gigantic chapter just about that one single topic.

So, here goes. There was one more group we never talked about in our earlier dissertation on coastal civilizations. And this one's a doozy. It's like a Gordian knot rapped in a can of worms inside a rat's nest at the bottom of a barrel of monkeys. In other words, it's pretty complicated.

Well, it's not that it's actually that complicated. Honestly it's pretty simple. But it comes across a little complicated because it's just so different from what most of us have been doing these last ten thousand years.

But I guess we can just start with introducing them. This civilization has been occupying a continuous length of forested coastline, stretching all the way from Alaska down to California. It's such a massive length of coastline that it fully extends over one sixth of the way across the entire hemisphere. So, because of their geographical location, it's very fitting to refer to this family of people as the North American Upper Pacific Coast Civilization, N.A.U.P.C.C., or simply with the stylized nickname Nauptsèh based on the acronym. And that would be said like their phonetics Nah-oop-tseh.

I suppose you might be able to refer to this region as the Potlatch civilization, since that's essentially the most salient feature which unites them, but they've never officially approved that term. And that would be like referring to our own Western civilization as the "County Fair civilization" without ever asking us. So, at least "Nauptsèh" is a fresh neutral nomenclature.

And I'm not exactly one for really caring about what's officially appropriate or not, but this culture is freakin' badass. And they're not just freakin' badass because they figured out how to make about a billion and one different things out of cedar trees; they're freakin' badass because they essentially solved civilization.

And also, they're still around these days. So, I can't just say whatever I want and get away with it.

Regardless however, the essential impetus for us needing to devise some sort of abbreviated new epithet for this civilization is simply because there's never really been one before. They've always been extremely independent and sovereign city-regions. And it's been like this for thousands and thousands of years at this point - all the way back to the Neolithic.

Furthermore, as we will see later, this ferocious independence has always been one of the most defining characteristic of this civilization - so much so, in fact, that we still don't have an official conventional moniker for them.

So, let us just keep that in mind as we proceed, that they have always been *extremely* serious about these values. And these values have thereby shaped them into one of the most progressively enlightened civilizations ever seen on this planet. *Truly* an extraordinary civilization.

Now, I've found over the years that every time I try to explain how their system works, the average person just doesn't quite get it. And that's why I have to have this whole, huge, final, separate, exclusive chapter on it. ... It's totally worth it though. Whenever you see how their standard of living is affected, it'll totally make sense. This culture is definitely a civilization worth talking about.

So, what I've generally found is the best way to convey this, is simply to describe what your current life would look like if we were living in a modern Nauptsèh community. And it doesn't even matter where you live right now. It works with anything - any level of technology, any-where in the world, any size community - it doesn't matter. It doesn't matter if you're in some cozy little village at the beginning of the Holocene, or if you're in some massive modern mega city that's visible from Venus. It's irrelevant. It scales up to anything.

But for our purposes, though, we will be discussing it in a very modern Western context.

Oh, and just to be perfectly clear about something, this IS an actual thing. This isn't like some *hypothetical* model or something. This is a living breathing system that's been around since the Neolithic. Well...
... technically it's been around since WAYYYY before that. But let's just keep things simple for the moment.

Alright. Enough of the intro.

Wait! ...No! ...There's more! ... I have one more thing to warn you about! There's gonna be a lot of talk in here about civics and taxes and logistics and bureaucracies and blah b'b'blah b'b'blah'blah. So, hope-fully I can make things pretty fun still. Hopefully I can make things pretty... *sexual* for you.

Okay, so first things first. We are now being transported into a parallel universe. Immediately upon arrival, we notice that your town looks pretty much the same still. But it's also kinda different. It's like we went back to the future or something. And even the low income areas look really nice now. Everything is just clean and well manicured.

Spoo'oo'oo'ooky.

We have now entered... ... The Nauptsèh Zone.

Alright. So, the first thing you need to know here is that you now make a lot more money. You're still working at the same job, but now you just get paid a lot more money - four hundred percent more to be exact. Yep. Your salary, your paycheck, your income, whatever it is - just take that value and multiply it by four. If you've been earning $50k a year, well now you're making $200k a year. If you've been earning $75k a year, well, not anymore. Now you're making $300k a year.

And there's no catch to that by the way. It's not like, "yeah, you're making four times as much, but now everything costs a thousand times more and you gotta pay all these extra fees and sh**." Nope. Everything costs exactly the same here. But now you just have a lot more discretionary income.

And that's just the easiest way to make the math simple. In the Nauptsèh world, your buying power is so amplified that it necessarily is the equivalent of you making about four times as much money. And, again, that's if you just leave all of your local prices exactly where they're currently sitting. We'll explain this later. But for now, just suffice it to say, you're making a lot more money.

Okay. So, what's next? What is it like having a run-in with the police? What's it like dealing with the cops in a Nauptsèh style city?

Okay. So. Let's say that you're just driving along in an 80s convertible with the top popped, listening to some hardcore hypa hypa, and getting a little too excited on the roadway. But then a cop sees this, and decides to pull up right behind you with his lights on.

However, on this particular day, you don't really feel like being pulled over. So, you just calm down a little, correct your driving, and then wave to the police officer to acknowledge your mistake. So, at this point, the officer can see that your driving is much improved, and that you're not swerving all around like a drunken sailor, and that your car hasn't been reported stolen or anything. So, he just decides to let you go with a warning. He never even finishes pulling you over. He just does some sort of signal back to you and lets you keep on driving.

Okay. Now. The reason why he does this... is because it's not his job to just try and catch you up, and raise money for his police department. His sole exclusive purpose for existence, in this capacity, is to help ensure the safety of the community for everyone. That's it. That's his mandate. That's his job. And that's also what he, as a fellow member of the community, wants.

Alright. Well. Then what happens when things get a little bit crazier? Let's spice things up a little.

And let's mix up the drivers too. Let's make the next one a female immigrant hailing anywhere from eastern Europe to the Philippines.

Alright. So. Lady luck is just drivin' along, minding her own business, swerving across multiple lanes of traffic without using her turn signal, flying down the highway with all four of her tires flat on her minivan and not even realizing it, and then trying to launch her car straight across an entire busy roadway like someone trying to shoot a bullet between two different freight trains going in the opposite direction and thereby causing me to crush down on my brakes so hard it felt like my car was gonna roll over forward onto itself.

Tryin'uh ASSASSINATE me out here!

Okay. So. Lady luck is just cruisin' along, doin' her thing, listing to some DJ Crazy Times, when all of the sudden a cop sees what she's doing and says "uuuuuuuuh, whaaat the fffffff..."

So, naturally, he springs into action and then pulls up right behind her with his lights on.

At this point, lady luck just reaches out of her window to wave him off. She's just not in the mood to be pulled over right now. And whether she's doing this because she knows it's a valid Nauptsèh rule... or she's just doing it of her own accord, we're not really certain.

However, on this particular day, our police officer *really does* feel the need to speak with this person. So, he continues to pursue her, and she continues to ignore him. Eventually however, she pulls into a grocery store parking lot. Then he just pulls in really swiftly right behind her. He needs to make sure that she isn't having some kind of medical emergency, or driving intoxicated or something.

Once they've had their little chitchat, he let's her know that he's gonna have to write a basic traffic report on this incident. So, he gives her the reference number and sends her on her way. He doesn't make her just sit there in her car forever while he's writing the full report up and taking a nap or something. Nope. All he needed to do was just simply make sure she wasn't impaired or something.

So, she's totally within her rights to behave ridiculously, if she wants to, but it's gonna have to be reported. That's the trade-off. And we'll talk more about this report in a minute, but suffice it to say... that it's just simply a way to *incentivize* her to drive more safely. That's it.

And the only time that this police officer would ever even *think* about intervening with someone is if they were *imminently* posing a threat to public safety. Other than that, he's almost just a cartoonishly Canadian caricature. He doesn't want to disrupt your day. He doesn't want any drama. He only wants to ensure the public's safety. That's it. And he doesn't even have the power to fine you.

He's almost like one of those security guards you see, where their vehicle was designed to look like a real cop car from a distance. So, when you see it, there's like that split second where your brain's like "oh sh**, oh sh**. It's a cop, it's a cop. Be cool, be cool." But then you notice it's just a security vehicle. Then your brain's like "HAAAAAA!! Security guard! It's just a security guard, you guys. What a joke."

And then you wanna go over there and be like "hey, what's up, chump? Wuddu yuh gonna do about it? Wuddu yuh gonna do about it? Ehhh-hhhh, securityyyyyy! HAH!"

But then you start walking over there and see that he actually seems like a pretty decent human being, so you're like "uhhh, never mind."

"At ease, sergeant! Keep up the good work!
... sir... ... sergeant!"

And then you salute him.

But you were never in the military, so your salute is all jacked up with your wrist all bent, like you're some kinda seal doing tricks. Which is

basically a way for military members to tell another one to go f***
themselves.

SALUTE!!!

Okay. So. Now. What's going on with this traffic report that our security... I mean our police officer has to file?

Well, the report doesn't come with any fine or anything. Like we mentioned earlier, he isn't trying to fund his police department by fining
people. That's not his job. His policing service is already paid for by
a transparently managed community funds pool which we will talk
about in a minute.

So, what's the whole point of this report then?

Well, it gets reviewed first and then entered into a decentralized database, much like our current credit reporting database, except that this
one is for reputation scores rather than credit scores - very similar
concepts, but slightly different. Additionally, the decentralized, distributed consensus nature of these databases is what gives them their
integrity, so that *nary* a shenanigan may occur.

Alrighty. So, what's the point of this database anyway?

Well, the main reason for this reputation reporting system is simply
to record any exceptional behavior on your part. If you did something
much more recently that's really naughty, then that could get recorded
in this database. Or if you did something at any point in your life
that's really positive, then that should be recorded to boost your reputation score. So, if you like volunteered or donated or something, then

that would go far in this system.

Also, these insurance companies in this Nauptsèh system are a lot like the cops here. They're not here to mess with anybody. These companies are *one hundred percent* competitive for your business. They have to be. It's not even an option. They have to do everything in their power to make you happy.

And we'll explain more about how this works in a minute. But for now, just know that this reputation system is mainly just there to keep these insurance companies in the loop. If you've been driving great for a few years, then they can do things like lower your rates to earn your loyalty.

However, if you're just driving around running your car into people all the time, then that's obviously costing them money. So, naturally, they have to enter this data into the actuary tables, which then calculates the probability of how much money your driving is likely to cost them in the future. And so, then they have to set your rates accordingly. It's just the math of it. And these companies have to be really good at this math. If they weren't, then they would end up going bankrupt from all of your accidents.

Furthermore, if you drive really well for a few years following your accident, then they can start reeling down your rates again. It's not a big deal. And they would even have an option to take a driver's improvement course. This would help them bring your rates down even faster.

Dealing with these insurance companies is a heck of a lot more convenient than dealing with a traditional police department. There's no fine to pay. There's no courtroom to get stuck in. There's no cage to lock you up in like some animal with rabies just, because you didn't pay some old parking ticket or something. None of that. Everything is handled seamlessly. And if you ever feel the need to get ahold of some-one, then you can do it at a time that's most convenient for *you*. Easy peasy.

Whereas a traditional police department, on the other hand, has pre-cisely *zero* incentive to make you happy. They're not one of these fair market companies that has to treat you "like a sir," like they're your butler.

And so that's the whole point behind these reputation reports. They just incentivize people to be responsible. They keep the community safe. They make sure everyone is treated fairly! And they spare your community a huge amount of pointless drama and wasted resources. The Nauptsèh figured this out thousands of years ago. ... And *all* you ever need to achieve this is just a little accountability and simple record keeping.

Okay. Sounds great. So, now that we've gotten all of that out of the way, why is it, exactly, that these insurance companies are so compet-itive? And why are they so insistent upon making you happy?

Well, this is where it starts to get a little interesting. You see, if you've ever had the great misfortune of having to crack open a modern pol-itical science textbook, then you may yet recall that much of our cur-rent world has these things called "revolving doors" or "iron triangles."

These relationships are basically just very close links between the various industries and the government bodies which regulate them. These bodies can then produce laws which make life disproportion-ately more difficult for the smaller competitors. This ends up having the effect of preventing fresh new talent from rising up and producing true, full-blown, one hundred percent competition. Rather, what you often end up with is... just a few old giants of industry completely monopolizing the competition.

Whenever you have extremely limited competition like this, companies simply do not have to try as hard to please you. That's what we have now. We have arbitrary regulations which are artificially limiting competition.

But the Nauptsèh system doesn't have these sorts of aforementioned "close relationships." To them, this would be an outrageous conflict of interest. Rather, the regulatory bodies in their system are selected neutrally.

We'll talk more later about how they're able to achieve such a high degree of neutrality. But the bottom of line is this, their culture takes these values of fair competition extremely seriously. If someone ever presented clear evidence that there were arbitrary standards being set in place for the sole purpose of limiting competition, then this evidence would be given with the utmost attention. Their courts, and their general assembly of community funds managers, would place top tier priority on these sorts of issues. In other words, they would have an extremely robust immune response to any and all anti-competitive shenanigans.

Their connection to competition was inextricable. That's basically why they invented the concept of totem poles. It was essentially a way of really showing off your brand name. It was like a giant, decked out sign for your family franchise.

This isn't to say, though, that there would be absolutely no limitations on industry. The Nauptsèh were extremely protective of their ecosystems. Their civilization essentially *invented* the concept of sustainability. They didn't mess around when it came to protecting their environment. ... While land tribes all over the planet were consistently hunting major species towards extinction, the Nauptsèh were *protecting* their resources and *respecting* their wildlife populations.

If a land owner was ever doing a bunch of damage to his or her holdings, then they would be admonished and socially rejected. And if the owner didn't cease this course immediately, then this would lead to them being boycotted and eventually bankrupt. And then *that* would lead to selling off their assets. And then the land would be returned to a responsible party.

A land owner would have to be clinically insane to abuse their land like that. Luckily though, the Nauptsèh didn't have to worry about insanity. They were getting nearly one hundred percent of their protein from fresh local seafood. They were about as close to an actual stable genius as humanly possible.

This is why they protected their resources in such a sophisticated manner.

There's this old myth out there that the merry old hunter-gatherers, of bygone yore, were all these super noble guardians of their environments. But that was not at all the actual truth of it. Hunter-gatherers were total opportunists. They were doing everything in their power to acquire calories. They didn't protect anything. And when civilization came along, it wasn't exactly a paragon of environmental protection.

So, if you wanna see a culture who actually *was* guarding something, then look no further than the Nauptsèh civilization.

It almost defies imagination how they were able to do this. They had full bore competition amongst businesses; their citizens lived like kings as a natural result of this; but they never let this harm their forests or waterways. And they accomplished all of these goals without a single authority figure.

That's right. There were no civilizational authorities who could force a Nauptsèh citizen to do anything. The only enforcement they ever needed was a person's reputation.

We'll talk more about this concept a little bit later, but suffice it to say... that no company would ever wanna be messing around with this. The simple fact that there were never any arbitrary limitations on competition meant that there was always a handy alternative if a company ever harmed their own reputation. So, just the threat of this alone would keep them honest.

And so it was, the Nauptsèh may have discovered one of the most profoundly unintuitive ironies in human history. It was competition itself which was preventing them overexploiting resources.

What?!

Yep. There was so much room for open competition that if a company was even remotely irresponsible, then you could drop 'm in an instant and switch your business. So, it was precisely the competition which was driving responsibility.

Woe. That's almost like a perpetual motion device.

It *worked* though.

It worked like a charm.

And it also helped to prevent smaller groups from being marginalized. Companies don't wanna split the market if they can help it. They prefer to service as many people as possible. So, if even one little group is none too pleased about something, then these companies are heavily incentivized to find an acceptable compromise.

And this is the type of environment that these insurance companies would exist in. I keep going back to the analogy of the butler, because that's the best example of how these companies are forced to accommodate you.

So, basically, everyone is incentivized to act responsibly - the drivers, the citizens, the insurance companies, the other companies, the community leaders, even the police officers. Especially the police officers.

And so, that brings up another important question. What precisely is incentivizing these police officers to be responsible? If the police officer doesn't have a quota system, then why not just put a mannequin in his driver seat... so it looks like he's trying to catch speeders... but he's really just down there in the woods going fishing all day? And why not go all super troopers and start messing around with teenagers who are obviously high on something?

"(((Littering and, uh... littering and, uh... littering and, uh)))"

Well, a police officer wouldn't wanna go to any one of those several extremes... ... unfortunately. ... If he got caught doing stuff like that, and it was a serious enough offense, then that could absolutely end up going on his reputation report. And if things ended up going that far, then many entities, such as employers and lenders and insurance companies, could see it. And if any of that information was highly relevant to them, then it could potentially affect the way in which they dealt with him. And if he got into some kind of legal situation, in the future, and it was just his word against somebody else's, then this could definitely the way in which they assessed him. It just wouldn't be worth it to push the envelope like this.

And of course, it wouldn't be the end of all the world... if you just got like, one little itty bitty markup every once in a while, but all of that stuff can add up, so you just wouldn't wanna push it and test your luck like that.

And, frankly, this isn't all that different from how things already operate.

The police department itself would be highly accountable also. It wouldn't be buried underneath mountains of bureaucracy. If a particular officer was doing stuff to people, then the victims of that mistreatment would file grievances. And then those grievances would be investigated - first internally, and then externally if necessary. So, again, it just wouldn't be worth it.

And not only that, but the court system would also be established with the most neutral methodology known to sociology. So, if something actually went into the court system, it would be reviewed with an air of fairness that was prudent and balanced.

So, suffice it to say, that the odds of a police officer actually doing something super ridiculous like this would be utterly and unequivocally negligible. Maybe if he had like a stroke or an aneurism or something. Maybe then he might act silly or babble something inappropriate. But other than that, everything would just be courteous like the town of Mayberry.

However, that's not to imply though that these police would just be glorified security guards. They wouldn't. If someone ever made things really dangerous for other people, then they would have to find a way to stop this hazard, even if that meant using forceful methods. So, they had the means to do what might be necessary, but they would avoid this if they could, because it was embarrassing and wasteful.

Their system was all about treating people like grown-ups. They dealt with one another like a rational species. And this was how they saw the wider world. They just assumed that other people were intelligent, ... because that's how they were. And what else would you expect from a population with so much brainfood.

Even if they got into a challenge over something, then they would defend their claims and titles in the presence of a council. And then the council's final decision was respected because of its merits.

They were always very strategic about how they approached things. If you put this into a modern context, then they would be all about the carrot cake as opposed to the bludgeoning stick. And they would be all about prevention as opposed to reaction.

For example, their police would never wage a "war on drugs" or whatever. They would just stop the war from ever happening in the first place. They would go back in time and kill the leader of the resistance as a child. And then they... ... ...wait... ... ...okay, I messed that up. They would go back in time and SAVE the leader of the resistance. Or better yet, they would just stay in their own time, and then their OWN leaders would protect their own children in their own communities.

The leaders of these communities would all be very heavily incentivized to protect these children. By doing so, they would prevent them from ever being predisposed to addictions in the first place.

There would also be a very a wide array of interests throughout society who would all have a vested interest in avoiding the calamitous aftermath of an unhealthy childhood. Thus, they would all be working together to make sure that nary a little kiddo ever felt left out or uncared for.

There are all sorts of creative and noninvasive ways to accomplish this. They could host special events throughout the summer and school year. They could work with families. They could have after-school programs. They could also do things like offer special discounts to families who have their kids in some sort of extracurricular activity. This way counselors could at least get a sense of whether or not a child is being cared for.

And they can also teach the children various precautions, like how certain substances will have a lasting effect on their brain development.

There's about a billion different ways to help these children - things that even I have no idea about. That's the power of crowd sourcing.

Oh, and contrary to some old-timey mythology, people don't turn into addicts because they tried a few tokes once. The susceptibility to addiction is imprinted in childhood.

We are profoundly shaped and molded by our childhood environment. Many animals are. Especially the clever ones and the mammals. And it just so happens that we are in both of those categories. So, we are extremely affected and influenced by our formative conditioning. It affects our long-term mood and causes echoing impressions. Thus, we are selected to be affected in this manner so that we can remember to repeat the most delightful patterns and then correspondingly avoid the bad ones.

Thus, if you or I were lucky enough to be cared for as children, then we were programmed with a sense of ineffable security. We were imprinted with a feeling that everything is just automatically safe for us. But these feelings are not an instinct we are born with. They have to be put there by our environment - not our genetics.

So, obviously, not everyone was imprinted with this sense of security. Some people got the complete opposite conditioning. To them, it feels like everything is gonna fall apart and then attack them at any moment, so they better be constantly trying to hold everything together while also bracing for impact. This is the exact opposite of feeling like everything's gonna be all fine and dandy.

And notice I said "feeling." This is all subconscious. They're almost never consciously aware of it. Because if you DO become consciously aware of it, then it slowly begins to dissipate.

But that awareness never comes for many people. So, they just keep on feeling stressed and never relaxing. They are far too subconsciously terrified to let their guard down. They don't even wanna know what calmness feels like. They're afraid it'll make them soft and break their focus. They're afraid it'll feel too good and then they'll want it.

And they're not wrong about that either. If they're still living in a toxic environment, then this feeling is ironically actually helping – it's enabling them to stay vigilant.

But this state of hyper vigilance is unsustainable. You can only clench your fists for so many hours. So, they start eating and drinking and smoking stuff, and then this helps them to feel relaxed without actually letting their guard down. So, it's almost like an ironic counterintuitive paradox. The chemicals make them feel like they're relaxing, but then the circuits in their brain are still just clenching.

So, that's the first of two paths to addiction.

And it's an ironic phenomenon if you think about it. People have been using these chemicals to outsource their abilities - their ability to relax, to be silly, to be honest, to be attracted to people. That's over half of your freakin' personality. And, honestly, if the environment is making us feel like we need to do this, then problem is not with us - it's the environment.

Alright. So, what's the second major cause of these addictions?

Well, the second cause is almost the same as the first one. In the same basic way that our childhood can make us feel stressed out all the time, it can also program us to be in pain constantly.

This is a very strange feature of our animal psychology - we can be encoded with a certain feeling arbitrarily. If our childhood was positively fantastic, then we can walk around feeling pretty wonderful. But if our upbringing was unfortunately quite the opposite, then we can end up feeling horrible no matter what's going on.

So, is this some kind of jacked up weakness in our biology? How can we be programmed to feel like garbage all the time? Isn't that a maladaptive quality?

Well, ironically, it isn't. If an animal grows up in a horrible environment, then we've needed to get away from that and start a new life. We've needed the motivation to build something better. So, we've been selected to be reminded of our childhood – all the good and the bad and the ugly. And now we *conveniently* get to feel like we're being tormented if we grew up in a toxic environment. This is nature's original mechanism for attempting to motivate us. And then this feeling will simply continue *until* we've escaped that environment.

Lucky us.

... or until we've just numbed it all out with some sort of addiction.

... or until we've mastered some sort of tranquility.

But this system is actually put there to help us escape all that. We're not supposed to be ignoring all of these feelings of agony. If there's an arrow in your leg, you're not supposed to be ignoring it. You're not supposed to be blocking out the arrow's sensation by using chemicals. You're not supposed to be sitting around just aspiring to nirvana so you can pretend like there's not an arrow sticking out of your leg. You're supposed to be doing something about it. You're supposed to be high on the endorphins of questing and the elation of hopefulness.

Nature puts every type of pain there as a signal. The sensation of a spear in your leg means "pull the damn spear out of your leg." The sensation of a knife in your soul means "pull your-damn-self out of that sh** hole." It's not telling us to start doping ourselves. It's not a sign from the transcendental universe that we need to be meditating. It's a sign from our epochs of evolution that we need to be doing something.

I'm not saying meditation is completely pointless. I'm just saying, often times it's only being used as a form of escapism. And a little bit of escapism is perfectly fine, every once a while - healthy even - but not when it's being used as a form of permanent evasion.

The only time that non-recreational meditation is really warranted is when you're in like inescapable torment or something - physical or otherwise. Like, if you have chronic pain or something and you've already done everything humanly possible, and now you're just ready to f***en end it, well, THAT is when meditation can become your superpower.

Pain is actually a pathway to the mastery of consciousness. It'll save you about two or three decades of meditation. ... If you wanna learn how to disassociate from your avatar, just expose yourself to torture.

... Just get one of those old, Bene Gesserrit, witch ladies to come over to your house and use that one mind-control voice on you to make you drop down to your knees and put your hand in a little devil box. ... And then you will become...

!!!!!!!~The~Kwissatz~Haderrach~!!!!!!!

It's kinda like that. Just do that.

A lot of ancient mystery schools used to do that actually. They would just subject themselves to torture. And it just forces you to disassociate from your avatar.

If you wanna get a sense of what this feels like without the pain component, just drink like a double shot of espresso and then try to meditate. ... But don't freak out if it feels like your brain is melting and you're having a heart attack. ... Just be nothing. ... Just... disassociate. ... Just... ... obliviate...

So, yeah, that's for sure a viable option if your agony is *inescapable*. And yeah, that's a sh** situation for sure. ... – without a doubt.

But other than that, nature has established these pain points for the purpose of motivating us. ... - not just for assuming that the problem is *inside of us* and that the *world* which is all around us is just fine and dandy. ... and then attempting to *avoid* all of this agony by just using

chemicals and escapism. ... No, the problem is not you, it's your surroundings. Pain equals pull your hand away. Pain equals escape. ... ...
... Pain equals... build... a better... world.

However... that being said, sometimes the situation still feels hopeless.
Or sometimes you're just a kid and you're unfortunately stuck with it.
... ... Well... this is when your elf's C.A.T. starts going A.P.E sh**. Your
cravings and aversions and trainings... start becoming your addictions
and phobias and eccentricities.

We see these depictions of torment in a lot of literature and music.
And it's also quite often depicted like a little tiny devil on your shoulder. But it's actually these A.P.E.ed out C.A.T. instincts that are feeding you this disturbing ideation. They are bombarding you with fear
and paranoia. They are instilling you with a sense of loathing and with
a host unnatural addictions like schadenfreude.

And, ironically, this most unfortunate psychological pathology was
actually selected by mother nature to protect us. When the world
that brought you up was excessively demented, then your C.A.T. layer,
which assists your inner E.L.F. core, just began to internalize it. It became a monster to defend itself from monsters. And now it rears its
ugly head when your life gets stressful again. ... It's always lurking just
outside in the shadows, like a paranoid vampire - perpetually waiting
for your elf-core's permission.

This is why things like alcohol cause so many bar fights. It diminishes
your executive ability to keep that monster from being activated.

This is why they used to call alcohol "spirits" back in the day. They
thought that the alcohol was the spirit of the plant matter. And they
thought that that spirit was attempting to possess you.

They also believed that these darker inclinations were a form of possession. They could instinctively sense that this darkness was not a feature of our own core personality. So, they just assumed it was coming from something external.

And, in a way, they were actually right about that. After all, this darkness is just the internalization of worldly evils which were perpetrated against us. And so, the darkness of the world becomes a part of us. It turns our C.A.T. layer into a paranoid vampire.

Then that vampire just wants to lash out at people. It wants to dominate, because it's paranoid, and because it wants to protect us. ... And it also likes to present us with various worst-case scenarios. This is why we dream of things that challenge us. Our subconscious is often attempting to resolve things by preemptively simulating them.

Yippee... :/

So, not only does this childhood mistreatment cause an animal to grow up feeling like garbage all the time, it also creates a monster in the background. And this, ironically, is how mother nature selected us - so that a victim could escape their environment.

We are supposed to be making our world better - not just numbing ourselves out with constant painkillers. ... Easier said than done, I know, but we are supposed to be fulfilling our purpose. This is where all the natural highs come from. And we are supposed to be high – I don't care *what* you're little, school program told you.

We are supposed to be having fun and nurturing our world - protecting the children, protecting our communities, protecting the environment, protecting our wildlife.

The truest aspects of our many layered identities are all aligned with this imperative.

But when our Neolithic ancestors started manufacturing these addictive substances, they just disrupted our natural defense mechanisms. And now we, as their modern descendants, just sit around doping ourselves all the time, and putting up with all sorts of nonsense. It's like we're all living in some sci-fi dystopia... and we're just taking our daily rations of our emotional suppressant drugs.

Oh, and when I say "doping" ourselves, I'm not just simply talking about the booze and stuff. I'm talking about anything that gives us an addictive dopamine hit. So, that could be all sorts of stuff - thumb sucking in little children, paw licking in animals, shopping, spending money, playing video games, getting social media attention, getting any attention, always being right, always dominating others, always watching television, having another person, having sex with another person, eating stuff, drinking stuff, smoking stuff, popping stuff, shooting up crack joints, you name it. It's all addiction.

Addiction is ANY type of behavior that brings us a moment of temporary pain relief which we pursue with a fervent compulsion despite its negative consequences.

So, that could be spending an excessive amount of money on things you're never gonna look at again. It can be burning up your critical study time to get attention on social media. It can be any little dopamine hit that we're willing to harm ourselves for.

That's not to say though that all addictions are necessarily deleterious. I think we can all agree that abandoning our tea time would be far too dreadful. I mean, what are we, barbarians? Don't be silly, darling. Ehmhmhmhmhm...

Oh, and, yes, caffeine is also producing a mild euphoria effect. It's actually helping us to deal with the "pain" of a stressful modern work environment. Oddly enough.

That's what addiction is all about. It's about blocking out our *pain*. And it's usually something lingering from a person's childhood.

There's this older misconception out there, that the junkies and the drug addicts are just these weaklings who always wanna get "high" all the time. But that's not exactly what's going on there. Taking something at a party is an attempt to get high. Taking something every day is an attempt to block agony.

And I hate to say it, but the simple fact remains, when you look around these street junkies, what you're usually looking at are not just some pathetic little weaklings, but rather they are the fragmented leftovers of unspeakable acts of child abuse. You couldn't even make movies about what some of these people have lived through. It would be too intensely horrifying. Audiences would get all mad that you had exposed them to that stuff. And yet, that was the actual lived experience of many of these people – something too horrifying to even speak about. ... They're not out there trying to get *high* off of something - they're out there trying not to feel *low* for a few seconds.

And there's another form of trauma that I've already sort of mentioned once just briefly. It's the passive form of trauma caused by child neglect. We tend to overlook it as something innocuous, but no mammal or pair bonding species was *ever* meant to be deserted like this.

I've said it once before, but it bears repeating, our children are not just some simple primitive insect larva. They can't just be abandoned to go off and hatch on their own somewhere, and then navigate their entire natural lives just purely based on preprogrammed instincts. Nooooo. They have to be nurtured and attended to and cared for, and then given lots and lot and lots of supportive attention. This is necessary to instill them with confidence, and a very grounded and self-reliant self-assurance.

... They never even think about it. They just automatically become confident. Their sense of who they are becomes invincible. Why else would their parents love them so much? Obviously, they must be a super awesome kid to be around. No doy.

Likewise, without a solid grounding, it doesn't matter what you look like. You could be the most outwardly attractive person in all the wide world and still not internalize it. Absent that initial imprinting, you'll just never feel good enough - ... at least not until you're finally apprised of this underlying mechanism.

But alas, not every child got that initial grounding, unfortunately. So, they lack a sense of comfort within their own skin. And this usually leaves them open to requiring approval from the other members of their peer group. And that's not a healthy way to live one's childhood.

Then we throw them in with all of these strangers in these massive public schooling systems. And it's a ton of random students with almost no supervision. And then these places are an absolute breeding ground for the most aggressive kids to dominate the discourse. ... So, what do you honestly think is gonna happen here? How will this affect an already insecure child?

Naturally, it's gonna make'm even more insecure. They're gonna be nervous and stressed out and anxious.

So, guess what's gonna happen when someone lets'm try a little dope for the first time. Yes sir, it's gonna make'm feel better. It's gonna turn off all those thoughts of insecurity. It's gonna quiet down those voices of total inadequacy. It's gonna help them with all those tensions from social anxiety. And if nothing else, it'll ironically give them a place to feel accepted.

Chronic pain... plus a source of dopamine... equals addiction.

And you don't have to be from a "bad" home environment for all of this stuff to happen to you. It only takes neglect. It's like those children's movies where there's a kid who has everything but he's always sad because his dad won't spend time with him. Well, that's the answer. It was right there all along.

People overlook this though. They think that kids need to grow up really tough or something. ... Well, it has to be said, this is not how you make a child tougher. This is how you make a child fragile.

If you actually wanna make a child tougher, then just take'm out to the country and go hiking and climbing and canoeing and stuff. Play sports with them. Work out together. Exercise as a team. That'll make'm a trillion times stronger than these poor, psychologically fractured, "toughened up" kids.

Neglect is often experienced on many levels throughout society. First it all begins within the home environment, and then it just continues in the school system. And if all those odds against you weren't quite bad enough, then it adds up one more time with critical nutrient deficiencies.

Many children have been growing up on a bum regime of empty calories, empty protein, deleterious substances and a substantial lack of brainfood. And this not only exacerbates the other problems, but it also starves your brain and causes depression.

And this isn't a parent blaming thing. Almost no one was ever even told this stuff. And what's worse, is that they've also been *misguided* by people who *honestly* believed they were doing the right thing.

It's like this thing that they're telling parents now, that they should leave their crying babies all alone at night.

Well, that "advice" is so bad it makes my eye twitch. That's like telling parents to feed their children weed-killer. The future is gonna look back on that with their jaws on the floor.

And yet, that's the actual world that we're currently living in. We live in a twilight world. And everything is backwards.

So, when you add up all these things on little children, it creates a toxic cycle of pain and suffering. Then this usually plagues their life for many decades. And then *this where you get your susceptibility to all addictions!*

So, this idea of trying to beat these substance addicted people into submission is just nonsensical. Their normal waking life is already torment.

And furthermore, trying to threaten an entire black market with punishments has proven utterly, catastrophically useless throughout history. It's just a massive waste of money that's enriching these antisocial syndicates. And it's like standing in front of a firehose punching at the water. It's like, no, dude, you need to turn the water off.

And so, that is *precisely* why the Nauptsèh system would *not* do this. It would be designed to protect the children and their families. And it would stop these susceptibilities to addiction *before* they ever even happened.

However, if anyone had fallen through the cracks already, then their community would just employ a method of holistically reviving them.

There would be a facility out of town designed to help these people. So, the features would need to be calming to the overall nervous system. That means aesthetics and scenery and music, along with good nutrition and caring people and empowering knowledge. ... And of course, you gotta bring'm those little petting zoos every once in a while.

From there, the formula is pretty simple. With the proper coastal nutrition, breath work, body movement, and hyperbaric conditions, you can fix just about *any* form of post-congenital brain damage a person can acquire.

And then the next thing that they need now more than anything is just the energy that they've been deprived of since their childhood. Whatever type of energy they were disconnected from, that is precisely the type of energy that they now need for healing. Didn't have protection? Well, now they need a refuge. Didn't have attention? Well, now they need someone to listen. Didn't have a decent caring guardian? Well, now they need a friend who wants the best for them. It's a pretty simple prescription for complementary kindness.

And then once you've started doing all that simple stuff, the only thing that's left is to start working with some brain wizards. They can guide you through your problems to detect where the origins lie. And then as soon as someone can acquire an adult's perspective of these origins, it slowly begins to shift that formative programming.

So, those are the steps. First you need that aesthetically healing environment. Then you need the tools for healing the brain tissue. Then you need the energy that they've been lacking. And finally, you need some wizards that are good with brain hacking. That's it. It's a beautiful, miraculous formula.

And when all is said and done, almost everyone can get their lives back. They can live a semi-normal life with dignity and happiness. And when you live in a supportive community where everyone's income is *four times larger*, then the prospect of achieving this happiness is *a bit* more attainable.

A sufficiently informed, modern, healing center can perform miracles. And they can do it on a tiny, shoestring budget, especially if they're using hyperbarics. It doesn't take an expensive, state-of-the-art facility.

It's kind of like one of those secret karate maneuvers where you just poke someone in the neck and then their body collapses. You can have an enormous impact if you just know the right technique.

Virtually all mental illness boils down to just three basic things - child mistreatment, bad nutrition, and toxic exposure - all of which are treatable. Our brains are quite resilient and incredibly adaptive. But they just need a little help with the proper inputs.

So, yes, the vast majority of people would be able to recover. And even the ones who couldn't, could still do other things. There are plenty of fulfilling engagements where they could earn some spending cash. They can do things like bagging up groceries, or being a waterboy for a foosball team."~

So, the point of all that is just this - their police would never be "warring" against their citizens. They wouldn't be wasting massive fortunes keeping nonviolent human beings in cages. They'd just be stopping all these things before they happened. They'd have formulas for health and prosperity, not just cycles of abuse and impoverishment.

So, like we mentioned earlier, their system was all about the carrot cakes rather than the bludgeoning sticks. If they wanted you to do something safely, then they'd simply find a way to just incentivize it. And they would never use a stick unless you forced them to. You'd have to be actively, manifestly violating someone else's sovereignty.

They've been sparing the rod and spoiling their population for a good long while at this point - thousands of years now to be imprecise. And this has always been enough to run amazingly smoothly – all they ever had to do was maintain a system of accountability. ... They had regular community meetings where recent news was disseminated, and of course their diet was, as mentioned, mostly brainfood, so they remembered all of this information pretty easily. And so, if someone had stolen some property or something, and it was proven at a meeting that they did this, then their entire community region would be informed of this infraction.

So, people just didn't steal sh**. Everyone would know that you had done this. And then no one would forget about it... *ever*.

And even if you were moving to a new community, then you'd have to get your leaders to officially vouch for you. So, your rep would always follow you forever - at least as far as your history of being an upstanding citizen was concerned.

So, people just didn't do stupid sh**. They saved all that stupid energy for the parties, which wasn't hard to do since they were basically always partying. The Nauptsèh pretty much invented partying twenty four seven.

... if, by "twenty four seven," you really mean is two hundred and forty seven days a year, son!!!

Okay, maybe it wasn't quite that much, but it was a lot. The Nauptsèh didn't mess around when it came to partying. Sometimes their parties would go on for WEEKS, SON!!!

Hi, their name was Rod and they liked to party. No, you don't party. Only they party."~

But most civilizations never rolled like this. They never maintained a network of accountable enclaves. Rather they'd just hit you with a stick if you did something bad. And so, that's the type of system that we inherited.

So, most people think that this is normal. And they've never been informed about these carrot civilizations. So, then a game they like to play when they hear about one of these incentive-based systems, is they try to find a way to poke a hole in it.

They've been playing this little game for generations now - thousands upon thousands of people. They keep trying to find a way to break the system, like a loophole or some kind of scenario that would force them to compromise their values. But they never find a way, because there is none. It's an organic reactive algorithm with distributed intelligence. If you think you found a way to game the system, then the system just adapts and solves the problem.

In other words, the Nauptsèh style, carrot-based system is not some robot. It never walks into a wall and then just keeps trying to walk forward. That's the sort of thing that a typical stick-based civilization does. Most of the world's other cultures were just acting like robots.

They would build up these massive pyramids, and then just waste a ton of money on useless vanity projects. They would tax their own economies to the point of self destruction. They would obsess over foreign conquests until they had bankrupted their own national treasuries. And they would manipulate their own native currencies, until they had annihilated your entire life's savings.

Then they'd overwork their land and drive it fallow. And they would mow down all their forests and then never replant them. They would ignore all sustainable practices, until everyone was starving to death. And then their elites would just ignore their populations, despite the fact that the people were the source of their prosperity.

And yet, the Nauptsèh never did any of that.

They never wasted money on any pyramids, because their citizens never wanted a giant pyramid. They never tried to conquer their regional neighbors, because their citizens never desired to rule their neighbors. And they never tried to tax their people oppressively, because there simply was no "they" to tax them anything. And they never destroyed the savings of any citizens, because you only became an elite by better tending to your people's wishes.

And one of those people's wishes...

... was to party.

So, let's have a party.

And let's have a contest.

Let's have a party contest. Let's have a party contest to see who can throw the best contest party.

We'll have two teams. Whichever team can throw the best party will win the final contest, and whatever other proceeds they can generate.

Alright. So, let's begin. Let's introduce our first team. The first team will be just ten individuals who will be working for me in this analogy. Let's call 'm team Deca.

First up on team Deca is Janice. Janice is a nice lady, but she spends most of her workday making artwork out of her cat pictures.

Then we have Todd. Todd looks like the mayor of Whoville - just a physically round human being, with a mustache, and a rosy little button nose. But don't let his appearance fool you. He'll try to drag you into some useless conversation about trivia so he can try to one-up you and make himself look smarter. Except he hasn't quite figured out one little fun fact yet, that there's this handy little thing we call an internet, and you can't just go around *bullshiting* people anymore.

Alright. So. Anyway. That's team Deca. OH wait! I forgot one. - ... *Jason.* That's right. Jason, Jason, Jason. How could I forget about Jason. Good ol' Jason. He convinced our project manager to buy some expensive new software, which turned out to be the biggest pile of sh** software you ever saw in your life. It was like some kinda April fools joke, except it was real. That f****n piece of sh** software couldn't even figure out what a white space character was, if you accidentally pressed the spacebar key after your search term.

"Duuuuuuuuh!!! What's this?! A space bar?! What am I supposed to do with this?! Duuuuuuuuuuuuh."

It was like it was some kinda ancient a** software, from like the 80s, that didn't even know what a f****n space bar was.

F****n Jason and his bullsh** software.

Yeah, and then something quite magical happened. Yes indeed. We found out that that this bullsh** a** software was from a company that was partially owned by little Jasy wasy's brother-in-law.

HMMMMMMMMMMMMM! Interesting how that works... JASON!!

That's right, ... Jason. You think we didn't know about that sh**, ... Jason?

We know everything... ... ... Jason.

Okay. So, that's team Deca. Jason's still on the team though. They all work under contract, so they basically gotta murder someone to get fired.

... ... Jason!!!

Alright. Now, let's introduce our second team. Our second team is a lot more interesting. They have about ten thousand people on their team. So, let's just call'm team Myriad.

Team Myriad doesn't work for me though. They're just members of the community. In fact, with ten thousand people on their team, they basically are the community.

Team Myriad's participation is completely voluntary. They can compete or they can abstain however they see fit. It's up to them. But if they do participate and they do a really good job at it, then they can get unlimited tips and proceeds and even personal recognition.

Team Deca on the other hand doesn't have a choice. They have to compete. And they're not allowed to receive any proceeds. They just get paid exactly the same amount no matter what happens. So, technically speaking, they can screw it up really badly or even do an okay job at it - it doesn't really matter to them.

Alright. So. What's this festival all about? What kind of festivities can we be expecting?

Well, it's mainly going to be like a traditional canoe festival. Both teams will come up with an original canoe design which will be manufactured for the festival. Then each team will write their own music, and then that music will be played throughout the festival. Then each team will host their own festival, and they'll coordinate all the logistics, like the venues and the food vendors.

So, team Deca will go first. They'll design a canoe based on whatever knowledge they may or may not possess of engineering. Hopefully it'll float. But don't get your hopes up. And even if it does float, it'll just end up looking like a bathtub, with like a red and blue stripe painted down the side.

Alright. Next up, team Deca will record us some music. Hopefully it won't make our ears bleed. It'll probably end up sounding like some stock synthesizer beat. And then they'll throw in a couple of sound effects for good measure. It'll be like donkey noises and laser sounds. Peew peew peew.

Alright. So. For their last trick, they're gonna organize the actual venue itself. It'll probably be hosted in the industrial sector. Jason's brother-in-law owns some property down there. And don't get your hopes up about food vendors. The only one that's actually guaranteed to be there will be whatever Janice's favorite food variety is. Maybe we'll get lucky and get a couple more. It all depends on what Jason's brother-in-law's got goin' on.

So, that's team Deca.

Next up we have team Myriad.

Most of the people in team Myriad are not going to participate. Naturally. But a few dozen will. These'll be people with backgrounds in woodworking and engineering and music and recording and planning and administration. In other words, it'll be people who actually know that they can do a great job at this.

Team Myriad's canoes are gonna be kick-ass. They're gonna look great, feel great, and handle great.

Their music is gonna be awesome too, like a music festival. You'll catch yourself just wanting to dance along to it. They'll probably have some really solid cover songs. And then they'll throw in some really catchy local flavor.

And then finally, you'll have the venue itself, which'll be immaculate. It's gonna be somewhere that's really aesthetic... and that's actually enjoyable to be in. And then the vendors'll be incredible too, with a ton of variety. So, there'll be aromas everywhere against the backdrop of the music and the festivities.

People are gonna love it. And they're gonna leave some pretty nice tips for Team Myriad. And then the vendors are gonna get some action also. And then the canoes and the music will be available, if anyone would like to purchase them.

And if any of the participating members on their team would like to promote something, then they can advertise that too in specific areas, like at a concert or a sports game. ... especially if they can advertise it aesthetically – like a promotion that actually looks good, and not just tacky.

And when all is said and done, they're gonna do pretty well for themselves. They're gonna end up doing better than even Jason did.

Alright. So, what just happened there? Why did team Myriad just trounce team Deca like that? Why did it turn out that way?

Well, for one, they way outnumber team Deca a thousand to one. Even the ones who ended up participating still outnumbered them by about three to one.

Secondly, they were people with a high amount of specialization in all of those areas.

Thirdly, they were also people with a *passion* for what they were doing. They *chose* to do it because they ENJOY these particular pastimes.

Fourthly, many of them also knew that they could make some good money at it. And there's nothing wrong with that. They produced an amazing value for the community, and then the community gave them value in return for that.

But, of course, the story is a little different for team Deca. They were just ten lone people with a very limited range of talents. And they also had no incentives to do a good job at it. However, they did enjoy taking their families to the festival that team Myriad put on. Jason even tried to sweet talk one of their coordinators there into going in on a franchise deal with him and his brother-in-law.

She respectfully declined.

Alright. So. What was the point of this whole festival competition? Why are we even talking about this?

Well, this is one of the best ways that I can help illustrate how the Nauptsèh regional affairs were managed. You see, in almost every other nation on Earth, we pick our administrators and our bureaucrats in much the same way that team Deca was chosen. They're not bad people - they're just... limited.

Well. Let me correct myself. MOST of them are not bad people.

I can't speak for Jason.

... ... Jaaason.

But in the Nauptsèh world, their community leaders are *organically* chosen, essentially like it was with team Myriad. They're mostly members of the local community who are exceptionally productive and unusually talented. They get stuff done. They make it happen. And they keep their consumers happy.

In any given population, there will always be about one fifth of the people who are responsible for literally about *four fifths* of all production. In other words, a pretty small amount of people are responsible for a very large amount of the total output. One version of this is known as the Pareto principle.

Oh, and when I say they're "getting stuff done," I don't mean that they're physically way faster or stronger. I'm not saying that they're out there just lifting huge boulders that are like five times larger. I'm saying that all of their heavy lifting is going on in their head. They're figuring things out. They're inventing things. They're coordinating. They're summoning courage. They're taking huge risks. They're failing, over and over and over. But then they're getting back up and getting stronger from it. Simply put, they're making things happen.

And that's the type of person who becomes a leader in a Nauptseh community. They are simply the people who are making things happen – the ones who are making all the good stuff that we love to have access to. ... – things like houses and clothing and restaurants... and things like music and entertainment and fashion. They're just people who have a talent for production.

And these leaders could also be anyone from basically anywhere. It didn't really matter what your family of origin or your station of birth was. Literally anyone could prove themselves if they had the tenacity. They didn't have to be from like a privileged bloodline or something.

We often take this concept for granted now, but this was extremely rare in the ancient world. Usually, you had to be a male of noble origins. But not in the Nauptseh civilization. You could be any type of person if you had the hustle.

And you may recall from earlier why this is so special. This is precisely the paradigm you want if you wish to select for intelligence. You want a system that never limits opportunity. You want a culture that never stifles competition.

And so, that is exactly what the Nauptsèh system featured. They never had any good ol' boy networks or secret handshakes. And they never had any iron triangles or revolving doorways. It was just pure, unalloyed, unadulterated, undiluted competition - at least to the extent that they respected the environment.

So, this was truly anyone's game. Any person from any origin could throw their hat in the ring. And just about any fairly talented family could eventually become a leader within their community, if they desired.

But you had to really want it, because it came with responsibilities. Yeah, it was fun to be rich and all, but being a leader also made you into a power servant. You were fully and unequivocally expected to serve your community. And I do mean "serve." That's not like a euphemism for, just like, "rule over them exploitatively while only *calling* yourself a servant." Nope. It was a huge responsibility.

In other words, they were not unlike the patrons of Renaissance Italy. And they were not unlike these great, local benefactors, who, for generations, have been donating things like parks, and other common amenities, to their local communities.

For the Nauptsèh, it was their duty and their privilege to take care of their communities. And I do mean "take care of them." They weren't just bossing people around. They had zero authority. They were *only* there to take care of the people.

Yes, you heard me right. They had ZERO authority. No one could tell a Nauptsèh citizen to do anything. Literally no one. There was no emperor, no king, no president, no governor, no mayor, no anything. None of that. Nauptsèh citizens were bound by no one. Or, to put it another way, they were all emperors and empresses.

These wealthy leaders were thereby bound by their reputations to do the best jobs possible at taking care of their communities. And, naturally, as the biggest producers in the region, they were the most highly qualified people for this job. They were the most extensively skilled individuals at keeping people happy.

So, they didn't just have to make products that made people happy - they also had to do all of these other public things too, like building roads and bridges. And they paid for all these projects just using their own money! So, that pretty much eliminated any fraud, waste, or abuse issues.

But this wasn't even a problem though... because they had tons of money. They would even throw these huge festivals throughout the year where they would pay for everybody's food. And they would even give out these expensive items to everyone, like a form of payment.

However, in a modern context, we would just call these people the "community funds managers" or something. They would all come together to work on the various projects for their local communities, and they would suss out all the details on how best to achieve this.

Usually this process would be simple. Most things just boil down to a very straight forward set of logistical decisions. ... But every once in a while, it's not so obvious - some things are not so clear or so objective.

Every community would have their very own set of conventions for systematically dealing with this kind of stuff. And sometimes they would even need the actual citizens themselves to officially vote on something.

Maybe they would like to build a bypass through a scenic area, and it would save them a lot of money, but then it would also be an eyesore.

So, what do you do in these situations? Do you just build the road so you can save yourself some money, or do you just scrap the whole idea so that you can preserve the natural beauty?

Humanity has never quite determined a perfect answer for this scenario. It usually just boils down to one agenda winning out over another. But the Nauptsèh system could never work like that. They always had to find a way to effectively balance out the interests of their local citizens.

So, one way to do this would be to compromise. They could come up with different solutions for the people to vote on. An overly simplified example of this would be the following – either putting in just a basic, inexpensive bypass; or spending a bit more money on some aesthetically complementary architecture; or maybe even just boring out an underground tunnel with a much higher price tag.

Alright. So, how do you vote on these options? Do you just vote on your favorite choice and then cross your fingers? What if the option that wins is still highly unpopular? What if like forty nine percent of your community is still really really unhappy with the final outcome? Is this a balanced way to run a community? Would the Nauptsèh system ever condone something like this?

Well, no. No, they would not. They would scarcely condone a system that was so divisive. This would never be seen as an appropriate way to run things, or as in accordance with their values and principles.

Rather they would use a balanced voting method. And the most simple, balanced voting method would be the so called "approval style Borda count" tally, but we can just call this "the balanced voting method."

Alrighty. So, how do you do a "balanced vote?"

Well, instead of just voting on a small number of highly polarizing options, you create a wide array of blended choices. Then these choices offer a spectrum of balance and compromise. And then everyone simply ranks each alternative by how much they prefer it.

So, let's just say that there are only ten basic options. We'll just keep it simple. Okay. Well. Then all you need to do is rank each option. So, you would just write a ten beside your first most favorite option; and then a nine beside your second most favorite option; then an eight beside your third most favorite option; and then a seven and a six and a five and a four and so on. That way each of these options gets ten and nine and eight points respectively. It's pretty straight forward.

And if any of these options are things you really don't like, then you can just write a zero beside these choices so that they don't get any points from you.

Alright. So. Part of the beauty of this system is that it's really really good at avoiding contentious and divisive options. So, for example, if like fifty one percent of the people clearly preferred something, but then like fourth nine percent were extremely opposed it, then that option would be avoided because of its contentiousness. Rather, the option that won would be the favorite amongst the things that nearly everyone could at least live with.

So, this system will find the balanced compromise and thus avoid the most divisiveness, contentious options. And that's exactly what you want in your local community. That is precisely what you need in the place where your kids are growing up. You don't want fighting and negativity. You want compromise and cooperation. You want orchestration and coordination. You want conversations and deliberations. You want congregations and salutations! You WANT...

What also helps communities to be more unified is to simply use incentives rather than forcing people.

When our societies use these forceful means to achieve stuff, it creates this constant battle over who gets to control it. Everybody wants the reins of power, and no one wants their neighbors to control them.

But if your system is built on incentives as opposed to just controlling people, then it removes this constant battle over power. People then tend to act really really cool because it comes with so many benefits. And there's just simply no control for people to fight over.

And of course, that's how the Nauptsèh civilization structured their system. No one could tell their citizens that they had to do anything. They just encouraged prosocial behavior through an incentive-based community structure. And that's basically why they partied constantly... instead of just squabbling over power all the time.

Simply having an incentive-based culture will eliminate most conflict. But then employing a method for balanced voting will eliminate even more of that.

However, if there are still some intractable deal breakers, even after incentive structures and vote balancing, then it's often a better inducement to simply separate geographically into a sort of alternating checkerboard pattern.

This way no one will ever be too distant from a group of like-minded people. And because of that, everyone will get to enjoy the different climates and regional zones. And furthermore, sports teams can also compete in a number of different regional rivalries that *actually mean something*.

I have a feeling this could be the beginning of a beautiful frenemyship.

And lo, be Portius amongst my neighboring community or be he not, tis no matter, for we shall meet... on the field of battle. And I shall lay him low.

Ready your foam enchanted swords, GOOD SIRS, for verily, on that morrow, twill be the undoing of the ill professed Spell Binder.

... and his tWISted appetites.

But anyway, this is why the Nauptsëh were never a single empire. They were always quite aware of the fact that different types of temperaments sometimes need their own communities. So, they divided themselves into regions. Then they divided those regions into districts. And they still maintained a broader regional identity, but they never surrendered their local autonomy.

And so, once you've added all of these different layers together, the potential for community conflict is virtually nonexistent – so, that's the incentives, and the voting, and the checkerboard arrangement.

But no matter how much community drama you've successfully eliminated, there will always be at least a modest necessity for a court system.

Part of the beauty of this checkerboard arrangement is that every single region can experiment with their own solutions, but the most effective model we know of, has already been discovered. It calls for something that is new, combined with something that is old, and then something which is distinctly Nauptsèh, combined with something that is borrowed from ancient Rome.

In the Nauptsèh world, the community leaders worked extremely hard to earn the trust of their fellow citizens. So, these leaders could also act as a sort of panel of judges for resolving civil disputes. But the main thing here to note is, that these leaders, were forced by their very nature, to be good faith guardians of their community.

Additionally, in the early Roman republic, they had this highly effective emergency response system. Sometimes there would be some huge emergency and they just didn't have time to sit around and talk about it, so they would select a single person to be their king for a while, and then this king would be given ultimate power to deal with the emergency.

But what was interesting about this emergency response system was that it forced them to exercise extreme caution. They had to choose a local person who was incredibly trustworthy. So, they would end up choosing someone like an old farmer guy or something.

And so, a modernized protocol would just combine all of these different tactics. The leaders would simply research a field of candidates, then they would vote for whichever ones they thought would make the best judges. And then by making their determination in this manner, using the balanced voting method, they would identify the most well-balanced candidates which most evenly represented everyone.

These judges would be constantly incentivized to strike a balance. Their reputation, as a fair and prudent arbiter, would be dependent upon this diligence. Auditors would be thoroughly monitoring their activity. And thus, their record of performing as an adjudicator would be publicly available.

The reason why they would be working so hard to be fair like this is because the plaintiffs and the defendants, alike, could *both* decide on who their judges would be. They would have to choose from a list of judges that they *both* agreed to.

But the idea here is that these Nauptsèh judges would have to be in constant competition for your faith and trust in them.

So, that is the unbeatable, unstoppable recipe for a community court system. You've got the fairly new conception of a balanced voting method; with the extremely ancient principle of competition; and then the convention of using a panel of trusted, community leaders; to which we've add the original prudence of the founding Romans.

There's no other court system on the planet that even comes close to this. There is seriously nothing else within a trillion light years of these principles and tenets. This system creates such an over the top, absurdly high standard for fairness and diligence that is not even funny. This is so high up in the stratosphere of true justice that you would have to wear an oxygen mask full of temporally inverted air molecules just to get into the courtroom.

The most intense aspect of this system would be its openness to competition. This one single feature alone could single handedly drive its fairness and its immunity to corruption right out of the solar system. ... As long as there was a sufficient market demand or it, these judges could do all sorts of things, in their pursuit of earning more people's trust in them. They could sign an official affidavit, or a notarized commitment, that they would pay an exorbitant penalty if a breach of faith was ever discovered. It all depends on how intense the competition for people's trust needed to be.

I mean, you could go on forever like this. As soon as you add competition to the equation, the sky's the limit. They could do things like... agree to have their assets periodically audited. Anything to best the competition. You name it, they could do it.

So, you would just have all these courtrooms out there, and they would all be ready to come and help you, in the most accommodating way possible. And since their whole entire existence is just to help you, then you wouldn't even need a lawyer half the time - you could just speak to a courtroom assistant like a regular human.

And if all *that* wasn't enough, well then, you've also got these businesses always competing, always working day and night to try and please you. ... ... Tell me *that's* not an epic society.

It's quite *astounding* what begins to slowly manifest, when you just release your competition from its artificial confines. It brings out the best in people's performances, and then they quickly begin to discover how much they actually enjoy it.

It doesn't get *aaaany* better than that, muh friend. I mean, really, it just doesn't - not in any futuristic scenarios; not in any technologically advanced utopias; or interstellar space colonies. This is the *universal* standard for accommodation right here.

Skrrt skrrt.

Alright. So. That being said... there's also one more little detail about these courtrooms... and it might alarm you by no small measure, before you see how all the details work... - these courtrooms wouldn't actually have any authority.

Wait, what?!

Yep. ... Just like everything else in the Nauptsèh world, no one could force you to do anything - not unless you had like stollen someone else's property or something.

So, technically speaking, if you were completely insane, then you would be more than welcome to just straight up ignore a judge's official ruling like that. But if you really did pull a stunt like this, then you might as well just pick up and go move to the amazon, to some remote little village in the middle of nowhere, because no one else is ever gonna wanna to do business with you ever again. And even then, they would even *still* probably send some official warnings down there... - down to that remote little village to let them know about you.

So, suffice it to say, no one would ever wanna be just scoffing at a judge's assessment like this. If they truly thought the judgement was made in error, then they could always just appeal the court's decision. But you would never just straight up flout a court's opinion like that.

Their rulings may not come with any teeth attached, but they can always pull the rug out from underneath you.

Alright, so... that pretty much explains the court system. The Nauptsèh never really took it to that level though. They never really needed to. Their incentives, and their arrangement into a checkerboard pattern, all but eliminated the internal drama.

And yet, most other cultures never quite figured this stuff though, at least not to this level. They just assumed that you had to fight over everything. They just assumed you had to choose between your values, like liberty or security. And they just assumed things, like, you could either take care of your people, or you could allow them to have their freedoms, but you couldn't do both of these.

And yet, the Nauptsèh unequivocally proved that this notion is a false dichotomy. They proved that you can have your cake and eat it, just as long as it's a carrot cake. ... All you ever really needed was just a culture based around incentives. ... And then *all* of your ideological extremities can achieve their end goals synergistically as opposed to adversarially.

So, like I said, they never had to choose between their liberty and their security. They always had both. ... – because, if you, as a fellow member of their community, were fairly successful in your own life, then they, as an overall ecosystem, were also benefiting from your productivity. But then, if you, as an able-bodied person, were for some reason struggling and not quite thriving, then they, as a fully interconnected ecosystem, were also wanting for your potential and carrying the burden. So, they made sure to do their best to help you prosper. They made sure to do their best to help you crush it.

And a lot of the magic of this system was in how their leadership was appointed... or... ... more accordingly... ... in how their leadership was *not* appointed. They were selected by a completely organic process - just like it was with team Myriad.

These were simply the local people who made stuff happen. They just naturally had a gift for logistical problem solving. And there will always be these people in every population. There will always be these exceptionally productive individuals out there.

So, the other citizens just let them do their thing and serve their markets, and then they purchased all their products from the ones who were doing the most to provide for their communities. So, they just harnessed all this natural productivity, and they just never tried to fight these organic rhythms. ... It's basically like they were just using tai chi for everything – working *with* the energy, and not against it.

And so... since these highly productive people were always competing their butts off all the time to try and make people happy at every possible opportunity, they were inherently obliged by their positions to be extremely accountable. Which is to say, if one of these regional *producers* ever did something unpopular, and they did this unpopular activity while they were performing in their role as a *community leader*, then they could still be affected as a *producer* by their actions as a *leader*. Which is to say, the people could just quickly withdraw their business from them.

Where else can you find a leader who can have their salary reduced like that? Talk about accountability. That's the crème de la crème of accountability. It's like these leaders were all working for street performer tips.

And people don't mess around when they're working for street performer tips.

They put on a show.

But they were also like those stores, in one of those big mega malls, if those stores were also supplying you with everything, while also paying to keep the mall going, and then just splitting up the costs amongst themselves, and never charging you some tax or fee just to be inside the mall there. They would be handling all of that stuff themselves so that you could just enjoy yourself. And they would even be trying to do a little bit extra, here and there, so that they could brag about it. They would be competing with one another to make you like them. They would be competing with one another to make you WANT to come and shop with them.

And that's essentially how the Nauptsèh ran their society. You could say that their stores were just paying for everything. And that means that *you*, as a regular citizen, paid for *nothing*. You, as a regular Joe Bag'o'Donuts, never paid taxes on anything - not on your property, not on your income, not on your house - not on your investments, not on your children, not on your spouse. Not on anything. The wealthiest local producers just paid for everything.

Alright. So, this actually has a pretty fascinating side effect that we've been building up to. Aside from simply ending all your taxes, it also makes the prices of everything *plummet*. It's like if everything was just suddenly on a fire sale.

And accordingly, the reason why this system reduces prices like this... is because every little product that you purchase, already has about *fifty billion* taxes in it.

Every single product that you pay for, lies at the center of a massive spider web. It's just a labyrinth of human labor and material resources. ... And indeed, this webbing is extremely resilient, because this webbing has to compete and remain efficient. But alas, this webbing is also covered in a mist of dew drops.

These droplets, are, each and every one of them, another heavy tax burden. And all of these taxes add up *big time*, and they just slowly cause the web to droop and sag a little. But then these taxes are so deeply buried within it that you never actually see them. ... So, you just think that you're paying for a product, but you're actually just paying for all of this extra weight in it. ... And yet, you never really notice, because of how strong the webbing is.

So, if you just simply get rid of those taxes, and then you replace them with a roughly ten percent markup, then the cost of every product will just crater.

And a ten percent contribution is nothing, by the way. Most businesses would crawl through glass to pay so little. That's just a fraction of their current total tax liability. And the accounting for this method alone is about a trillion times easier.

But if you're keeping much more of your money, and then every single product just suddenly costs less, it's the same as if you're making a lot more money. It's the equivalent of having your income essentially quadrupled.

Yes, quadruple. Increasing by at least four hundred percent. Multiplying by four. Doubling from what you're making now, and then doubling that again.

That's exactly what happens when you take our astronomically complicated modern tax codes and you replace them with a point-of-sale markup.

Okay, soooooo... wait. Hang on. I thought we had to pay those taxes for a reason? How can the Nauptsëh leadership afford to pay for everything if they're only using about ten percent of their business income?

Well, the full version of that answer is insanely complicated, but you can boil it all down to a pretty simple axiom. The reason that their leaders can afford to pay for everything like this is because their leaders are on team Myriad, not team Deca.

Their entire existence is about doing stuff with maximum efficiency. They just simply can't afford to have team Deca's Janice sitting around all day not doing anything. And they really can't afford to have team Deca's Todd just breaking everything all the time. All he does is waddle around most of the time, quite unmetaphorically breaking over HALF the stuff he touches. Then he just leaves it there and waddles away, because he knows he's under contract and you can't touch'im.

And then there's Jason. If you thought Janice and Todd were inefficient, then just forget about Jason. Jason's brother-in-law scams would be costing team Myriad a fortune if they were able to spend other people's money the way team Deca does. But of course they can't. Team Myriad can only ever spend their own money - not yours. So, they just need everyone to put in an honest day's effort and work efficiently.

And it's not that big a deal, honestly, but team Deca has virtually no incentive to work as efficiently as team Myriad does - so, they don't. ... Some of the new guys do - at least for a while. They show up with a bunch of fresh idealism and spunky hustle. But then they just acclimatize after a while and start "pacing themselves."

Thus, when it comes to how efficiently team Myriad can get the roads built, it's not even a contest. Team Deca doesn't even register. And it's not even their fault though - they're just hopelessly disincentivized to be as efficient. That's why they'll often bid out their own contracts for things. They'll just get portions of team Myriad to build the roads and stuff. But even then, Jason still loves to get involved with these projects.

... if you know what I'm sayin'.

Wink, wink.

Wink, wink, it's not his money, wink, wink. So, why should he care, winkitty wink wink.

But it's not just about team Deca and their shabbiness - it's also about team Myriad and their dazzling dapper shnazziness.

By now, you may have already heard of things like crowdsourcing, distributed intelligence, emergent behavior, self-organizing complex systems, neural networks, and even quantum optimization algorithms. These innovations are being lauded right now for their power to find solutions and optimize efficiency.

And yet, this is precisely what team Myriad has already been doing. Team Myriad is all of these things. It is a self-organizing, organic, neural network with emergent innovations and crowdsourcing optimization.

And if that wasn't *already* mind blowing enough, the Nauptsèh already figured this out... *thousands of years ago.* They've been space-age since the stone-age.

And so, that is why their system is so efficient. That is why they've never needed taxes.

So, once again, without all of these taxes on your citizens, their buying power just explodes through the ceiling. And then they never have to question how much an extra side of avocado costs. They just get it if they want it.

So, it's little wonder the Nauptsèh loved to party so much. They partied all year long. ... especially in the winter. Their leaders threw these awesome community potlatch festivals. And they gave away massive amounts of wealth to every single citizen.

It wasn't indiscriminate gift giving though. They knew exactly who you were and how much they intended for you. But they always went all out and fully gave away the maximum.

Often times they would just giveaway their entire fortunes, all the way down to the literal shirts on their backs. In our current stick-based civilizations, this would be insanity. This would leave you penniless and bankrupt and essentially destitute. But in their incentive-based civilization, you were never more wealthy, because a major source of wealth was your reputation.

As soon as these leaders were done gift giving, their stock was solid platinum, and their popularity was on fire. So, the people were more than happy to remain loyal to them as consumers. And before you even knew it, these leaders would all be wealthy all over again.

The leaders would help the people, and then the people would help the leaders. It was all a perfect circle of symbiosis.

So, when it comes to building roads and stuff, they don't just wanna lazily build some basic, bare minimum thing for you to drive on - they wanna build something that you'll love to drive on. They wanna take every opportunity available to do stuff that you'll love them for.

And there are lots of opportunities for that. They could help take things to next level on any number of projects. It could be caring for the needy; parks and recreation; roadways; infrastructure; regional defense; special projects; community outreach; children's organizations;

you name it - anything for the bragging rights - anything for your loyalty.

Just like the judges are always competing to inspire your faith in them, the wealthiest local producers are all competing for you also. They want your trust and faith and patronage, and they are willing to go the distance to achieve this outcome.

They can easily afford it, by the way. No one is paying taxes on anything in this civilization. You yourself are not paying taxes; they are not paying taxes; no one is paying taxes. All of the basic necessities are already being taken care of because of the standardized, point-of-sale markup. So, everything else is just a bonus, and they can easily afford these projects without it even affecting them.

That is the amazing power of competitive crowdsourcing. Public projects end up costing way less money. So, then there's all this extra wealth for the common citizens. But then there's also these extra funds for all of the community leaders. So, they just airdrop all of this cash on their communities.

Things gets pretty darn interesting when you start crowdsourcing like this. It's like announcing a schoolwide competition for an open, voluntary science project. Then these teams of little brainiac whiz kids start forming up naturally. And then they usually invent some project that actually qualifies for an actual patent or something.

But if you try to just get some regular, ordinary group of students to do this same project, and you can only insist that they aspire to a minimal standard, then that is exactly what you'll get. This command-and-control type of management lacks the power of competitive crowdsourcing.

That's okay though, because no one actually needs a command-and-control type civilization. Incentivized, competitive crowdsourcing can take care of everything. Just incentivize all of the preferences that your local community wants, and then an army of entrepreneurs will take care of everything. Not only will they bring you all the products you want, but they will also use these profits to pay for your community projects.

So, these guys won't be messing around at all with any fraud, waste, and abuse nonsense - not when it's *their own money* on the line. They will be achieving the absolute pinnacle of civic efficiency. They will be attempting to impress you, but they will also be working hard to do it cost-effectively.

And of course, this efficiency means that their civics will inherently cost less. So, they just simply won't need as much money to afford these collective amenities. And thus, the prices on all of their products can be cut down dramatically. And furthermore, no one will be taxing you anything when the producers are already paying for everything.

In other words...

... you're rich, bitch!

So, this is how the Nauptseh achieved their wealth expansion. They just simply found a way to exist more efficiently.

... Hence a civilization where everyone is making four times as much money.

And, again, this isn't just some hocus-pocus, semantical shell game. If you're driving along in the winter, with your windows down and your parking brake on, well, then you're going to be wasting your fuel, just fighting against all that friction and blowing out your heater. So, the Nauptsèh simply rolled up their windows and stopped driving around with the parking brake on. It was never about some arbitrary, semantical shell game. It was just natural, logistical efficiency.

But we often don't observe what's right in front of us, because of our biases and our fear conditioning. We've just spent thousands upon thousands of recent Holocene years, being consistently conditioned, that the only thing standing between us and total chaos is our stick-based civilization. So, of course we never even *looked* at these other alternatives.

We've just been bottled up inside of this stick-based paradigm, trying to solve an unsolvable problem.

Fear will do that. It'll keep you running from things no more nefarious than a weirdly bundled up pile of clothing in a dark room. ... or like an empty old pizza box being blown across the roadway on a really dark night, ... like just skittering along in the darkness all sporadically and jerkily, ... like it's some fucked up rectangular animal with like a billion little legs under it.

But really though, there has been so much completely uncalled-for stupidity in the world, and it's just absurd how freakin' much of it is caused by panic. People have been getting in shipwrecks for centuries, right off the coast, but then instead of just swimming over to the shoreline, they just freak out like a maniac and start panicking. Then they just assume that they don't know how to swim. So, they just start flailing around all wildly, or they just freeze up completely and start sinking. But *even* regular, everyday land animals, like freakin' deer and porcupines know how to swim, and they're not even amphibious like we are, so we've got no excuse.

Or another one is rip currents. Rip currents are these lazy little flow channels that move slowly away from the beach, like a gentle little river. And then they get back there behind the wave break, and they either just stop right there abruptly, or they just wrap around in a big horizontal circle and make their way back to the beach. They are completely harmless. And if you don't wanna be in one, then all you have to do is move slightly to the left or right a little. Easy-peasy. But then people have all just heard these totally bogus scare stories already, about some mythological "undertow" that just "pulls people under" or something, and then they find themselves in an actual rip current in real life, and they just totally freak out and start panicking.

Yeah, there's no such thing as currents that just randomly pull people down. The only place where you could actually find something sort of like that is if you were scuba diving, way way out on the horizon, really far down underwater, in an area where there's a major underwater cliff face. Sometimes along these sheer faces there can be currents that run horizontally or vertically, and if you don't wanna be in the current, then all you have to do is just move back away from the wall a little and you'll be out of it. Easy-peasy.

And then people go and watch these nonsense hollywood movies about these like demon possessed sharks or whatever, and then they think that these sharks are all out there, just swimming around, trying to hunt people down all the time. No. Sorry. Sharks only exist for specific purposes, and hunting humans down is not one of them.

Sharks are freakin' pu**ies, honestly. *A shark is not built* for going after things like wildebeest and buffalo. They're just built for going after things like little fish and stuff, and they are *terrified* of getting injured.

So, if you just poke'm with a stick or something, or if you just smack'm in the nose or in the eyes or in the gill area, they just freak out like a chicken and start crying.

I wish *land* predators were that easy to intimidate - just smack a lion in the nose and he gets all hurt and runs away.

And out of the hundreds of different species of sharks in the world, pretty much all of'm prefer stay the heck away from us, except for just a few that sometimes get humans wearing black wet suits mixed up with seals. I don't know why everyone has to keep wearing these jet-black wet suits all the time though. There's all sorts of other stuff that we can wear - stuff that makes it abundantly clear we're not a seal snack. And yet, no one does this. People just keep tempting fate all the time and wearing these solid black swim suits. And sharks STILL almost never tag us by accident. That's pretty freakin' impressive.

But, yeah, there's all sorts of different stuff we can do. For example, a person can just simply wear a swim suit that has some really big and bold, black and white stripes on it. Wearing a wet suit like this makes you look like a pilot fish or like a sea krait or an orca - all things that sharks don't wanna mess with. There is a reason why evolution has painted these big, bold, black and white stripes on things. Sometimes this is nature's way of just sort of breaking up your appearance a little, to create a visually disconcerting profile, however, notwithstanding any usefulness in a snowy or bleached environment, this is generally nature's way of saying, "yeah, you're not gonna have a fun time if you mess with me, son." ... ... especially in the ocean.

Especially with orcas. Sharks are absolutely terrified of orcas. If they even catch a glimpse of an orca, they freak out like a like pearl clutcher and start swimming away as fast as possible. And they don't even *think* about coming back for at least a month or more. They are... TERRified of orcas. As they should be. Orcas are freakin' lethal. Thank God they respect us. And we should respect them too and take care of their environment.

So, anyway, if you and your fancy new black and white surf board just simply *kinda look* like an orca, then this is just *way* too unnerving for a tiny little shark brain.

Sometimes they like to get up close to us and just check us out a little, because they're honestly just curious about what type of predator we are, since we're obviously not trying to swim away from them like a seal would, but if you even vaguely resemble an orca, they are not curious about anything. They're only curious about how fast they can get the hell outta there.

And a lot of oceangoers are already fully aware of this, but then they *still* don't even worry about it. They don't wanna look like they're just some dorky, hamburglar, nightmare-before-christmas guy, just swimming around out there in a prisoner costume. ... We human beings are about a million times more concerned about what our fellow human beings are thinking, than we are about what any shark is thinking. ... Shows you what our actual priorities are.

But, yeah, so, anyway, we spend all sorts time being afraid of stuff that isn't even an issue. Some people are afraid of relying on crowdsourcing for public services. They feel like it wouldn't guarantee them a reliable source of public utilities. Or they're just afraid it would somehow place our entire society "at the mercy" of the producers. And yet, competitive crowdsourcing has precisely the opposite effect. It is unhyperbolically about a billion times more reliable than anything team Deca could ever come up with. This arrangement, of open competition for public services, unequivocally places these producers at YOUR mercy - not the other way.

And to be perfectly honest here, not only is the idea of relying on competitive crowdsourcing for your basic public services *not* a scary thing, the *rEALLY* scary thing is actually *not* doing this. The scary thing is having all of your stuff being arbitrarily managed by team Deca, with their extremely limited pool of collective talent. THAT... is the scary thing.

And without any of that team Deca, artificial, administrative authority interfering, Jason doesn't have any power, whatsoever, to just come flying in, like the hand of God, and keep arbitrarily protecting his brother-in-law from the natural consequences of trying to scam people. No way. Every producer, and every entrepreneur, is subject to *your* good graces. Period. You say "jump," and they say "how high?" or their reputation becomes tainted, and it's not long before some other competitor just swoops in and replaces them.

They don't mess around when it comes to customer feedback. They know that every single person who bothers to voice their opinion about something... is usually just the tip of a much larger iceberg. We consumers... are *incredibly* powerful.

It's sort of like a scene in a movie where there's some really tough crime-fighter guy who ends up getting sent to prison for some reason, and then a bunch of dirty rotten criminals try to attack him in there, and then he just like, beats the living hell out of the first guy who tries to come at 'im, to make an example.

And then he's like... ...

"None of you seem to understand... ... I'm not in here with YOU... ... ... you're in here... WITH ME!!!"

Well... ...

... we're not in here with the producers... ...
... the producers are in here... WITH US!!!!

And when I go swimming in the ocean... ...

I'm not in there with the SHARKS... ...

... the SHARKS ARE...

Actually though, a competitive crowdsourcing system isn't all confrontational like that. It is precisely the opposite, in fact. Thankfully. It creates a culture of aligned common interest. And it breeds a spirit of symbiosis and appreciation.

Citizens don't have to threaten their regional bureaucrats. They don't say to their local "representatives," "look, chump, you better do this or else." They just say to their local producers, "pretty please with a cherry on top." Then their local producers reply back, as fast as humanly possible, "much obliged, is there anything else I can help you with?"

And not only that, but the consumers will not just abandon their local producers at the drop of a hat, if they've already built a long-standing relationship with them. So, when some hot new competitor just shows up with some shiny new product, the old producer will still have time to adapt, and this makes life a heck of a lot easier on all of the people who work for them.

So, yeah. This system is like leaving cookies for Santa, where there's a mutual exchange taking place, and where, inherently, both parties are very much appreciative of one another. It's not a stalemate, like a Mexican standoff, that causes endless anxiety.

That anxiety is instead just a consequence of having team Deca manage your community. And team Deca can be anyone, by the way. Absolutely anyone. Any group of leaders who can force you to do stuff. So, this could be royalty and their palace advisors. This could be your Greco-Roman representatives and their delegated bureaucrats. This could be the Senatus Populusque Romanus and the emperor's praetorians. Or this could even be the head of a crime family and his troop of ruthless, wise guy enforcers. Just any group of people with enforcers that make you do stuff.

So, beginning around two and half thousand years ago, Western civilization started acknowledging that the normal, everyday citizen needed to be represented by their leadership. You couldn't just have some king or queen telling everyone what to do all the time.

So, the Greeks could clearly see this kingly power, which their hereditary monarchs had wielded, and they realized that they could destroy it, and thereupon rid their lands of its malice.

But the hearts of men are easily corrupted, and the kingly power has a will of its own.

So, they divided up its power into multiple pieces, and they gifted it to three elvish elders, seven dwarves, and nine men. Or maybe it was all men. I forget. It was so long ago. But the point is this, none amongst these men could resist its power. According to them, one does not simply relinquish the kingly power.

It was very dear to them... ...

It was their own... ... ...

It was... ... ...

... their precious.

Their precious was a burden though. You could see it in their eyes... how it weighed on them... how it poisoned them... how it gave them these really dark circles around their eyes, even when they had just woken up from a nap and you would think that they should be rested by now. That's because the power of the kings can never be truly wielded by any man without consequence. There is only one true lord of the kings, and he... does not... share... power!

Or so I'm told.

But I felt really bad for these men - always bound to this burden and suffering. So, I offered to carry it for them every once in a while. But there was always this creepy little opium junky just lurking around in the shadows, and every time I offered to help 'em, he would just come running out of nowhere, calling them "master" over and over and accusing me of trying to steal it.

Then he would go sneaking around at night like a freakin' psycho and start throwing away all of our food rations. Then he would sprinkle the crumbs all over me while I was sleeping and then accuse me of being fat in the morning.

He was like, "tHe faT One dId iT!"

And I was like, "no, I DIDN'T!"

"... He's a LIAR!"

"... He's trying to poison you against me, Mister Plato!"

"... I promise! ... ... It's me... Mister Plato sir... your Samokles."

But it was too late for Mister Plato Sir. The power of the kings had already corrupted his heart. Mister Plato d'Athens believed that this power should be harnessed, not gotten rid of. He believed that the citizens would simply destroy themselves if you gave them too much freedom.

And Plato d'Athens was not alone in this regard. Pretty much every Greek aristocrat shared this sentiment. They thought that the citizens should still be commanded by this royal kingly power system, but that this power should instead be divided, and then apportioned out accordingly to every realm, and distributed amongst a group of many, smaller, little, district, kings. And then these smaller, little, district, kings would be chosen by all the land-owning men of that realm from amongst a group of local aristocrats.

That was their idea of giving "power to the people" - letting them choose from amongst a group of local aristocrats whom they wanted their little, local king to be. And if that wasn't antithetical enough, these kings could still subject them to just about anything. And they each had *years* to inflict their damaged before a new set of kings could be selected.

And they also didn't call themselves titles like "king" anymore. Instead, they started using titles like, "representatives," ... because in theory they were supposed to be representing you, ... although in practice they had *no such* obligation.

Thus, as soon as you had selected your "representative," they could exercise all of this power you just gave them, in pretty much any way
they wanted. They could promise you just about anything to solicit
your support for them - then they could send you off marching into
battle within the space of a fortnight.

"Hey, I thought you were gonna lower my taxes. Why am I marching
in the snow with a sword in my hand?"

"Is this the line for lower taxes?"

"Why is it way out here... AHHH THEY'RE SHOOTING AT US!!!"

So, they never actually got rid of the concept of royalty. They just divided it up and started calling these little, district kings "representatives." And then they claimed that this was all so very necessary
because the citizens themselves were not quite smart enough to be
making their own decisions. But then they still wanted YOU to be
the one who decided who gets all the power. ... Soooooooo... we don't
trust *you*, and your *decision making*, but we still want YOU to be the
one who decides who gets all the power. ... ... Hmmmmmmmmm... ... ... Interesting...

Yes, that was their logic. ... And that was the *best* idea that Western
civilization *ever* came up with.

Luckily though, one of the New World civilizations had already solved
this problem... ... several *thousand* years earlier. And it's a little bit
ironic too, honestly, because Western civilization actually had a three
thousand year head start on them. That's why explorers from the Old
World already had things like gun powder and iron by the time they

got there.

And yet, it was the Nauptsèh of the New World who perfected it. They created an unequivocal form of representation, and they did so... without it causing a mob mentality. And if any leader within their civilization was ever causing a bunch of problems, then they only had *weeks*, as opposed to *years*, to do their damage. So, they solved it. They perfected sociology.

Mwah! ~ Bellissimo!! ~ EUREKA!!!

They did it!

The fellowship of the Nauptsèh was victorious. They had cast their kingly powers into the fires of Mordor. They had destroyed its lidless eye enwreathed in flame... ... ... rrrAAA!!!

And anyone could have done this just like they did. ... Literally... Any civilization could have done this. All you ever really needed was just a series of delineated enclaves, and then those enclaves could effectively function like a network of communities. That way, everyone remains accountable to someone – everyone remains connected to some broader social network.

Oh, and spoiler alert, this is *precisely* what our species is evolved to do. We are *evolved* to keep track of people's faces. We are designed to retain a memory of their reputations. We are adapted to spread this news throughout our peer groups. And we are bred to establish these communities whereby everyone remains accountable. ... It's in our makeup.

And this is why we're generally more suspicious of out-of-towners. No one knows a thing about their history yet. ... And this is also why we're drawn to certain people with a healthy social circle. The fact that they have friends means that someone is always keeping them accountable. ... So, why should our civilizations be any different.

So, like I said, it's pretty a straight forward process to convert your ancient or modern civilization to a carrot system. *And it even happens on its own... if you simply let it.*

But there were people like Mister Plato who claimed that this was impossible. They said that if the citizens were ever given too much freedom then the world would end in chaos. And yet, that is precisely the antithetical opposite of what the Nauptsèh experienced. The world that the Nauptsèh inhabited could only scarcely have been more peaceful and prosperous.

And that was in large part due to the wisdom of their leaders. There leaders were a source of stability. You see, their leaders were not just there to be their servants - they were also a source of wisdom with exceptional insight. So, the people would always listen to their counsel. Then they would often largely agree, and thereby follow their guidance.

Centuries ago, this phenomenon was known as "elf counsel" – or, in the language of the people who had spoken it, this was simply known as "ælf ræd." This expression was inspired by the wood elves, who's counsel was a source of instructive wisdom. So, if a man had thus obtained this elvish virtue, then he may have earned the title Ælfræd for his elf counsel.

Today we simply know this name as Alfred.

The name Alfred just means counselor of elvish wisdom. And that is precisely what the eldest Nauptsèh leaders were. They were servants of their local communities, but they were *also* their counselors.

So, they were essentially like a loyal, elderly butler. And although technically it was *them* who were supposed to be doing what *you* said, in actual practice it was usually *you* who was following *their* advice.

And this is what you get from an incentive-based, crowdsourcing society - you become wealthy, because of the efficiency with which the public projects are being paid for, but then you also get these Alfreds as your leaders. So, not only are you many times more personally wealthy, but you also have your Alfreds looking out for you, ... Master Wayne. ... So, does that mean you get to go around fighting bad guys and seeking vigilante justice in the dark of night?

Well...

... yes.

That's exactly what that means.

But you'll have to look around to find some crime worth fighting, because your Nauptsèh city won't really have that much. This system just prevents these things from ever even happening in the first place. It just handles all these issues and takes care of everyone. So, no one ever needs to steal or plunder anything. Everything is already provided for - assistance for the needy, police, fire, utilities, healthcare, education, infrastructure, defense, community outreach, festivals, carnivals - you name it.

So, if you *really* wanna stop a bunch of crime from happening, then you'll be better off just switching over to a carrot system. And if anyone tries to give you any flak about it, then *that'll* be your chance to rain down justice with your fists on someone. Or not. It's up to you. Just listen to your heart. And listen to Alfred. Alfred knows what's up.

A civilization organized by a bunch of Alfreds is a pretty chill one.

Things like police and fire would be paid for by the leadership. Same thing with defense and infrastructure. These sorts of secondary services would be accountable to the leadership. And then this leadership would be accountable to the people. So, these secondary services are still accountable. And these leaders who are employing their services are simply acting like a mediator.

For example, if there were people driving their cars really recklessly, in a neighborhood where there were lots of little children playing, then the citizens could raise this issue to the local police department. But if this particular police department didn't seem to be doing anything about it, then these citizens could just bypass them altogether, and go directly to the leadership themselves.

Then, at this point, these leaders would have to spring into action. They would have to go down there, rather quickly, and find out what was going on with this. And if it turned out that that police department actually wasn't doing their job correctly, then these leaders would have to take this very seriously. But if it turned out that there was only a very simple misunderstanding taking place - like maybe the police department was already doing a secret stakeout or something - then the leadership would have to assure them, and even stake their reputation on it, that the situation was being ameliorated.

And the key phrase here is that the leadership would "have to." It wouldn't be an option. Ignoring the concerns of these citizens would be like telling a loyal customer to get lost. It just simply wouldn't happen. Because not only would you be losing them as a customer, but you would also be invoking their wrath as a citizen. And a citizen that has been scorned like this is a citizen that can do some serious damage.

So, what kind of damage can a citizen do? ... Well, for starters, they can leave some really bad reviews, and even report them to the regional, consumer protection agencies. ... They can also attempt to get some publicity on it. And if that attempt to get some publicity on it doesn't quite work to their satisfaction, then they can also file a lawsuit, and thereupon have the matter investigated by the court system. Or they could even do a class action lawsuit. Or they could even get some helpful assistance from a consumer protection agency. Then the courts would have the ability to really dig into the details. And if the judges ever found that any of these leaders were in any way slacking, then this would be a serious local scandal. ... and then this would certainly garner them that attention they were looking for.

This situation would be an absolute nightmare for these leaders. This would make them look like a bunch of chumps. And it would harm them on a private business level, ... in addition to all of the other drama it was causing. This is the number one worst possible scenario for a local producer. Damage to your reputation is brutal. It takes a lot of hard work to recover from that. And that is precisely why these local producers would just avoid this like the plague to begin with - by ensuring that all of these local services were performing admirably.

Likewise, if a group of very adamant, local citizens ever wanted something that was just functionally impossible, then their leadership would have to address this... and actually explain to all of these people why what they were asking for was just fundamentally undoable. They would have to treat them like adults, and provide an answer that was satisfactory. They would have to actually address their concerns, in the most earnest way possible.

So, if you wanted to go around and start looking for injustices to battle, especially in the dark of the nighttime, then you're gonna have to find another city, unfortunately. ... ... Or you're gonna have to focus your efforts on little kids with *overpowered* paintball guns. ... You're gonna have to go teach those little sons of bitches a lesson. ... You're gonna have to go show'm what injustice is really all about.

But this discussion also brings up another good point. When you're living in a society that's almost entirely based around people's professional reputations, then the *journalists* of this society are *extremely* influential. They can enormously affect a producer by the things they choose to say about them. Luckily though, the media is just as obliged as anyone to behave themselves. Even *they* can be subjected to a lawsuit, just like any other business. And any attempts to actually change the news or editorialize it would very likely result in a lawsuit, whereby their actions would be exposed during legal discovery. So, it just wouldn't be worth it to try and spin the news like that. They would be bound and compelled to the most accurate reporting possible.

Furthermore, you also couldn't just clandestinely establish some sort of giant, cartel monopoly. Like, you couldn't, for example, just like pay off every single media organization to keep quiet about something. As soon as you tried to do this, a competitor would just expose you. Then *they* would soak up all of *your* customers, as the hero who had exposed you. You would just be handing them all of your market share on a silver platter. ... Then you would lose your reputation, all for nothing.

So, corruption never prospers in this system.

You can't control the playing field in the Nauptsèh system. You just simply cannot block the competition. There's no mechanism to do so. There's no hammer. There's no "power." It's like playing a game of Whack-A-Mole without a mallet. So, it's just a game of Mole. You just stand there in front of the machine and moles come out. And you always lose. ... if you're a monopoly man. Otherwise, the moles just bring you stuff. And they're all a bunch of little Alfreds. So, the game is just Alfred-Mole, and you have to, like, buy little prizes from 'm.

I would play that game.

"Look, babe! I won you a giant panda! ... Well... technically I just paid for it. ... But it still cost a lot less than impossible darts or something."

So... anyway. The Nauptsèh of the Neolithic destroyed their kingly power. They cast it into the fires of Mount Shasta. And thus, there was no more kingly power to force people to do stuff. There was no more giant mallet to whack the moles down with. There was no more arbitrary authority to empower corruption. And then all of these little moles just started popping up everywhere and offering the people things they really wanted. And then the ones who were doing the best job at it were the Alfreds. So, it then became the natural expectation, that, in addition to being the producers and the local suppliers, these Alfreds would also offer their guidance, and their sponsorship of public amenities. And, of course, the Alfreds were so efficient at this process that these projects only cost them just a tiny little fraction of what it used to cost. And what that all means for us is this, that there was a *massive* amount of wealth that was just *never* absurdly wasted or inefficiently squandered. So, then the people became effectively RICH, BITCH!!

That's definitely a game that I would like to play.

And some people think that this is weird, that you would have all of your local producers just sponsoring your overall community like that. And the reason that they think this is weird... is because it just kind of seems strange... that you would *also* be judging a local business... by how well they were taking care of your civic requirements. So, you're not just simply shopping and thinking about the products you want, you are *also* now paying attention to what that seller's reputation is.

However, I would definitely counter this assertion of weirdness, and I would posit... that this is *actually* the *only* way of living which is legitimately NOT weird.

So, what do I mean by that? Well, it is extremely freakin' weird that we *freely* give our hard-earned money to many of these large-scale producers who we are ideologically opposed to. We just go out and buy some product, from a well-known producer, and then we'll see them in the news at some point, and just decide we don't like 'm anymore. But then we'll STIL go out and buy their products again! We'll just continue to give them our money... *deSPITE* the fact we don't like them. ... ... Well... THAT... is freakin' weird. ... And pretty much everyone does it... on one level or another.

And that's what living in a stick-based culture does. It disassociates everything and scrambles it all up. So, no, the way that the Nauptseh have been doing this is *not* weird. In fact, it's been the absolute *least* weird way of doing things for thousands upon thousands of years at this point.

Alrighty, so, anyway. Hopefully that explains things like police and firefighters and charity and stuff - the Alfreds would employ them as contractors – thus, they would always get the best for their communi-

663 - BARTROLOMEW MC INNCEL

ties. And if a leader ever failed to perform these duties properly, then they would organically be defunded and effectively boycotted and potentially replaced by someone. ... That is *exponentially* more accountable than any of these stick-based societies.

So, this idea of a leader actually failing to perform their duties properly, to the satisfaction of their community, was almost unheard of.

Stick-based leaders make excuses... - ... carrot-based leaders make things happen.

But, anyway, what about some of these other things, like utilities and healthcare and education and whatnot?

Well, utilities are a pretty straight forward subject. ... Woe, I just felt myself get sleepy with how boring that sentence was. Le'me shake my head a little. UHBUHB UHBUHB UHB UHBUHB... ... Ugh.. ... alright. So, utilities. Alright! So... There is an ever-present threat of competition amongst these local utility providers, and there are simply no protections for any tomfoolery shenanigans. So, your utilities would be as ridiculously affordable as scientifically possible. It wouldn't even be an issue. No one would even think about it. Parents would just gladly allow their children to leave the doors open and heat the wilderness.

Okay... well... ... maybe it wouldn't go *that* far. But it would still be pretty chill.

Alrighty then. Well, then let's talk about healthcare. ... everyone's favorite subject. Our current systems are a joke. They're just some weird old amalgamation of team-Deca inefficiencies, combined with an archaic early system from the bad ol' days, of when "the company-stores" had total monopolies - the days when your employer would only pay you in some like weird, fake, monopoly money... so you could only spend your paycheck at that same exact company that just paid you.

Well, that is where our current systems of healthcare originally came from - either that or some massively inefficient team-Deca program. The whole thing started out on a really bad foundation, sorta like Venice, and instead of just going back and fixing it, they just kept on adding layers and building on top of it. Then they were forced to incorporate all of these extra, random things like building floats and flood controls and tidal suppressors. And now it's just one enormous cluster ffffffffffffffffff... fricassee.

Lucky for us though, healthcare never had to be like that, and a Nauptsèh healthcare system would easily fix this. With their full-bore crowd-sourcing competition, and no brother-in-law shenanigans to get in the way, the most state-of-the-art modern healthcare is literally always entirely available to absolutely everyone.

And it's affordable at every single income level, to be sure. Not that this would even be an issue when everyone's already making *four times* as much money as us. People wouldn't even be thinking about it. ... And for the very few who were, there would be inherently be some helpful assistance.

Sometimes people just have really bad luck for whatever reason. It happens. Or sometimes they're just starting out in life, and they don't really have any family who can help 'm. Well, people like this would get assistance. And the program would also be designed to assist them in achieving a better lot in life. Or, in some cases, people are just physically AND mentally compromised - provably incompetent, in other words. These people would just inherently get assistance, if they had no other options. And the costs for all of these programs would be almost imperceptibly small.

And last but not least, there is also one more important detail about these assistance programs. The people who were receiving this assistance would still need to choose which providers and hospitals and clinics they wanted. Then these companies would be given vouchers discretely. And so, these providers would still have to compete for their business. They wouldn't be treating them any differently because they were on assistance. And this of course preserves the element of competition, which itself drives up efficiency and reduces prices.

And to that end, every able-bodied person receiving assistance would be provided with things like counseling, for education and job placement. They would even be given help with rehabilitation if necessary. They would have every opportunity to become successful. No one would be left behind - because doing so is not just immoral, it's also wasteful. And a carrot-based society is anything but wasteful. It can't afford to be. Everything is connected. The people's success is the leaders' success, and vice versa.

These sorts of full-spectrum, holistic assistance programs definitely get results. Certain communities have already been using these for many generations at this point. And nearly everyone they help goes on to become successful.

And that pretty much sums it up for the Nauptsèh-style healthcare system. It would be extremely competitive and affordable. And then those who actually needed assistance would accordingly be given it.

But there is just one more little point about these assistance programs.

Now, I know it might come as a bit of a shock to most people, but not everyone who applies for these assistance programs is being honest. ... ... I know, I know. It's terrible. But some people actually do it. ... They lie!!

So, of course, any functional assistance program needs to periodically check for fraudulent activity.

My team and I developed a test for this years ago. First, you simply have the person that you suspect of being a benefits cheat come down and meet you at the office. Then you have them sit in a chair while you attach a bunch of lie detector electrodes all over their body. Then you have 'em place their right hand up in the air, like they're making a sworn statement, and then their left hand on a stack of religious recruitment pamphlets.

Then, as soon as you've got 'm in place like this, just strap 'm in the chair so that their arms and their legs are immobilized. Strap their left hand on the pamphlets, and their right hand in the upright and swearing position. And be sure to get their elbows too when you do it. That's usually how they squirm out of it.

Alright. We're done. The suspect is now in truth mode. It's time for some answers.

And don't worry if they try to get all rowdy on you. There's a specialized hood in the back that you can pull down over their head as a spit guard.

At this point, they'll probably be all up in a fuss. They'll be yammering on about how they were "totally willing to comply" when they came in there but now they're not saying sh** and blah'blah'blah'blah'blah.

Well, next I want you to grab a big wet sponge and just squish it down over their head. Then secure the sponge in place with a helmet-style skullcap electrode. The skullcap electrode should have an archaic looking chin strap, and bunch of wires coming out of it.

There should also be an oversized electric lever in the room, right where they can see it. And it should be like cartoonish large in its proportions, and like painted red or something.

Then just look'm in eye and be like, "I'm only gonna ask you this once... and if you don't start tellin' me what I wanna hear... then we're gonna have ourselves a little sizzle party."

"... Are you... or are you not... a low down... dirty rotten... filthy grubbin'... ever-lovin'... *BENEFITS CHEAT?!?!*"

"DON'T YOU LIE TO ME!!!!"

That's the way we used to do it. ... And I still recommend it, honestly. ... But I guess *technically* you can't do it anymore. ... Everyone turned into a bunch of little cry babies all the sudden. ... Now you have to use, like, some little sissy method.

I'm gonna miss it though. It was *great* at getting answers. ... My favorite was when like, some old, elderly guy would come in there, and get all upset about the way we were treating him, and he'd be like *rambling* about all of the wars he's fought in and blah blah blah, and you would just grab'm by the hair and be like "shut the *fUCK UP!!!!!!*"

30

19 ¾ 2 -- SHEEP DOGS FOR BEING THE SHEEP,

... NOT FOR CHASING THEM

I could never get a girlfriend back in high school, so I had to date a shopping cart in a square dress.

She was quite the lovely dancer though. ... except when I would bash her into people's ankles on accident.

Oops.

My bad.

Frowny face.

I still feel bad about that.

It's not my fault though.

Human girls just didn't wanna go out with me for some reason. I would pick'm up for a date. And they would seem like they were enjoying it. Then I would turn on some really heavy death metal. And I'd start singing, and yelling along, until my eyes were bugging out and completely bloodshot.

Then they would just randomly get all hormonal and decide to go home early.

Pfffff. It was like, why would you even go on a date with me? ... if you didn't even wanna have dinner. ... you stupid bitch.

Whatever.

It didn't even matter. I was too busy anyway, working on my inventions. I was working on a commercial that was unavoidable. It was gonna be made like those warning strips on the road - the ones where it's just a bunch of bumpy little ridges, and it sounds like you're driving over a money printer? B'r'r'r'r'r'r...

Well, I realized that if you changed the spacing on those, you could change the frequency, and then you could make a bunch of sound effects and audio tracks. And it would sound like someone was talking to you in a really gravelly voice. And every single car would *have to* drive over it.

"u'u'u'u'u'u'" "H'e'l'l'o'" "u'u'u'u'u'u'" 'A'r'e' 'y'o'u' 'o'r' 's'o'm'e'o'n'e' 'y'o'u' 'k'n'o'w' 'h'a'v'i'n'g' 't'r'o'u'b'l'e' 'g'e't't'i'n'g' 'p'r'e'g'n'a'n't'?' 'u'u'u'u'u'u' 'A'r'e' 'y'o'u' 'h'o'p'i'n'g' 'f'o'r' 'a' 'm'i'r'a'c'l'e'?' 'u'u'u'u'u'u' 'W'e'l'l,' 't'o'd'a'y's' 'y'o'u'r' 'l'u'c'k'y' 'd'a'y' 'u'u'u'u'u'u'
'I' 'a'm' 'p'l'e'a's'e'd' 't'o' 'a'n'n'o'u'n'c'e' 't'h'e' 'v'e'r'y' 'l'a's't' 'f'e'r't'i'l'i't'y' 'd'r'u'g'

'y'o'u'l'l' 'e'v'e'r' 'n'e'e'd' 'u'u'u'u'u'u' 'B'a'r't' 'M'c'I'n'n'c'e'l's 'p'a't'e'n't'e'd' 'F'e'r't'i'l'i'z'e'r' 'N'i'n'e' 'T'h'o'u's'a'n'd' 'u'u'u'u'u'u'''

Just rattling your car to hell...

"u'u'u'u'u'u"'H'e'l'l'o'"'u'u'u'u'u'u..."

You would have a special pass where you could take a detour around the road commercial. ... once you've purchased the product.

"Yoe, I've been waitin' in line like five minutes ovuh heayuh to avoid the commercial. Can we speed things up a bit, chief?"

The idea never got off the ground though. No pun intended. My girlfriend told me the cars were gonna hate it. ... as a fellow four wheeled object herself.

Alright.

Oh, hey, you know what? Speaking of high school. I actually had a chemistry teacher that was friends with the real-life Breaking Bad guy. Yep, the real-life guy. I know this because she told us this whole crazy story about this friend of hers, and that was like years before there was ever a show about it.

And what's weird is that you can't even find this information if you try to search for it. If you try to look it up and research it, it just shows you this completely different guy who wasn't even from the right decade. And he didn't even do like one little fraction of what the real guy did. So, I don't know what that's all about, but I still need to go back there and track down my chemistry teacher and get the scoop.

But anyway, he was definitely a real person. He did exist. But I guess he just ghosted after all the drama. I don't know. And I don't know why I'm telling you this. I guess I feel like I can trust you now. ... now that I shared with you the truth about my high school girlfriend.

Alright.

High school is an important time in a child's life. Or I guess you're a teenager. But it's an important time none the less. And that's why any self-respecting carrot-based civilization would do an extremely good job at it.

Yeah, that's right, we're *still* talking about the Nauptsèh. It turns out it's a massive topic. So, I had to break this beast up a bit.

So, we finished up with healthcare in the last chapter, and now we have to cover education. But before we get to that, I have to share a little bit more of my backstory.

When I was a wee little school laddie, I was initially going to school in a low-income *private* school. It wasn't some crazy, elite, private school - just to be clear. But because it was the only school I had ever gone to, I had no idea how much better it was than public school.

Our private school was so good that I didn't even understand what the purpose of a test was. To us, a test was just this boring waste of time. Every single student knew every single subject. We all got perfect scores, and we all knew every single detail. So, it was like, "why are we even doing this boring test thing?"

That was how good our private school was. There was no such thing as a student who was getting left behind or struggling or not understanding something. Everyone knew everything. And we all knew it. And it was never because we were little geniuses or something. We weren't. It was only because our instructors were teaching us *properly*.

There's no way that I could quickly encapsulate why a private school is so much better. It would take an entire chapter. But it basically had a lot to do with how we were being taught to understand things. Our instructors were actually teaching us the hows and the whys of everything. And when you do that, you can store any type of concept in your memory like a landscape. And you can see precisely where everything is located within the landscape. You can see how everything fits together, and how all the little pieces change one another. And this of course makes it effortless to remember things.

Learning for us was like watching an entertaining movie, that was dynamic and emotional and engaging and whimsical - not just sitting there drooling in a mind-numbing lecture, frantically taking notes trying to keep track of all these disconnected fragments. No way. Our learning was actually interesting. Our teachers actually cared about what they were talking about.

Everything that sticks in our mind, basically does so, because of emotion.

And so, this is why people can't hardly think outside the box anymore. Public schools were never designed to produce thinkers. They were designed to produce factory workers. The institution was originally spawned out of what would eventually become zee Germany. Zat iz vy zee Germans have such a square head now, and zat iz vy zey are so good at following zee orders. No offense, zee Germans. I know it's not your fault. And I only mention this because that is the original pedigree of our public schooling system. And then the other nations saw what they were doing and just started copying it.

And honestly, it's pretty sad, because most people have only ever known the public schooling system. It's like they've spent their whole entire life in a cave, and they have no idea what the sunlight even looks like. I saw it first-hand. After my initial education in a private school, I was airdropped into the public schooling system, like being thrown out of the back of a cargo plane, butt naked over the arctic ocean without a parachute.

If I were to say that being catapulted into the public schooling system was horrific, it would probably be a vast and failing understatement. It was unreal. I just kept thinking, it must be some grossly anomalous oversight. I was like, "how are all these students just being left behind like this?"

But, yeah, within the first of couple weeks I already felt like I was the one that should be teaching the classes. My teacher came over to me and asked if I needed any help with like some crazy simple arithmetic, and I was like "uuuuuuuuuuh, no, I think can handle it."

They could see how well I was doing, so they actually had me tested for the "gifted" program. They made me do all those IQ tests and stuff. But then they ended up deciding that I didn't qualify.

... because my dick was too huge.

They were looking for something a little less substantial. And they were looking for people who were just autistically able to know things automatically - basically like savants. ... so they could shuttle those children off to the secret government training program for making child super soldiers.

I didn't get in though. Or I didn't last anyway. They said I was "taking too long to answer the questions."

Whatever.

I didn't wanna be a psychic, child, super soldier anyway.

The only thing I actually did like though, about the public schools, was all of the inappropriately dressed female teachers. I had this one lady teacher that was always wearing these really tight pants. Every once in a while, I would see her from the backside, then my little kid's brain would be like "wow... ... I don't... understand... what this... is... ... exactly... ... but... ... I want my whole life to be about it."

And most people just assume, about the public schools, that their quality is affected by their location. But that's like splitting hairs when comparing them to a private school. There's no contest. None at all. They might make it *look* close, on paper, but it's not. It's a completely different universe.

And I'm only taking a moment to talk about this, because of how important it is. I went to some of the most hoity-toity public schools growing up, and I also went to some of the worst ones. One time I ended up in this seriously messed up inner city school. It was freakin' just like in the movies. ... absolute insanity. There were like shootings and stabbings and fist fights all the time. And of course there were all kinds of drugs everywhere. It was just unalloyed depravity.

There would be like some fight going on in the hallway, and then some brand new teacher, who didn't know how things worked yet, would just go marching out there to try to break things up. But then some student would just like punch'm in the face. And then before you knew it, they'd just be lying on the ground, just getting kicked a million times by a bunch of students.

That's a weird thing to see.

You don't forget a thing like that.

Yep.

That was a weird year.

I had to become a pimp to survive.

That's why I don't take no lip from no anthropologist!

YUH HEAYUH?!?!

Alright. But what was even more insane was the academic quality. It was almost identical to that of the "nice" schools. Yep. Same curriculum. Same books. Same instructors. Different paint job. There might've been like a fifteen percent difference. That's it. The instructors at the nice schools might have done like a slightly better job. But that's it. It was still the same exact instructors, from the same exact universities, with the same exact curriculums, and even the same exact textbooks. It was all standardized.

And the real reason you often see such an abysmal performance disparity is because of the campus environments - not the instructors or the textbooks. Dangerous environments are stressful; stressful environments shut down your thinking; and if you can't think, you can't do school work.

This is why police and military always train so much. They're getting ready for a stressful situation - a situation that'll shut down their thinking. So, they transfer all of their knowledge into muscle memory.

Well, it's really no different for children. They need *peace* so that they can think. They need *support* so they can live their lives with confidence. It's just a fundamental, biological necessity. And of course, the classroom environment is no exception.

Passion wakes your mind up, but stress will shut it down. And so, students need teachers who will protect them.

... especially if they're a student who can multiply really large numbers in her head like a calculator... and who can move things with her mind and make them fly and stuff; and who also has a family who just sits around all day watching television; and a dad who always buys poorly concealed stolen car parts for his used car dealership; and a headmistress principle at 'r school who's always terrorizing all the students with a choky, and who used to compete in the Olympics, in shot put, javelin, and hammer throw; and who also caught a student slithering like a serpent into the school kitchen once to eat HER chocolate cake... like none other than a vicious... sneak thief. "DO YOU CONFESS?!"

You know... in case... *that* happens... ....children need to be protected.

Alright, moving on. So, that's why certain public schools have a worse performance record than many others. It's not so much the quality of the classroom instruction as much as it is the nature of their home and school environments.

But even the nicest public schools just *do not* compare to the private ones. They were never meant to. They were only ever designed to produce test takers and instruction followers, not students who can actually understand things.

Oh, and one very overlooked component of these environments is their aesthetics. Most public schools look like cr**. They look like some kind of minimum security children's prison - all these kids out there, just fenced in playing hopscotch. And then everything's like these orange and yellow puke colors. It's like, I don't know if they're actually trying to demoralize us, buuuuut...

And even when you're in your thirties, and you're like finally starting to get that nostalgic endearment for all of the stuff of your childhood, you even still go back to your public school again, and look at, and you're like, "ugh, this *still* looks like sh**."

Well, I've never seen that kind of soul-sucking architecture on a private school. Sure, they might be fairly quaint sometimes, if they're in a poor area, but they're never just disgusting to have to look at.

It's like my mama used to say, "life is like a box of nonsense. But just because you're poor doesn't mean you have to be dirty." ... She was always making everything really beautiful. And she was always doing so on an extremely limited budget. Well, this same exact philosophy can be applied to architecture. Aesthetics are powerful. They can liberate our souls, or they can crush them. We *need* aesthetic buildings. We *need* artistic talent.

So, for this and other reasons, a traditional, Nauptsèh-style community uses private schools. They are available to absolutely everyone, regardless of income - not that income would even be an issue in a Nauptsèh community.

And even if there were some struggling families, they would still just pick whichever schools they wanted, and then the schools would be anonymously compensated. It would be just like with healthcare - even people on assistance would still have a choice in the matter, and then the schools would have to compete for their approval, just the same.

So, not only would this system provide every single child with an absolutely delightful education, it would also achieve this outcome with a lot more efficiency. Children who are educated in this matter not only vastly outperform their public school counterparts, they also tend to complete this process in a much shorter time period. In fact, it's not that uncommon... for these kids to do things... like earn their very first master's degree... before they're even old enough to drive a car.

And they're never gonna tell you about this either. They don't want anyone to know our little secret. They just want you to be an order-following rule-abiding peasant. ... ... Not really. They have no idea. They have no idea how bad the public schools are. And neither do the kids in the public schooling system. To them it's just "normal." They're like that proverb about the two little fish, just swimming along, and then the older fish swims up to 'm and says, "howdy, boys! How's the water?" And then the two little fish just look at each other like, "what's water?"

"What is this 'water' you speak of?"

Well, that's how the publicly educated kids are. They have no frame of reference to compare their experiences to. And that sucks. It causes people to develop an adversarial relationship with critical thinking. They never develop that confidence of being able to work things out from first principles. Rather, the world just feels scary and confusing and complicated, so you better just hold on to someone who looks like they know what they're doing.

And so, children just learn to fall back on other instincts, like groupthink and tribalism and conformity. They start doing things and acting differently just because others are - even when it's senseless and pointless. They start doing stuff like watching professional sports, and always talking about it, and walking up to you in the break room and asking you if you saw the big goal match last week... and all of the game units they scored.

... And you're like "uuuuuuuuuuh..."

Alright, alright, alright. That last part wasn't serious. There's nothing wrong with professional sports... or the people who watch them. Pro sports are AWESOME! And the people who watch them are AWESOME! Except for the people who root for the OTHER team... ... Those people are BAStards!!

Okay... what the hell was I talking about? Oh yeah, critical thinking. So, children just pick up this habit of groupthink and never really trusting their own abilities - not unless it falls within some sanctioned rules or dictates. Of course, as we previously discussed in an earlier chapter, this is already a normal instinct for our species, but the public schooling system turns it up to eleven.

... Especially in people like anthropologists. They get brought up, and extensively conditioned, to do all sorts of sciency stuff, but then they do it within a box of pre-approval. And they're afraid to venture out of that approval zone. So, they'll dig up all this data and do their science, but then they'll do it within a very delineated sandbox. It's almost like some MK Ultra stuff - like they've been programmed to be terrified to leave the town that they're living in.

Man, this education section is going off in all kinds of directions.

Alright. So. There you have it. From kindergarten to college, and from medical school to trade school, everything is extremely affordable, and everything is extremely high quality.

And in the very rare instances of people who actually need assistance, it would certainly be available. And just as it was with healthcare, these people would *still* pick which local school they wanted, which of course would drive up efficiency and maximize quality.

So, there would be nothing ever thwarting your potential, and your community would thereby gain your skills and talents.

Oh yeah, I forgot. Kids these days don't even know what some of the most famous classic movies are. This is unacceptable. Therefore, one of their most important required classes will just be sitting there watching movies all year.

And their homework will be to watch more movies.

This is non-negotiable.

Yes, the classroom will be shaped like a theater.

Yes, there will be concessions and refreshments. Don't even ask.

Okay. So, that pretty much sums it up for your basic services. That's your police and your fire and your defense contractors, and that's your healthcare and your education and your anything else we've failed to mention.

When a civilization can just estrange itself completely from power, then it organically produces a bunch of Alfreds. Then these Alfreds end up taking care of everything. In fact, they *compete* with one another for the fame of patronage. And then that fame inspires trust and garners loyalty. And then that loyalty helps them sponsor further projects. Everything is balanced and reciprocal. And everything is mutually beneficial. In other words, everything is synergistic.

It's an eloquent system, to be certain. And it's been humming right along for literally thousands of years now. And it's been standing up to scrutiny for generations.

A lot of scholars have been attempting to find its weaknesses. They've been endeavoring to discover some sort of sneaky little loophole that could possibly break it. But that of course, has *never even come close to happening*. There is *no scenario*, within this system, that a traditional, team-Deca, stick-based civilization could ever *even dream* of handling better.

And that's a very important point right there. When you're looking at the Nauptsèh system, you have to compare it to every other system out there. You can't just detach it... and then attempt to compare it... to some perfect golden standard. No such *perfect* standard has ever existed - not even close.

So, as we round off this discussion here, on the home stretch, let's just keep in mind... that the world is messy... and that there's no other system on the planet... that actually functions quite as good as this. A lot of people don't quite know this though. They're still suffering from the illusion that their current system is competent. But I assure you, without going into a trillion light years of detail, the more and more you learn about your current system, the more and more "concerned" you will continue to find yourself. Just remember what we talked about with team-Deca.

Aaaaal"hal'hal"hal'hal"righty then...

One very common question about this Nauptsèh system, however, is what would happen if certain people were being antisocial? What if there was an antisocial business owner, and he was never giving anything back to the community? Or what if he was always causing trouble and disrupting every deliberation at the assembly of community leaders?

Or worse yet, what if this crazy mother f***** was actually scamming people? Or what if he was just ignoring critical safety regulations?

What then?

How would a system with no compulsion, no authority, and no official sanction to force any peaceful citizen to do anything ever presume to handle this?

Well, for starters, in the world that we currently live in, you often see things like certificates of inspection on the doors of restaurants. This is a way of clearly depicting that this restaurant is maintaining the current safety standards.

Well, a somewhat similar sort of certificate could also be used for a company's *reputation* rating. If a business was consistently making their community funds contributions, then they would have a generally good reputation rating. And they would show this to the general public by simply posting their reputation certificate. It would be featured at their "point of sale" locations - likely in both physical and digital formats.

So, if a company was simply *refusing* to make these payments, then they could quickly lose their status and their positive reputation rating. They would no longer have an updated reputation certificate; and they would also lose their cryptographic certificates on their websites; and they could even get notices on sign posts, erected just outside of their property; and the media could even run stories on it, if the situation had gotten bad enough.

The Nauptsèh used to do this by using a "shame pole." They would carve a really embarrassing and critical totem pole, and they would place it just outside the producer's property. And suffice it to say, you *never* wanted to get a shame pole outside of your property.

This was far more effective than using force on people. A damaged reputation is worse than almost any monetary penalty. If you've ever had occasion to have your reputation damaged, then you may have also felt like it would be impossible to ever recover from. Well, contrary to how it may feel, it actually IS possible to recover from... but it's extremely difficult. Psychologically speaking, it takes about *ten times* as much energy to recover from a bad reputation. Whatever you did wrong, you have to do *ten times* as much good stuff. That's a whole lot of good stuff.

So, although it may seem... like using force... is a great way to do things, it is actually *reputation*... which yields the greatest amount of order.

So, if you didn't pay your portion as a local producer, then this of course would affect your reputation record. And if you continued to not contribute, even after a grace period, then this would be additionally noted, and thus made known to the general public.

In a normal Nauptsèh society, this sort of negative publicity could be a death sentence. Most people will no longer want do business with you anymore. And that's because most people like having *roads* to drive on. And they also like having parks and recreational areas. And they *really* like having agencies which keep their community safe. ... So, if you're not even helping them to pay for this, then why in the everlovin' world would they ever wanna do business with you?

And this isn't just for ordinary customers - this also goes for other types of entities. Why in the world would they want associate with you if you're not even helping them? ... This could be utility companies, wholesale suppliers, insurance companies - all sorts of places. They may no longer be incentivized to offer you a good-citizen's discount. And if this sort of antisocial, non-contributing, irresponsible activity continues, then they might just wanna cut their ties completely.

Then there would also be the certifying agencies. These agencies ensure that the consumable products are meeting safety standards. And these agencies are paid through the community funds pool. So, if you as a local producer are just no longer contributing to that, then why in the everlovin' heckfire tarnations would they ever wanna help you? They probably wouldn't. And then your customers wouldn't even know if your products were meeting safety standards.

Society is a team sport. When people work together as a team, they achieve things that isolated individuals couldn't even dream of. So, if you're not gonna play the game or help society, then they don't have to help you with their teamwork. There's no need to *force* compliance by using a stick on people. I've said it once before and I'll say it again, they might not have any teeth behind their regulations, but they can sure as heck pull the rug out from underneath you.

It's all about free association.

We could probably end this answer right here, right at this point, but this book is all about the overkill. So, let's just drill down two more layers and get a little ridiculous.

Alrighty then. Well. What if by some miracle these social disincentives just weren't working? What if there was a company that just absolutely would not respond to this?

Well, honestly, it wouldn't really matter. They would be an isolated island at that point, and it wouldn't really matter *what* they were up to. You could just let them do their own thing off in the wilderness.

But let's just say, for argument's sake, that this company was still causing problems for whatever reason. Let's just say that there was a guy name Lloyd, and he was really really trying to impress this girl, Mary Samsonite. Lloyd had recently come into some money - *a lot* of money - so, he wanted to start a business to make it look like he was really smart and stuff. Then he talked his friend Harry into going into business with him. And then they bought a tract of land together from these local guys, Tucker and Dale, who themselves had probably gotten it from like an archaeologist or something."~

Alright. So, let's just say that Harry had an idea. Let's just say that Harry wanted to start selling some Rocky Mountain hand warmers. Okay. So, they start up their business and get things going, but then they try to make their community contribution payments using IOUs instead of real money.

And let's just say that this situation continues to escalate, and then it eventually gets to the point where the community has to cut them off completely. But Lloyd is really stubborn. And Mary already told him that there's definitely a one in a million chance that they could be together. So, he's not giving up. And Harry's starting to fall for Mary too. So, he's not giving up either.

Alrighty. So, this of course has reached the point of absurdity in this example, but let's just say that it happened anyway. Okay. So, what happens now?

Well, C&D Industries is gonna have to start shipping in *all* their own resources. But let's just say that this shipping was really disruptive. It could be blocking up traffic or even wearing down the roadways or something.

Okay. Well, at this point, the community leaders would be well within their rights to deny them access. The leaders, after all, are the ones who paid for these roads, so they own these public commons on behalf of the community. They can vote amongst themselves to block C&D industries. And, to that end, they can also deny them access to their airways and their underground. Of course, they would have to make sure that the public sentiment was almost completely behind them, but pragmatically speaking, they could *technically* exercise this option if it was *absolutely* necessary.

This of course would be an insane situation, but that's how stuff could happen if things got loco.

And yet, C&D Industries ain't quittin'. They decided to dig a tunnel so that they could ship in their industrial resources *beneath* the community. And technically speaking, this would actually be trespassing still, through people's underground property, but the community could also just ignore this, if it wasn't really bothering anybody.

But then just when you thought this couldn't get any dumber, Lloyd came up with a plan to totally redeem himself.".~ He realized he could save a ton of money if he just started hiring little orphans, and he started paying them with play money instead of real money – of course, they would have to be like, blind and in wheelchairs and stuff, but it could work.

But, of course, word would get out about this eventually. And the community would have to do something about these children. Now, they could try to just approach these little children, and let them know what was happening to them; or they could also just run an awareness campaign, wherever most of C&D industry's market was, and thus effectively produce a major boycott.

But let's just pretend like these options weren't exactly available, for whatever reason. Maybe Lloyd had established his very own orphanage so that he could just isolate all of these children; and maybe it wasn't exactly a very legal or simple process to just get in there and warn them.

Well, if these poor little children were actually being kept like this, then this would be analogous to human trafficking... and the community would be well within their rights to just go in there and free them. A carrot-based society is not a pacifist one. They just simply reserve these tactics as a last resort, defensive measure.

And that about does it for this example. There is no degree of dumbness that this system cannot handle. Any amount of tomfoolery just results in a corresponding degree of disassociation. And that has *al-*

ways been more than enough to maintain order and civility.

And what's more is this, it also doesn't matter if these producers are selling their products *locally* or not - the community can *still* keep them honest, regardless.

These companies would still be required to make their ten percent contributions to their local communities. Then those funds would be divided up proportionally amongst the producers who actually *are* selling to the local consumers. And then those producers, who are selling to the *local* consumers, would spend those funds, along with their *own* ten percent contributions. So, the consumers in the *local* community can *still* ensure that these funds are being spent well, regardless of the fact that they're coming from a producer who isn't even selling to them.

Okey dokey. Well, that's all well and good if you've only got like one or two little companies acting shady, but what exactly is a poor mother to do... if there's an *entire* black market out there?

Well... you do nothing. It doesn't matter. A black market is irrelevant. It wouldn't last. Almost no one wants a black market product if they can just as easily get it at a normal place. And I'll tell you why. Black market products have no accountability.

If you had a whole entire network of these underground companies like this, and they just couldn't care less about their reputation ratings, then there would be *no guarantee,* from any third-party agency, that their products were ever safe or reliable. And if you ever bought a product that was just straight up defective, or even dangerous for that matter, then there would be no official form of consumer protection.

And there's just one more little reason for why a market such as this would never prosper - no one wants a product that's inherently sketchy. If a producer has the skills to create something of quality, then they take pride in that, and they want it recognized. The absolute *LAST* thing that they want for their products is to have the stigma of the black market attached to them. So, of course they're gonna run their business responsibly. And everyone really wants these better products, so the market will inevitably gravitate towards these higher quality producers.

There was a really big scare back in the day that these black market industries were gonna destroy everything. When I was just a wee little child laddie, you could still get these things called cassette tapes. And what would happen was, whenever there was a new album that was coming out, there was this one guy over in Cincinnati Ohio that would just go down and buy it, and then make like fifty copies and pass 'm along. And then all of *those* people would make like fifty more copies and repeat the process, and before you knew it, we would ALL have the new album for almost nothing. And then no recording artists were ever able to make any money ever again; and then the entire music industry collapsed; and now there's no such thing as music anymore; and now no one even knows what the word "music" means.

Or at least that's what they said was gonna happen. They said the sky was gonna fall and the music industry was gonna collapse and our society was gonna descend into chaos. They *swore* we would be fighting in the streets over the last remaining cans of pickled pig's feet.

Sooooooooo, how did we survive all of this? How are we even alive to talk about it?

Well, I don't know about the other survivors, but I myself was hiding under piles of corpses. That's why the marauders never found me.

But then I came out of hiding a couple of years later, and I found out that the apocalypse had never even happened. Yeah. *It never happened.* There was never gonna be some giant apocalypse. The entertainment industry was never in any trouble. Civilization was never in trouble. Everything they told us was a bunch of *bullsh***.

So, why is that? Why didn't everything we hold dear become a smoldering toxic wasteland from people using cassette tape recorders?

Well, it turns out, people don't wanna buy sh** products. They don't want some cheap, ugly, embarrassing, low-quality, pirated, mix tape. They want the real deal. They want that factory fresh sound with the original album covers. They want beauty; they want texture; they want quality. They want that fancy new album smell when they open it up.

And here's what's so incredible about a truly open market - when a market is legitimately free and competitively open like this, then everything becomes a *race to the top*. Talented entrepreneurs strive for excellence; then the market seeks out that excellence; then this forces their entire industry to compete at that level.

If even just one single person started raising their standard of quality like this, and then they offered a better product which also included things like a warranty and such, and then they sponsored a bunch of projects throughout their local community, and if somehow no one else was already doing this stuff before them, for whatever reason, well, then the entire consumer population would come rushing to their doorstep. And then every single, major league competitor would have to rise up to that level.

This is why there would never be a black market. ... at least not for any type of "normal" products.

And this is why every single, major league competitor would always strive so hard for accountability. It's not just about avoiding a bad reputation. It's also about that race for higher quality.

This race up to the top is unavoidable, by the way. As long as property rights are protected, and as long as a healthy and open market is unmolested by power, then this attraction to an elevated standard is just like gravity – it is unavoidable. Thus, if a more socially responsible business *can* exist, then it *will* exist, and everyone will want it, and most assuredly, it *will* become the new standard. ... So, it's really not a question of *if* this will happen... ... beeeeeeeeecause'it'will.

In fact, there's actually nothing more addictive to us than this sort of positive social recognition. That's why producers become so addicted to it whenever they start sponsoring their communities. ... As soon as they get that attention, from the fame of their patronage, the elation just goes straight to their head... and becomes totally irresistible.

You could see this effect in the Nauptsëh whenever they would have their big blowout parties. Their leaders would get so ecstatic, at the prospect of being admired by their entire community, that they would just give away almost everything. Sometimes they would even give away their actual clothing.

"Screw it! ... Who wants this kick ass shirt I'm wearing?!"

"I said, who wants this KICK ASS SHIRT?!?! YeaaaahhhhhHHHH!!!!!!!!"

"Now, who wants THESE PANTS?!?!?!"

It was cool though – even though they did gave away nearly everything, they would build it all back within several months or so. They were actually never more wealthy than after these parties. That's because wealth is also stored in people's love for you.

So, once again, this system has no incentives for unscrupulous markets, because nobody wants their reputation damaged; and because this system also drives up both quality and integrity; and if all *that* wasn't enough, it also turns out that we're addicted to it.

Soooooooooooooooooo... no... There will be no black markets. There will be no mass chaos. There will be no environmental degradation. Entirely the opposite in fact. There will only be the owners of bakeries twirling their shirts around their heads like a helicopter.

Alright. So, if you just take the seed of abandoning kingly power, and then you immerse it within the soil of protecting people's property rights, well then what grows from that will far exceed all former achievements.

And that brings us to our third and final question. How exactly are their leaders supposed to "run" things? And how in general should they handle disagreements?

Well, no one actually runs a Nauptsèh society. The leadership themselves only dictate within their own individual companies. Rather, the conventions and the rules of their society are all established organically. It is up to the people to decide what they want for themselves - and the leaders are just there to fulfill their wishes.

The people can decide things however they see fit. They can let their common sense determine policy, and they can vote on other matters of local preference. They can construct a constitution to reflect their values. And they can enforce these rules upon themselves however.

So, the big difference with this Nauptsèh system... is that if any population, for whatever reason, actually wanted to enforce some extremely draconian rules upon themselves, then they could impose this will upon themselves however. But this yoke would be a *self-inflicted* one - not one of an oppressive tyranny.

So, technically, this wouldn't actually be a departure from the traditional Nauptsèh system. Yeah, you might have a community where the police have been forceful - like, maybe they've been forced to just go around and constantly arrest people, who have their ankles showing, *ooh la la*, but this would be your *own population* just imposing this will upon themselves - not an ilk of super powerful "representatives" ... with their *own* agendas.

So, in other words, it's basically like this community is just indulging in its own private naughty time, and they're just telling their local police force to keep on spanking them. So, it's like, yeah, technically you're being assaulted by a person in a police uniform, but they're also just doing what you told them to do, so they're inherently going to be very very *careful* about it. ... ... That's the difference.

And, as you yourself well know by now, you can always have them stop at any moment by just using the safe word. As you yourself know. From experience. In other words, the community itself could always change its mind about these policies. They don't need some violent revolution or a political blood bath. They don't need some endless, impossible battle with a bunch of bureaucratic agendas. They just let their leaders know through some consensus, and then their leaders change these rules like yesterday immediately.

And of course, these leaders just exist to serve their people. They're not there to be setting their own agendas or ever dictating policy.

They are almost entirely just a bunch of wise old Alfreds, who all are experts at catering to you. And if they ever have their own unique ideas about stuff, then they can always just explain them to try and convince you.

But, as is always the case with any leadership, there will almost invariably be a realm of foggy grey area. The Alfreds can't just sit there and ask you like million little questions about all your different preferences on everything. So, they have to do their best to try and guess sometimes. They have to do their best to try and interpolate.

And so, this is where the process gets a wee bit nebulous. The Alfreds have to discuss things amongst one another. They have to consider an array of proposals brought forth by their project managers. They have to determine what the best way to impress their community is. And if they cannot, for whatever reason, just agree on an optimal course of action, then they can always get it straight from the people themselves by invoking the balanced voting method.

That last point right there is incredibly important. The leaders don't *ever* want to upset the people, because the market could always take it out on them with a few targeted boycotts. And since every single decision that they make will forever be recorded in a publicly available archive, the people can always just look this stuff up, and see exactly who it was who was the most responsible for something, and to what degree. So, they would definitely wanna keep the people happy.

Of course, this process will never be perfect, since you can never quite please every single person simultaneously. But they can always just employ things, like customer polling or the balanced voting method, to try and remain as accurate as possible. And then this way they can optimize everyone's satisfaction.

And, of course, there is also no room for any shenanigans within this paradigm. There is just simply no way to game this system. It's too accountable, and it's too transparent, and it's too adjudicative. It's too responsive, and it's too accommodating, and there is just simply is no tax trough to be exploited here. Therefore, this system can *only ever serve the people.*

The end.

Alright. So, there you have it. That is the North American Upper Pacific Coast Civilization system of social organization within a contextually extrapolated nutshell.

... so that Sally can sell her sea shells down by the sea shore.

The Nauptsèh sold all kinds of stuff down by the sea shore. Nearly all of their dietary protein came directly from aquatic sources. And THAT is the entire point of this chapter. *That...* is the entire point of even mentioning this. The Nauptsèh were not just *amongst* the most impressive of all global civilizations – they *were* the most impressive.

Chief amongst them, you might say.

The Minoans, The Phoenicians, and The Nauptsèh.

While other civilizations came and went, and while other earthly empires rose and crumbled, the Nauptsèh simply rose, and then just continued. For thousands upon thousands of years... they just simply continued.

No wonder they liked to party so much.

Just about everything they did was truly brilliant. Their diet was nearly perfect. Their societies were almost flawless. And literally everything they did was completely *sustainable*.

They never did slash and burn agriculture. They just managed their regional forests to enhance their natural food production. They also did this synergistic enhancement with the underwater kelp forests. And they also shuttered their seasonal harvest of critical salmon extraction, to be protective of, what we would call, their genetic diversity.

They also had these incredibly beautiful, little, tidal pools called clam gardens. They were just these pools that were filled with clams and other shellfish. ... When the tide was at its peak, it would just spill over into them and replenish them with nutrients. And so, by doing it this way, it would protect them from their natural predators, like starfish.

In the same way that they never tried to force *one another* to do stuff, they *also* never tried to force their habitat. They just found a clever way to share its rhythms. And their management actually made their environment *healthier*.

Talk about a win-win situation.

And they also had these different types of dog breeds. One of them in particular was their so called "wooly dogs." They were so called because of their super woolliness. They were basically like a slightly larger version of a Pomeranian.

Often times you'll see a Pomeranian with their little backside britches trimmed, and you can see from their exposed little undercoat just how extremely soft and wooly it is. Every time I have to trim a Pomeranian's little britches like that and I see that fluffy woolliness that's coming off of'm, I always think to myself, "someone should be making like little socks out of this stuff." And apparently a bunch of the Nauptsèh were also thinking that exact same thing. They would trim their wooly undercoats, when the weather was getting warmer, and then they would use that fluffy wooliness to make some really warm fabrics.

Genius!!

Yep, those people were definitely some dog lovers.

Sometimes I'll be out walking my dogs somewhere and then a family with a bunch of little kids will suddenly notice 'm, and then they'll just start burying their little fingers into their fluffiness. And you can tell that every party involved is just absolutely loving it. ... Well, that's basically what was going on in a lot of these Nauptsèh households.

It was sort of like how the ancient Romans were all crazy about their Pomeranians. They would have them carved into stone effigies like a family portrait. And they would write these really touching little poems about their personalities. ... Well, it was basically the same exact thing with a lot of these Nauptsèh families.

Yes indeed, they really did love their little doggy critters. And I think it might've had something to do with the fact that they themselves were actually werewolves. ... Yep. Some of them were apparently werewolves it turns out - one community in particular.

Their most famous Nauptsèh werewolf was a young laddie named Jacob. Young Jacob became the first ever human being, that we know of, to barter a truce between werewolves and vampire teenagers. And he actually did it so that he could keep on being in this really unusual love triangle, between him and his wheelchair dad's human friend's teenage daughter... and that girl's one-hundred-and-four-year-old man, teenage man boyfriend vampire.

It was pretty sexy.

And the whole entire thing was caught on camera. Yeah, I know. How crazy is that? It's like a five part documentary.

And one time, some of these Pomeranian loving Roman vampires tried to come *all the way out there* to North America and start some shit. ... And it was like ... *whaaaaaaaaaaaaaaaat?!*

It was nuts. It was a faceoff between TWO groups of people that BOTH love *Pomeranians!!!* D8

And they *didn't even KNOW* it!!!!

Holy... ... ... mother...

Talk about the Twilight zone.

Alright.

Hashtag'Team'Ashley.

Okay...

Moving on.

Hashtag'Team'Jacob'for'Civilization.

Okay.

Alright.

Moving on.

Hashtag'team'all'the'chicks.

ALRIGHT!

For real this time.

Okay, so... Clearly the Nauptsèh were some pretty sharp folks up there. They invented things like intellectual property rights, even way back before these other empires like Rome were even a thing.

They would announce that someone in their regional area held the intellectual property rights to some original piece of artwork, or some invention or something, and then that would be disseminated throughout the entire community, and then everyone would just have an awareness that so-and-so owned such-and-such.

Wow... talk about brain power. They would just remember all of this stuff about people's property rights, and their reputations, and then all of these different inventions that they were creating out of cedar trees, and then all of these little details about sustainably cultivating food throughout their environment... and just, like, a billion little details about everything. They were some serious freakin' brainiacs.

And that is what we as a Homo chinnichin were built for. We were built to keep track of all these little details. We were built to eat a coastal, blue zone diet. We were built to live in a society which is based upon reputation. The Nauptsèh were simply living almost PRE-CISELY how a human being was evolved to.

They even had to battle with our age-old nemesis - overpopulation. And this would unfortunately cause a little bit of warfare from time to time. And some people might try to even criticize them for this, but if we take a closer look at their situation, we can see how this overpopulation problem was just logistically unavoidable, given their circumstances.

The Nauptsèh were in a tight spot. ... literally. They were actually sandwiched between the mountains and the ocean. And so, they occupied a relatively narrow stretch of coastline, from Alaska down to California.

And they couldn't exactly limit their population size either. I mean, you can *try*, but without our modern birth control, your success rate is never going to be perfect. So, for that and other reasons, it just was just simply not a viable option.

Trying to limit your population without the advent of modern amenities is basically like trying to balance an egg on its end. It's like, yeah, you can do it, but it's not a very stable position. You're balancing on the peak of a mountain, not the trough of a valley. Just the slightest little breeze can knock you over.

First of all, it would take an absolutely obscene amount of discipline for a preindustrial civilization to just artificially restrain their natural birthrate. And if just a few too many people were accidentally not doing it right, then they might get sick of being denied their fair share of carrot rewards, just for having an extra kid because of an accident. Then they would break off and form a whole new separate community. Then this breakaway community would just completely disregard this whole, not-having-too-many-kids rule, and then the whole entire venture would just start to unravel.

But let's just give'm the benefit of a doubt though. Let's just say... that they were *somehow* able to consistently do this. ... Alright. Well... then now what? How do they keep their neighbors from trying to attack them?

Well, the most obvious way is to go all Sparta. You just create a professional military that no one wants to mess with. That ought'a do it.

Oh. Wait. No. That would not ought'a do it. A military like that would be extremely expensive. And this would cause their population to feel a sense of resentment towards their neighbors. And then the neighbors could just do the same thing and make their own military. And then the wars that they *would* inevitably have would just be that much more insane.

And I do mean "expensive." The Spartans were *only* able to do that, because they had like ten helot slaves for every Spartan. So, that's insanely expensive.

Alright, well, that's a nonstarter. ... Okay... Well... then why don't they just trade with their neighbors and create an economic incentive for peace? Would that work?

Well, if their neighbors actually had something that they themselves did not, like some sort of scarce or limited commodity, then, yeah, that could work. The Nauptsèh really respected a person's property rights, so it was very against their nature to just invade someone and take their stuff. So, yeah, that could work. ... if their neighbors actually had a limited commodity like that. But alas, they did not. All of their different regions had pretty much the same exact selection of resources.

Alright, well. Sparta and trade negotiations is out. What else could you do?

Well, you could invoke the ancient practice of professional hostages. Which is to say, you could exchange a bunch of teenagers between the regions. And then the teenagers could essentially act like an insurance policy against invasion. Soooooooo, that might work. But that's an awful lot of teenagers.

You couldn't just have, like, a single lone teenager from every single Alfred family get exchanged like this. You would need to have way'way'way WAAAY more than that. ... because if you didn't, then some random, freakin', crazy a**, Caesar type, who didn't even care about these Alfred families, could just rise up and attack the neighbors anyway. So, you would need a lot, a lot, a lot, a lot of teenagers.

And then this would only work if most of these regions were about the same size. Your neighbors would have to be like, roughly about the same, demographic size as you. ... But, alas, this condition did not exist for them. The size of their different cultural regions was super divergent - it was enormously disproportion in many cases. So, there was just *no way* that a smaller-to-medium size region, could just absorb an absolutely massive amount of teenagers from a larger one. And since their world was essentially just two-dimensional – just situated along a really long stretch of linear coastline - the idea of actually forming an alliance, between a group of many, smaller, little, separate regions, was just simply not a practical solution.

So, here again, this strategy fails. And that would be a really strange and awkward way to live for them. They would have to just embody this *enormous* amount of self-discipline... so that they could keep their own population from overproducing... while simultaneously taking care of all their neighbor's teenagers. It would certainly go against their natural instincts, that's for sure.

And then what if all these teenagers just decided to rebel or something? What then? ... ... Well... it would be a shit show. That's what then.

Or what if these teenagers just decided to assimilate instead? ... ... Well... then I guess they wouldn't be hostages anymore, would they?

This strategy is not a super viable one. ... at least not for the Nauptsėh civilization.

So, Sparta won't work; trade won't work; and the use of professional hostages won't work either. ... Alright. Well... then what else could they do? Could they figure out a way to increase their carrying capacity?

Well, yeah, they *could*, but that's only gonna make things a whole lot worse in the long run. You're still gonna hit some hard, new limit at some point, ... and then the carnage is just gonna be that much more extensive. So, finding new ways to increase your carrying capacity is just simply not a viable solution.

Okay, well, then what if they just expanded, and started moving into another geographical region? ... Would THAT work?!

Weeeeeeeell... no. ... It would not. ... There actually WAS no other geographical region. The continent was full up. ... - no vacancy. There were people almost literally everywhere. There were people down living on the sides of cliff faces, all the way down in the crazy hot deserts of the American outback. And there were people up living in these tiny little snow houses, all the way up in the freezing cold expanses of the arctic circle. And these people were not just living in these places because they *wanted* to - they were living in these places because they *had* to. The continent was occupied. There was essentially nowhere for them to go that wasn't already someone else territory.

The Americas were not just some vast and open wilderness that was waiting to be settled. There were hunter-gatherer groups living everywhere. And they had already been living at the hunter-gatherer carrying capacity for a very large number of generations. ... And they were also some pretty sketched out and volatile individuals, quite frankly, from their obvious lack of necessary coastal nutrition. So, they didn't exactly tend *to take very kindly to it* when a bunch of random strangers started showing up in their territory.

So... ... you couldn't just simply expel your extra people into the continent like that. All you would be successfully doing is simply exporting your internal crisis elsewhere.

... Although, they might've actually tried this strategy already, a really long time ago. There is an ancient linguistic connection between the Dinéh Navajo and the Nauptsèh. So, who knows. Maybe they did. But either way, this quickly began to start drying up, as a viable option, sometime around the early Neolithic.

Speaking of which... if you ever wanna party with some people with a wicked sense of humor, the Dinéh Navajo are just as bad as the frikkin' Irish. ... They have this, like, extra, natural talent for just being crazy freakin' funny when they wanna be. ...

They even have it in their ancient spirituality that there should always be comedy in the world. ... For real. Whenever one of their little babies finally *laughs* for the first time, they always throw'm a really festive, sort of first-laughter baby shower, ... because that laughter is very sacred. ... ...

Now, THAT is how you make a religion, my friend. ... ... - a whole entire *planet* of other religions out there, and only *one of'm* got it right apparently.

Well, actually... on second thought... they also have swastikas in their religion... so, it's like, uuuuuuuuuh... why the hell do they have sWASTIKAS in their religion??? I thought they were on the side of the United States. ... I thought it was *them* who won the war with the Navajo Code Talkers. ... And now these old Navajos are just walking around with a bunch of sWASTIKAS on?!?!

These old grandpa Navajos don't give a fffU##, *son!*

Well... ... either that, or it's an ancient symbol in their culture. ... ... I'm still not sure which one yet.

I wanted to join their comedy religion once, and get like, a really big tattoo on my chest, to prove my conviction to it. ... but then they showed me what the symbol was, and I was like, "uuuuuuuuuuuuuuh... ... ... wait... ... ... Is there, like... ... any *other* symbol I could use??"

And they were like, "no, it's gotta be this one." ... ... So, I was like, "uggggggghhh... ... fINE... ... then just give me, like... ... a little tiny smaller version on my forehead then."

I get so many dirty looks these days.

But I'm like, "no, no, no, you don't understand... this is an... ancient... Navajo..."

Alright, alright, not really. But that really is their ancient symbol though. So, don't blame me for it. It's them you wanna look at.

It's these freakin' grandpas, man. They're the ones who started it. They're the ones always talkin' about their prophesies. ... They say that their ancestors came up through three distinct realms - the black realm, the *blue* realm, and the yellow fire realm - and in time, they will eventually ascend into the "white" realm... and it will be the "best" realm. ... - the final solution to all of their problems. ... - their final white solution and superior fourth realm prophesy.

Well, I may be a little rusty on my World War II German, but I seem to recall another guy who was a pretty big fan of some *final "fourth realm" prophesy*. ... ... And then he wrote about it in a book called his "struggles" or something. ... ...

Well, guess who else is always whINING about their "struggles."

"waaaa, waa, waaaa... I don't have any electricity or running water... waaaAaaAaaa..." **D,x**

"The government took my land and tried to genocide me... wyaAAaAaaAaAA..."

D,x

Their """"struggles.""" ... ... *PFFFFF!!!*

And they're always yammering about their *CLAN* loyalty all the time.

It's terrifying!

It is *ab*solutely terrifying.

We better keep an eye on these old Navajos.

Anyway. I didn't mean to frighten you guys with that. ... ... I apologize. ... I sincerely apologize. ... Let's just... let's just try to not think about it, okay? ... ... Alright... ... Alright, as far as we can tell from the anthropology though, it appears that many of these Dineh Navajo were once a more northerly people, along the central Alaskan coastline. ... but then they just moved, because of overpopulation pressures.

And so, the point of all that is just this, you can't just simply stop your overpopulation problems. You can't just go all Sparta to prevent future conflicts. ... You can't just simply trade... when there's nothing to actually trade with. And you can't just expect to absorb a bunch of teenagers, or simply move your extra people into the continent.

Alrighty... Well... I guess we're screwed then.

I guess we're gonna have to do what our ancestors have had to do... for the last...

TEN...

MILLION...

YEEEEEAAAAAARS...

That's right. It's gonna be a neighbor fight.

All it takes is just one single, desperate or dodgy neighbor to start a blood feud. And we, as a very social and intelligent species, have always had an instinct to take revenge on people. So, if someone just starts some sh** and assaults your people, then every single fiber of our chimp brain wants revenge on them. All it takes is just that *one single neighbor* that's been pushed to the point of desperation.

Like, for example, one group just lived on an island. And it was small. So, they always felt a little bit precarious about their position in the world. And so, they tried to overcompensate for it by being aggressive.

And then another belligerent group was precisely the opposite. Their empire was massive. So, sometimes they just attacked their neighbors, essentially because it was easy, basically. But what I also find really fascinating about this, though, is how they did it, because they were never really attempting to actually conquer anyone. Rather, they were just responding to their own, internal, population pressures. They weren't, like, trying to just absorb the entire planet. ... like every other empire.

It's just fascinating, though, to see how such a peace-loving, party-throwing, property rights respecting, non-authoritarian civilization had to deal with this issue. Because, quite frankly, it sucks. ... It's really easy for us, honestly, to just sit here, on our comfortable modern high horses, and then just judge some other ancient historical period, using our own enormous advantage of all this hindsight, but it really is *quite* a different thing to actually be there. ... and to be faced with those difficult decisions... right there in the heat of the moment.

But, once again, trying to actually stabilize the birth rate is like trying to keep an egg up on its pointy end. Not everyone will be able do it. Accidents will always happen. And then what? There would have to be some sort of penalty for it. And this would make your entire civilization extremely stressful.

And then you would have to hope that none of your neighboring civilizations ever attacked you, because *then* you wouldn't have the people to fight them off with. ... - your neighbors will *still* have lots and lots of people if they're not actually following this stabilization plan. So, for this and other reasons, throughout all of ancient history, the idea of actually *trying* to reduce your overall birthrate was just always seen as tantamount to suicide.

So, yet again, this problem of overpopulation was just inevitable. They were going to have to confront it, one way or another.

So, how did they?

Well, they basically created a mixture of the Spartan style helot conscription, combined with the tactic of keeping their neighbors as professional hostages.

Was this a fantastic thing to do for them? ... ... No. ... No, it wasn't. Would this be considered defensible by today's standards? ... ... No. ... No it would not. But hopefully I've been successful by now, in making the point... that they weren't exactly spoiled for many options.

Okay. So. What did it look like then?

Well, once you've lost some loved ones that you really care about, and you've lost them in a battle with your neighbors, then it kind of changes the way you fundamentally see them. It makes it really really hard to perceive them as human. To you they start to look more like a monster.

So, when they would defeat another region in a battle, they would select a bunch of people to keep as hostages. Then they would take them all back to their own communities to come and live amongst their people, like a regular citizen almost. ... They would live in the *same* houses; eat the *same* food; dine at the *same* tables; and even do the *same* work. And they even got paid for it.

So, not only were these people a potential insurance policy, they were also taking on some of the workload so that some of your men could have enough time to train for battle. It was basically like a way of ensuring that your neighbors were the ones paying for your military expenses.

And they weren't like some massive, overwhelming, labor force, by the way - just like, holding up the entire economy and making people rich and stuff. Nope. No way. Not even close. ... They were only just a minor, little augmentation to essentially compensate for those military expenditures. That's it.

And they weren't out there just working for free either. They weren't slaves. At least, not in the truest sense of the word. You'll see that a lot in some of the academic literature - they'll keep referring to these people as "slaves." But I find this mildly dishonest and intellectually lazy. I've never heard of a slave that eats, sleeps, lives, works, dines, and basically does everything right along side their host community. ... or one that is always at liberty to save up and purchase their citizenship.

That's right. If they wanted to save up their money and purchase their citizenship, then that was always a right that they were afforded. And sometimes their hosts would even just grant it to'em anyway. ... So, all things considered, I think that the terms "conscripts" or "indentured servants" would be much more appropriate.

And that about does it for this whole, overpopulation discussion. I just had to share that little history there, in the spirit of full disclosure. And I definitely had to make that little point right there about the hostages - they were never this massive portion of the overall labor force. And they were never this driving engine of the whole entire economy - not like in other civilizations. Not even close. Every able-bodied citizen in the Nauptsėh civilization was entirely responsible for their own advancement.

Oh yeah. And for anyone who doesn't really know about this sort of darker side of human history yet, this type of *super accommodating* indentured servitude was virtually *unheard of* in the ancient world. Nowhere *on the planet* were there *ever* involuntary workers being so *ridiculously* well accommodated like this – just like any other ordinary citizen. ... I mean, by modern standards, involuntary servitude doesn't exactly pass the muster. Obviously. But by *ancient* standards, this was like some *absurdly* humanitarian treatment.

The ancient world was a shit show. And so is the modern world. But it was even worse back then. Most of the world's empires ran on slavery - brutal, dehumanizing, ruthless slavery. And there was SOOO much more than the average person even knows about. So so SOOO much more. It was freakin' *everywhere*. ... for *thousands* of years.

Everyone alive is descended from a group of ancient slaves at some point - multiple times over. Our ancestors have all been conquered and subjugated, over and over and over. ... all the way back to the Paleolithic.

I mean, take mine for example. The Irish were being rounded up *for centuries* and sold into slavery. The whole entire reason that the city of Dublin was originally founded was to function as a slave port.

I always wanted to go back there and be like, "hey! What the hell?! Why are you trying to enslave us?! We're *white people*! ... just like you! You can't *enslave us*!"

"... Other people... maybe. But *not us*!"

They've been f***ing with us forever though. I mean, the whole term, "beyond the pale," just originally meant, as insane as an Irish person - beyond the lands of "civilized" control. ... ... And then the Irish have been, *on more than one occasion*, said to be some sort of "in between breed... somewhere between humans and primates." ... ... And it's even been said that, "the Irish women have been given special dispensation from the pope to wear the thick ends of their legs on the bottom."

'the fuck?!

'fuck's that supposed to mean?!?!?!

Fuck it, that's a sweet burn. ... I'll give you that one... ... ... you medieval jackasses.

But that's the kinda shit we've been having to put up with for *centuries* now. And they've just been talking all this shit and enslaving us, over and over and over. ... even as recently as the American colonial period. A lot of us were rounded and then forcibly sold into indentured servitude. And not like fancy, professional hostage, indentured servitude. We're talkin' sloppy, *brutal*, American colonial indentured servitude.

But if you look it up and you ever start to read about it, then you'll notice that they will be *veeeeeeeery very* careful to draw a *veeeery* clear distinction about what happened during this era. They'll be like, "yeah, this is what happened to the Irish, but there's no way that you can compare it to the institution of full-on, irrevocable, chattel slavery, and blah blah blah blah blah."

And that is EXACTLY correct. There is noooo way... that you can equivocate... the general social practice of indentured servitude... to the brutal institution of full-fledged, chattel slavery. There is just *no* way. ... It's apples and oranges to the nth degree.

So, they need to quit referring to the Pacific Northwest situation as "slavery." ... because it wasn't. It wasn't anything like that. Vaguely reminiscent? Maybe. But the same thing? No. No way. Not even close. And, quite frankly, it really causes people to get distracted... away from all of the other accomplishments of this *mind*-blowing civilization.

And besides, whenever we use the wrong word for something, it just destroys what the original word means. It doesn't make the *ordinary thing* seem like it's the exaggeration - it makes the *exaggeration* seem like it's the ordinary thing.

Like, if suddenly the name "Goddess" was just getting really, super popular for some reason, and then like, there was just all of these women out there with the name Goddess all the sudden, and like, all of them were just these, like, really really ordinary women, and then like, a few of'm were even just these short little barrel chested women... then it wouldn't, like, *elevate* all of these women *up* to the level of a goddess. It just pulls the word "*goddess*" down to the level of these women. And then when people would hear the word "goddess," and they would be like, "yeah, sure... I've seen your '*goddess*'." ... Then they would make that one jacking-off hand motion with the sound effect.

Pl'i't'i't'i't'i't!

Words mean things!

Okay?!

And words have consequences!

Can a barrel chested woman still be hot?! ... Yeah, sure, why not. ...

But is she the most universally desirable tier of womanhood... in all the whole wide world??? ... Uuuuuuuuuhhh... maybe for Neanderthals. Sure. But I mean, c'mon, we can't ALL be the top tier. That's why it's *a tier!* ...

We can't just go around using the wrong words for stuff! And we can't just go around pretending like there's a meaning behind something... when there's *not.*

We can't just keep watching professional sports and acting like there's anything interesting about it!!

... bunch of grown ass adults... sitting around watching a child's game... acting like there's something important going on. ... like anything that ever happens in a sports game would ever matter.

"Oh my goodness!!! He kicked a ball in a hole!!! It's so amazing!!!"

Fuck outta here.

Oh, don't worry, I know you're one of 'm. I know you're a sports fan too. I know how common this shit is. It doesn't matter though. It's still a joke.

And everyone just sits around at work all day talking about it constantly. "Oh yeah! Did you see what kinda season he was having?! Did you see how he did that thing that every child does?!"

"Did you see?! :D "

All serious about it. ... like someday you're gonna get a lifetime achievement award. And like one of the team members is gonna come out and give it to you in person. ... at this big banquet ceremony. And when you're shaking his hand and getting your picture taken together, he's gonna look over at you, and just for a moment, your eyes are gonna come together, and he's gonna give you that look, like, "thank you. Thank you for having my back out there. Thank you for always standing by my side... ... in spirit... while you were sitting at home eating stuff."

What a joke.

Professional sports is a joke.

And so are you.

You're a waste of a human being.

How do you live with yourself?

How do you look yourself in the mirror every day?

How do you...

Alright, alright, alright... ... Let's just relax. Let's just relax for a moment. ... and take ten deep breaths... and have a sports time out. We can fix you. It's not hopeless.

I've been watching too many sports games lately, trying to understand it, and now I'm having trouble controlling my emotions. I was hoping if I kept yelling at my TV... for four hours a day, I would eventually start to see the point in it.

But now I just feel angry and confused.

Maybe I wasn't meant to be a sports fan. Maybe I was just meant to be a ref.

Yeah! Maybe some of us are just refs. Maybe I should go down to the bar pub and start cheering for the ref.

I'll just dress like a ref and start yelling at people.

Blowing my whistle.

Trying to eject people from the bar.

"YOU'RE OUTTA HERE!!! YOU'RE OUTTA HERE!!!"

"That was WAY OUTTA bounds!!!!"

I would make a good ref.

I would make some call on the field, and then everyone would get all mad about it, and I'd just be like, "who caaaaares?!?! ... It's a chILD'S game!!" Pl'i't'i't'i't'i't!

We need to fix this though. We can't just keep going on this way. ... just staring at these screens, acting like there's anything more substantive than paint drying going on.

We need to fix this. ... once and for all. We need a new sport. We need one that's actually fun to watch.

And it'll have like every single sport combined into one. ... That way you can get all your sports watching done in a single sitting. We'll call it like "every-ball." ... or "omni-ball." Yeah, that one sounds good.

Screw it. Let's call it "allyball." ... just to make it confusing. And let's make it BADASS! Let's have like chicks in bikinis. ... and I guess, like, dudes in little thongs. ... if that's your thing. But they'll be on opposite sides of the arena. ... so you don't accidentally look at the thing you don't wanna look at. ... unless the jumbotron flashes by it really quickly... and you have to recoil your face all dramatically.

I'm gonna be up there in the jumbotron, just flashing it on purpose.

I'll wait 'til some like really important play is about to happen, and then just flash it.

"AAAAAAAHHHH!!! ... dAMNIT!!!"

Alright. So, we're gonna have like every sport combined into one. And it'll be like, it starts with baseball. And you'll have a baseball bat, and you'll be standing at a batter's plate. Then they'll pitch a ball to you. And instead of just a baseball, it'll be like four different types of sports balls. And one'll be like a rugby ball that's not even shaped right.

Then you'll just hafta hit as many as you can and take off running. But it won't be running. You'll just be ice skating. And the whole entire arena will be an ice rink.

And you'll just be skating and dribbling a basketball as fast as you can. And you can try to shoot the ball, and make it into the net behind the base, but if you lose the ball for any reason you hafta start hopscotching the rest way. And the pitcher just gets to hit golf balls at you the whole time. ... even while you're batting and hopscotching. So, there's like multiple pitchers. And the pitcher's mound is all huge. It's like the pitcher's island. And there's a moat around it. ... so you can't just get all angry and rush'm. Or you'll have to swim across the moat. But they can defend it like a castle.

... - trying to swim in your hockey gear.

... in ice water.

It's alright. We're adapted for it. They'll be fine. ... And once you get to each base, there'll be like, a warm blanket and cocoa waiting for you. And once they nurse you back to health, you'll have to play ping-pong to get past first base. ... And then second base is foosball. ... And third base is just a roulette table. ... You have to gamble your way past third base.

Alright, so... the pitchers can try to get you out the old fashion way - trying to get three strikes while you're batting – but *now*, they can *also* get you out the new way - trying to get a bowling strike while you're hopscotching.

There'll be like four different bowling allies forking out - one for each base. ... And each one'll have an arcade in the back for some reason... with a buncha games in it that no one would ever play, unless they had a fight with their family and they're back there just crying by themselves.

But then I guess this like nullifies the whole point of even batting in the first place. ... - those balls don't even enter into the equation anymore. ... ... It's just, like, an act of aggression to hit the pitcher.

... – see if you can injure one of 'm before you take off skating.

And as soon as you take off skating, the catcher just switches sides. Normally he's on *their* team – trying to help strike you out - but then like, as soon as you take off skating, he just suddenly turns against 'm... and starts hitting golf balls at'm. ... trying to lay down some cover fire for you. ... And there'll be like a ref there, monitoring his emotions... ... making sure he has authentically switched sides in his heart.

31

19 ¾ 3 -- BARKOLOGICAL CHILDREN

I was in the military once. But I had to get out. They put this Down syndrome sniper on our unit and he kept shooting our own people.

It was heartbreaking.

He even shot our commanding officer.

And we weren't even in the field.

He was just giving us speech about morale or something.

I saw that and I was like, "*screw this*, I'm outta here!"

They were trying to enhance our diversity. ... trying to like, give more people a shot at stuff. ... Well... now you see what happens. Diversity is bullshit.

There should be *no* diversity. Everyone should be identical. And all the identical looking males should be exactly the same height. We should be like China.

Alright, alright, alright, that was a low blow. That was a low blow. I didn't mean that. I haven't eaten in like twelve hours. So, I'm a little... you know. I overslept by four hours and I'm starving.

You know what's funny though, if you sit and watch a camera feed of just a bunch of people walking through a Chinese airport for an hour, you see that's it's all just the same exact stereotypes of all the same exact people you already know from your own life.

You've got big ol' heavy set Todd, from your work, who's always trying to reel you into conversations about nothing... so that he can show you how much smarter than you he is *not*. You've got the punk rockers and chach'skies. You've got all the different types of dad archetypes, just trying to get by. ... just trying to make it in a world.

It's just funny to me. We always manifest the same exact archetypes no matter where we are. ... as long as we're in like a normalish middle class lifestyle.

I knew this guy named Chaun Wei once. When I met him I was like "whaaaaaaaaat?! Your name is Chohn Wayne?!" ... "That's awesome."

And he just looked at me completely blankly.

I was like, "like John Wayne... the actor..." Still all excited. ... and smiling. :D

723   -   BARTROLOMEW MC INNCEL

Then he was like "Chown??? Weeeeeiiiiiiiiii???" All perplexed.

I was like "yeah, the famous cowboy actor."

"CowwwwwwWWW??? boiiiiiiiiiiiiiiiiiYYY???"

Chaun Wei was cool though. He was all tall and skinny like Ichabod Crane. And his body was like, cartoonishly arched forward like he was a C-shape. He looked like he would make the perfect goth, ... if he wasn't in his tidy little science clothes.

I was like "yeah, you should go look him up. You have the same name as him."

And he just looked at me like I was speaking nonsense. ... like he probably has the same exact name as like a quarter billion other people already.

What if he did though? ... – just goes and looks it up... makes his whole entire life about it. ... Starts dressing like a cowboy goth. ... Then his life starts to spiral.

His parents - "... ... *Why you do dis?!?!*"

Chaun Wei - defiantly at the edge of tears - "I'm Chaun Wayne now!! You unduhstan?!?!"

All in his goth clothes. ... like a black cowboy hat and mascara.

Still crying.

Alright, what the hell's this chapter about? I'm not even gonna relate that to anything. This book has gone on for far too long now. And now I've had to break it up into THREE parts. And, no, I didn't get that Chaun Wei thing from the movie. That... ALSO... happened to me. ... in really real life. ... like literally or whatevs.

Okay. So. Why are we STILL talking about this? ... in a *third* chapter? What in the world is still so important that we *still* need to be talking about it... sTILL?

Well, I'll tell you what, Mister Impatient. It's the future of the human race... as we journey through *outer space*!! Is that important enough for yuh?

... huh?

HUH?!

I SAID ANSWER ME WHEN I'M TALKING TO YOU!!!

YOU THINK I WORK ALL DAY... SO I CAN COME HOME... AND WRITE A BOOK... AND LISTEN TO BELLYACHE ABOUT...

Alright...

But first... we have a terrestrial question.

We just got done talking about the Nauptsèh system, and it brings up an important quandary. That is, how do you make friends with your neighbors when you have nothing of significant consequence that you really want from each other?

That question is especially relevant as it pertains to minority groups. How exactly do you protect a minority group, especially if you're living in a carrot-based society? After all, a carrot-based system is non-authoritarian. ... unless you *deliberately* want it to be. So, there's no ilk of omnipotent overlords to force you to cohabitate if you don't want to.

Okay, well, then how do we protect them? Or how do we *not* protect them - maybe they're like some crazy out-of-towners that're trying to pay blind kids with play money. How do we handle all of this?

Well, the system is *by the people*... and *for the people*. So, if you want the system to do anything, it's mostly gotta come from the local people.

For example, if you want the system within a community to treat everyone equally, then this generally needs to be the opinion of the majority of the people. And then they can go about officializing this however they want to. They can have a local vote and make it official, and they can write it into code as a standard to live by. Whatever makes'm feel all warm and tingly.

As long as most of the people fall at least somewhere along spectrum, between indifference and ambivalence, then it's just simply not an issue for these communities. It only becomes an issue when there is a sizable opposition for some reason.

And if there IS an opposition, for whatever reason, and that opposition is proportionally appreciable, then that essentially leads to a sort of bifurcation or enclaving effect. In other words, the two distinct sub-groups just sort of reassociate... into two distinct sub-communities.

And this is exactly what people have always done throughout the whole of human history. ... up until very very recently, that is. People have always just sort of split off and formed their own distinct sub-communities, where they were essentially much more comfortable with who their neighbors were. ... It wasn't until much more recently that we started trying to force people to live on top of each other... before they were ready for it.

And forcing people on top of one another has never ended well. It has always been better for people to just observe one another from a comfortable distance while getting to know one another. And they will, eventually, if you just let'm.

We are instinctively attracted to a friends group of people who share our interests, and just people, in general, who have a spectrum of complementary temperaments, regardless of their backgrounds. But this process never gets a chance to happen if you just start smashing different cultures right on top of one another. Doing so just creates a tension that keeps on lingering for generations.

And that's because human beings have always had a habit of reserving their most potent and unalloyed hatred for their closest neighbors. It's just human nature. We're either fine and okay with a particular neighbor, or we absolutely hate them, even more than we hate any other thing in the entire world. We'll just hate them with even more passion than our worst mortal enemies. And that's because these are the people that we have to look at most often.

It's like, when you have some friend that you've been really really close to for a while, so you decide to become roommates together. And then all of the sudden, you just slowly start to discover how much they annoy you.

"Dude! Why the hell did you put the dog bowls on the second shelf of the refrigerator?! That's right over my exposed watermelon!!"

"Uh, well, why didn't you put a cover on your watermelon? ... ... Seems obvious."

"Well... it also seems obvious not to put the crumbly splashy dog food over my watermelon."

So, it is always incredibly important to establish a proper harmony within the various associative spheres of our relative proximity.

We can't just go smashing people together if they don't want it. However, if you *do* give people a chance to get to know each other, then they pretty much always end up discovering that we are just all the *same* array of the same exact archetypes in different packages. Everywhere has a big Todd who thinks he knows it all. Everywhere has the punks, and the jocks, and the nerds, and the this, and the that, and the other thing.

My first best friend when I was really little was my buddy Matt. He was the son of a preacher man. And his dad just always had a neck brace on for some reason. Like, every time I saw 'm. But we were friends though. And we were friends because we were the two fastest kids in school. ... even faster than Percy. ... ... especially faster than Percy.

But there was something different about Matt, and I never even noticed. Yes, it's true, there was something very very different about 'im.

It turned out, in fact, that my buddy Matt was actually *black*.

That's right... Matt... ... *was black*.

He still is, in fact. Decided to stick with it.

But what was funny... is that I never even noticed it as a kid. And that's because I was never taught to notice. The first time I saw a black man, I was like three years old. I was waddling around through a store, on my own two feet, and I looked up and saw a man that was as black as midnight. ... not just regular black. This dude was like *super* black. It's like, there's black, and then there's purple, and then there's *beyond* that. He was beyond that.

And then my eyes got as big as saucers. So, I reached up and tugged on my dad's bandana. He used to tie this red bandana around his belt loop... so I could find 'im in the crowd and hold onto it. So, I tugged on the bandana and whispered to him. I thought the guy was injured or something. And my dad was like "oh! Oooooooh... oh, uhhhh, yeah, uhhhhhh... some people are just uhhhhh... a different color." And I was "ooooooooooooooooooooooh..."

And that's all I ever needed to know. And then I just forgot about it. And my little kid's brain never even thought about it after that. It was like, "different eyes; different hair; different tones; stuff is different; moving on."

And so, I never even realized that my buddy Matt was actually *black*. ... at least, not until the public school system started showing us all these really freakin' traumatizing movies as little kids. And then all of the sudden I just noticed.

And then I also noticed something else. I noticed that all of the older kids in our school were divided. They were all segregated apart... based on color. And this completely blew my little kid's mind. ... because I thought, "surely they're not doing that on purpose. That would make about as much sense as a bunch of kids just deciding to be friends because their shirts were all flannel or something, and they were like, 'well, we have nothing else in common, but we're still gonna be the flannel friends, ... because screw logic'."

So, I thought it must be a coincidence. But then I was completely mind blown by that, because I was like, "what in the world are the odds... that all of those friends... would ALSO be the same shade of color?!?!"

That was the first time I ever contemplated probabilities. "How is this even... pOSSible?!"

But it was a fascinating experience, though, in hindsight. And I'm not sharing it to be like, "ooh, look at me. I had a little black friend. I'm so special." I'm just sharing it because it's important.

And it doesn't make me better than you or anything. It's not a competition to see who had the first little black friend. No one's keeping score. ... I mean, I *am* winning... but that's not the point. I'm only sharing this because it allowed me to witness two distinct principles first hand. And that is, one, that we all can get along... if we're just given the *opportunity*... to see the humanity in one another; and, two, that if you try to force people together, especially when they're ill prepared to do so, then they just simply won't want to do it, ... and it'll only make things worse.

The reason that Matt and I had a chance to get to know each other is because our parents were perfectly comfortable with living around one another. It was our *parents* who enabled it... of their own accord -

not some grand, overarching, administrative authority.

This principle is universal actually. You can't just force things to come to you. You can't just force a *cat* to come to you. You have to start working on something really really important, that you really *don't* want stepped on, and *then* they'll come to you.

And you can't just force a *girlfriend* to come to you. You have to start sharing your hopes and dreams, that you *really* don't want stepped on, and *theeEEEeeen...*

There's an old adage... that if you wanna get people to do something, then you have to make it *easy*. And there's another old adage... that if you give a person a "why," then they can bear any "how."

"What the heck is an 'adage'," you say. Well, I do not know... And it shall remain that way. But the point is this - if you want people to do something, then it needs to be fun and easy, and there needs to be a positive incentive at the end of it. There needs to be a carrot - not a stick. Forcing people to do stuff *never* results in a high quality output. It only stresses people out. And that shuts down their thinking. So, it's carrots for the win, if you can dig it.

And leave the kids alone, for f*** sake. Just let 'm be kids for a while. Quit dumping a bunch of dark ages bullsh** into their heads. ... I was perfectly content when I was four years old to just see people for who they were and what their character was. ... and also to assume that the Earth was hollow for some reason.

... I was always trying to open up the globes in our house... because I wanted to see the "inner world where we lived." But then my dad was like, "no, we live on the outside." And for some reason I thought idea that was horrifying.

I must've gotten my sense of cosmology from ancient Greece in a former life. ... must'uh been walkin' around the marketplace and heard ol' Pythagoras saying something about a little mini sun at the center of the planet. ... And I was like, "oh snap... das wuss up!" ... And then that just became my truth forever. ... ... until my dad shot his mouth off.

Stranger things have happened.

(•)

pəuəddeɥ ə^eɥ s6u!ɥ+ ɾ36ueɾ+S

Wait!!! ... They say that you can get in there through the north pole!! **D:**

What if that's where *Santa lives*?!?!?!

D :

But yeah, you gotta let kids just be kids for a while. You gotta just let'm be children and protect their innocence. ... Sure, they might be imagining something crazy, ... like humans are actually living in the upside down world, but they will also see the world through the eyes of innocence. ... It's a perspective which will always help us to rediscover what we've long since forgotten.

So, ... back to the original question. How do you protect a minority group in a carrot-based society? ... Well, you start by getting them a safe space - not by throwing them into the lion's den.

If you, along with a sizable portion of the population, want to help them, well, then you just help them where they're at. Or you find them a new home and a haven. You, don't just thrust them into a situation where they're surrounded by wolves. How fun do you think that is for them? Talk about stressful.

And another thing, if you're actually forcing some other larger population to unwillingly do something that they really don't want to, then you are literally holding them hostage. You're saying "no, I don't care what you want. I'm forcing you to do this regardless." ... How humanitarian is that - just forcing your opinions on other people? That's not exactly leading by example... if your ultimate goal is enlightenment.

It's like those bumper stickers that say, "Be Kind."

It's like, really?! Your objective is kindness, and then you imagine that you're going to accomplish this by just giving me an order? ... ... Shouldn't it say... ... "Let's be Kind?"

But anyway, if you really want a community to accept another one, then you have to show those people what they're missing. They need to *see* what they haven't been *seeing*.

At one point in this latest modern century, there were groups within Western Civilization that were really not being treated so well. And the average person never saw a problem with this, because the average person never saw it in *any capacity*. They had no idea what was going on. All they knew was what they were being told about it. And what they were being told about it... was just some rosy little narrative... about how everything was all fine and dandy.

But things were not all fine and dandy, and the majority of Western citizens had no idea about it. That is, the majority of Western citizens had no idea about it... *until* someone showed them. Some clever entrepreneurs made this fandangled new contraption called a personal camera. They miniaturized the technology and made it affordable to normal folks. And then all of the sudden people could just start taking pictures of things. And then a flock of new photographers were now taking pictures of everything. And some of these new photographers were, now famously, taking pictures these other people's hardships. They were showing the actual reality of what these people had been enduring. And when the majority of Western citizens could finally SEE this reality, well, then *that* is when their opinions started shifting.

A lot more people started caring about these minority groups. And it was happening because they were finally *seeing* the images.

Now, to be fair, you can also paint a picture by just using your words, ... and you can additionally enlighten the masses by just employing your vernacular, but this ability has not been so common, as it also tends to take a lot of know-how. And so, photography has generally been a lot more effective.

And so, that is why ignorance and darkness have so perplexingly persisted. People just didn't know stuff, ... because they weren't being informed about it. So, when the pictures of these terrible conditions started flowing in from these photographers, well then, the world, as it had long since been before this, was just no longer viable.

So, once again, how do you protect the minorities who are just trying to live their lives peacefully? Well, you have to make people *want* to protect them. And you do this by simply informing them. ... You just show 'em these poignant pictures of other people's humanity. ... You just provide them with a whole perspective that's fair and balanced. ... Because, we are *all* just these regular people... who are, each and every one of us, struggling with something. ... And we are all just these pair-bonding mammals, who, at our core, are just aspiring to be happy.

And this isn't just the case as it pertains to a carrot-based system - this is also the unavoidable reality as it pertains to *EVERY* civilization. ... It doesn't matter how many human rights you've got enshrined on some fancy piece official, sacred, parchment paper; and it doesn't matter how many paramilitary enforcers you've got employed on your domestic police force - if the majority of your population does not agree with these "laws," well, then they just simply won't acknowledge them, and then they'll just find some way to judicially rationalize it. No amount of forcefulness, or coerciveness, or authoritarianism, or *stick*-based legalities, can *ever* protect you... if the majority of the population does not agree with it. ... Indeed, history is *replete* with such examples. ... But tis nary... like a sir... to repeat such humbuggery.

There is simply no substitute for a culture of conscience.

So, if your family was in a very small minority, and if you had the basic choice of these two systems, then you would *definitely* want the carrots and the incentive-based Alfreds, because at least in *these* societies... everything is tailored for *positivity*. It actually *wants* its population to be empowered - not just obsessing over power struggles all the time, and consumed with endless, existential anxiety. ... So, truly, which of these two groups of people would you rather be surrounded by - the ones who are being kept *happy*, or the ones who are being kept agitated?

Well, I think that the Nauptsèh already answered this question for us *quite* conclusively, thank you very much - they *even treated* their most *egregious mortal enemies* with the *same exact respect* as any citizen. ... ... W'o'o'o'o'e'e'e... ... ... That... is... insane. ... ... I don't think it could get *any* more clear... or absurdly *obvious*... just exactly which of these two civilizations you would rather be living in... – *especially* if you were a vulnerable minority.

It has been said... that you can tell a lot about a people... by the way they treat the most vulnerable amongst them. ... Well... ... there you go. We've got thousands upon thousands of years of recorded history, and it *clearly* shows that the carrots have been absolutely dominating at this.

And this isn't just for some ethnographic or philosophical minority, by the way. This is for everyone. The most vulnerable minority in any society is the people who have been condemned to face the justice system. And you never know when some weird twist of fate is going to thrust you into that category.

Like, what if you were just driving down the interstate in your big truck, and then you accidentally ran into somebody's aunty Donna? ... and then you didn't even realize it? ... so, you just drove off and left her there?

"That is called... a hit and run."

Well, what then?

What if you committed a hit and run?

And then what if you got caught? And what if the prosecutors wanted justice? What kind of world would you rather be living in if this just suddenly happened to you - if you just suddenly became this most vulnerable minority?

Would you rather be in a world that was obsessed with things like fairness and honesty and rehabilitation and second chances?

Or would you rather be in a system of sticks and hammers - a system that just sees everything as a nail?

It's a fair question.

And then, of course, you also have the *other* types of vulnerable minorities – the people who are just down on their luck for whatever reason, and as a consequence, the people who need some sort of assistance.

A lot of people live in fear that they might end up like this. They see someone panhandling on the side of the road, and a cold shiver runs down their spine. They're like, "f***, I hope I never end up like that."

As if any of us would actually do that.

As if we'd be like, "damn... I guess the day has finally come. I guess I finally gotta go beg in traffic now... ... "

"... SH**!!!"

Then we're all out there with our sign.

Yeah right.

No... ... we would not do that. And we would never *have to*... if we were living in an incentive-based, carrot-style society. ... And this isn't just the case because of all of those other reasons, this is *also* the un-avoidable eventuality *specifically* because people are so afraid of it - this system will *always* find a solution to the things that people are afraid of. ...

... No one likes to think that there's no safety net. No one likes to imagine that there's no support for them. And so, that is *precisely* why this system would provide a solution to that. This system is *byyyyy* the people... and *fooooor* the people... quite literally... so if you *want* something... and it isn't like, one of those types of wishes that only *a genie* could grant... then the producers will *create* it for you. ... ... aaaaaaaaaand you'will'have'it. ...

... - period, point, punto, *finito*.

It's basically like magic.

So, if you *want* a social safety net, then there will *be* a social safety net. And there will *be* a social safety net, because you *want* a social safety net. ... We all do. It's common sense. No one wants to think that they're just one little stroke of bad luck away from oblivion.

Oh, and speaking of a social safety net. There is also nothing more consistently reliable than the Alfreds. They are required, by their very existence, to be supportive and dependable – it's not even an option

– it's obligatory. ... They're not just some fly-by-night, unaccountable, "representative," who can promise you a bunch of big fancy stuff... but then never actually deliver on it... and then just sit and make excuses every time. ... Nope... No way... Not even close... The Alfreds are your *butlers*... for an entire *lifetime*... They can *only* survive in their professions if they *actually* take care of you. That's it.

World history, as we know it, would've been radically and profoundly different... if the rest of the world had been using these carrot-based systems. There's no political faction, party fighting; no endless destructive battles over who gets to hold the reins of power; no political assassinations; no succession crises with everybody trying to kill each other; no civil wars; no dissolution of an entire empire just because one single dude died; no wild, schizophrenic swings in official policy every time some brand new set of political leaders gets appointed; no crazy, catastrophic *collapses* in the *economy*, every time some brand new set of unnatural regulations gets created; no massive inefficiencies in the allocation of societal resources for defensive applications, which has always left a civilization far more open to an outside invasion; no excessively disproportionate disempowerment of the civilian population, which has always left a civilization far more *inviting* to an outside invasion; and no more *massive*, apocalyptic armageddons just because *Orlando Bloom* had the audACITY to go and kidnap Helen of *TrOY* like that. ... ... No! ... ... No way. ... ... None uh that. ... ... It would'a been just smoo'oo'oo'ooth sailin', ... for the most part. ...

... which is kinda what you want where your little kids are growing up. ... Am'a right?

I like my drama shaken into situations of my *own* choosing - not just stirred into my life with unsolicited profusion, and making me wonder if I should be stockpiling beans and rice, and some sort of medieval weaponry, that'll still be functioning as a viable, marauder

deterrent, long after the last of the real ammo has gone extinct. ... ...
Nyyyyope... ... Nyyyyo thank you. ... ... I do mind if I *don't*, thank you
very much. ... ... Let's save that drama for your mama cookin' pasta,
boyuh.

There's this old saying out there, in politics, that goes, "there's nothing
more expensive than free." ... ... Well... ... ... there's also nothing
more unreliable than a guarantee... if it's coming from a coercive civilization. ... ... Indubitably, there is just nothing more reliably *unreliable*
than a stick-based civilization as a whole. These empires rise and fall
like clockwork. ... - just repeating the same old, doomed and cursed formulas, as if it'll somehow just be different this time. ... ...

... Isn't that the definition of insanity?

... according to people who haven't actually bothered to look up the
definition of insanity yet?

I think it is.

... and the people use the word "poignant" like fifteen times a sentence.
... ... like it's some sort of tell for sophistication. ... and you get points
every time you use it. ... ... "Oh my gaud, it was such a poignant
experience. And I love how Tina and I just have such a poignant
relationship. And then I talked to my mom yesterday, and we just
had this like, conversation that was sO poignant. And like, you'll
never even guess what we poignantly talked about. Well, I don't really
poignantly remember per se, poignantly, but like, we just made so
many poignant points about the poignancy of being poignant..." ...
aaaa*AAAAA!A!A!A!H!H!H!H!H GET A NEW WORD!!!!!!!!!* **Ox**

Ughhh...

This is why we can't have nice words.

However! ... Notwithstanding... These axioms are indeed as yet withstanding, as it currently stands...

Rule by the sword, die by the sword.

Rule by reputation, thrive by reputation.

Reputation is how you govern amongst sovereigns.

Coercion is how you govern amongst conscripts.

And so, verily, my dearest Pacha... I pray thee... how now?"~ ... Be'est thou a soveREIGN... ... or a soverAIN'T? ... ...

Ehh???...

Pacha?!?!? ; \

Never mind. ... But again, there is simply nothing more consistently inconsistent than a supposed guarantee from a coercive civilization. ... The only thing that's actually *guaranteed* to happen is disfunction.

So, if it seems a little scary to have a system that doesn't *forcibly* tax people, then just remember that the actual logic is precisely the opposite. It is actually the *stick*-based civilizations which are highly unreliable. They may *seem* like they're giving you assurances, but really, they are *intrinsically* just a source of volatility.

So, there you have it. Every type of vulnerable minority is taken care of in a carrot-style, Alfred civilization. Period.

And so, this brings us back to the original question - how do you make friends with your neighbors when you have nothing of material value that either of you want from each other?

Well... let me ask you this. Let me answer that question with another question.

What if we were flying along through space, and we found this really cool moon planet where everything glowed at night? And what if we found this like, really special mineral there that was insanely valuable? And what if like, this mineral was just, unobtainable anywhere else in the known universe? But what if like, every time we tried to mine this unobtainable mineral - let's just call it... unobtainius - a bunch of giant, blue, cat people started shooting at us... - arrows dipped in a neurotoxin that would stop your heart instantly? ... "*They are very hard to kill...*"

I'm just coming up with this all on my own right now.

This is good.

Okay, yeah, and what if like, all of these giant, blue, cat people had these like, really long ponytails... and like, all of their really long ponytails had these like, little squiggly ten'acle things inside of'm... and like, every time you touched these little squiggly things together they formed a *hehlooh*!?!

Alright. ... So, now, let me ask you this. What would there be to actually stop us from just wiping these people out completely? Who cares? Why should we care? They're in our way, right?

Okay, well, maybe they have some like, critical, biological significance or something. Fine. We won't kill'm. But can we at least like, kick'm off their land, or just like, put'm in reservations or something? Why not? Who cares? Why shouldn't we? Why would it even matter?

Well, how 'bout this... What if some of these giant, blue, cat people were actually living alongside the beaches? And what if all of these giant, blue, *beach* people had these really cool whale friends that they liked? And what if they would always swim out to them and meet them whenever they were visiting? And what if they *loved* each other?

What if they really, truly loved these whale creatures? ... - they didn't want anything from them; they didn't have any contingencies for them; it wasn't any sort of like, mutually, inter-dependent relationship or anything; ... they just *plain* and simply loved them without expectation. ... for exactly who they were. ... just like we love our own critters. ... our dogs and our cats and our animals.

Alright... well... would that make any difference?

Would it matter that they possessed this *literal* super power? ... this power to just *love* another being? ... this power to just care for another species that is wholly unrelated? ... this power to just, want the best for an innocent child... despite anything else their parents or their culture may have done?

Does this matter? Does this *actual*, real-life, super power sort of change things at all? Because it IS an actual superpower. Think about it. We're not just these blind, biological robots, running around pressing copy and paste on ourselves. We have a power. We can *care* about something... that isn't even our species. ... that isn't even our biological relative. ... ... *That...* is a gift... *That...* is a miracle... *That...* is a superpower. ... And we ALL have it! ... How sick is that?! ... - each and every one of us have an ACTual... ... ... superpower. ... ... Think about it.

So, if we end up just hurting these other cat people, on this moon planet, then we are *damaging* an actual spark of real-life magic. It's like we're just smashing this little pixie fairy that was brilliant and sparkling and beautiful. ... – this little pixie fairy that had the power to legitimately love you.

I mean, think about it. How fascinating is that... that there are oooother beings... who can look at yooooour little children... and just be hopelessly disarmed by their love for them... even though you are a completely different culture or species? ... And they could even raise them as their own if they've been orphaned. ... ... *"monkey booooy!!!"*

Seriously.

How insane is that?

And we ourselves can do this also. ... obviously. ... I mean, we're basically just defenseless in the presence of a puppy. ... And that is just mind boggling.

And if you're still not exactly sure yet what I'm talking about, then just go look at some videos of little toddlers playing with Pomeranians...

… … It's exCRUciating.

And, quite frankly, this isn't just some hypothetical, immaterial, rhetorical flourish – this is, under no uncertain terms, the glue... which unites... the cosmos. This is our single best hope for universal solidarity. … - this *actual*, preternatural magic which resides within *all* of us.

And maybe that's what the big, blue, cat people are talking about when they say, "I see you."

So, that's my best answer. That's why I say don't be a dick to the big, blue, cat people. They are the keepers of the flame... the same as we are. … And we are *all* these incredible progenitors of unlimited creativity.

And *even* beyond that, just hurting another being is really dirty. Bringing tragedy to something that was innocent is just not a good experience. And you would always have to live with that in your consciousness. … even if no one else would ever know the difference.

… That… … is what an actual, real-life hell is.

Imagine being immortal and then having to live with that in your memory... all of that regret... of what you've done... for centuries... just that knowing of how you hurt people… … that some really beautiful things were brought to ruin... and it was all your fault.

That would feel like eternity. … like, a really really protracted stint in eternity. … … … Yeesh.

… until the day would finally come when you could re*dee'ee'ee'ee'ee'ee'eem* yourself!! ... ... Can I get uh ayyMAYUHN?!?!

Why would we want that? ... – all of that dirtiness in our soul... It's funner to just be chill. And that's probably why the aliens haven't wiped us out yet.

Yeah... that's right... YOU heard me!! That's probably why the *aliens* haven't wiped us out yet... ... YET. ... They figured out this big-blue-cat-people thing already.

Because maybe WE are the big, blue, cat people... so to speak. Maybe WE are the ones standing in the way of someone else's resources. Ever think about !?THAT?!

Of course you have. That's like half of every sci fi movie ever made. And that's the reason that a lot of people think we're alone out there. They just assume that these other races would all be a bunch of prim-itive jackasses like us, and they would've invaded already if they were out there. They don't imagine that some highly advanced, spacefaring, blue people might be a lot more like the Nauptsèh.

Maybe they're just biding their time - hoping that we'll eventually evolve our humanity. And maybe then, and only then, will our tech-nology finally reach the point... where we can just look at them as reg-ular people... ... ... instead of deities or devils.

Why don't you put that in your pipe and smoke it?

And while you're filling up your cheeks with that conception, why don't we *also* take a puff of yet another cogitation?

Why don't we do a thought experiment? Let's just run through this for a minute. What would've happened... if one of the Nauptsèh regions would've sent out some emissaries to all of the other regions, and then discussed with them this basic philosophy, that we all possess this magic within us, which is definitely worth protecting? ... and that maybe we should seek a higher calling? ... and that maybe we should spare ourselves the horrors of warfare?

And what if they tried to convince them that we should all do this for the children regardless? ... because anytime we do something really dark to someone, even if we feel they deserve it, it is always going to leak out and get onto other things, and it is always going sadden something innocent... it is always going to devastate something tangential?

):

After all, the children don't deserve to be affected like this. None of our stupid bullsh** is ever their fault. And to the extent that they haven't been too affected yet, they're still just happy, little, munchkin people who wanna be friends with everybody.

And one could even argue for their sovereignty. They're not just *clones* of their parents, and therefore destined to agree with them. Not at all. They have an instinct seek out their own way, and a free will to choose it. And they even have some *brand new* genetic material... that neither their parents nor their ancestors ever possessed.

And those children might even discover that they agree with YOU more... – even more than their own genetic relatives. ... especially if *your* culture is shining brighter. ... So, even if they grew up hearing bad things about you, *your* actions can still speak louder than *those* words.

Alright. So, let's just say that nearly everyone agreed to this, amongst the various, Nauptsèh, civilizational regions. Alrighty. Well, then how would we go about implementing this? How would we establish peace using this as our framework?

Well, one way to do it is with a citizens' exchange program. You could have some students and some young people go and apprentice in the other one's regions. Or you could also have some emissaries and their families go and relocate to one another's communities. This would give these regions a chance to get to know each other. ... And this is sort of like the effect of sharing poignant photos of people. (((... aaaa*AAAAA!A!A!A!H!H!H!H!H GET A NEW WORD!!!!!!!!* **Ox**)))

And not only that, but these emissaries could also serve another purpose. They could essentially be one another's eyes and ears across the regions. And this would allow them to trust but verify that no one is building an invasion force.

Alright, so, let's just say that this worked. Okey dokey. Well, then how are they supposed to go about stabilizing their populations... which of course is to prevent their overpopulation wars?

Well, they could always just incentivize it. For example, they could have like tiers of different social incentives and financial benefits, and the people could get these privileges based on how conservative they were being with the number of children they were having - bearing in mind the caveat that any and all benefits would be revoked if there was ever any evidence that something *inhumane* was being done in order to maintain these lower child numbers.

But that could work.

You could make incentives such as discounts and priority service privileges. And you could honor these people in public for their personal sacrifices. After all, by, in any way reducing their number of children, they would be helping to prevent that future warfare.

And maybe you could have like, even more benefits... for the people that just resolved to not have *any* kids. They could get like, grants for assisting with research, or for doing some sort of other socially beneficial initiatives or something.

Then they could also get other things too, like free food and free tickets and free vacations and stuff. And they could get like, bottomless paleo nachos and free refills on kava beer, along with courtside tickets to any allyball game. And you could let them throw out the first pitch... behind a sheet of safety glass.

So, yeah, this could work. But it would definitely take commitment. None of this master plan would be inherently easy.

And let's talk about why.

Well, first of all, we already alluded to the first issue. These incentives to limit your children could encourage some unethical behavior. We wouldn't want any accidental parents to get any dark ideas about this, and then just go all Sparta on a little baby. That would just be transferring the horrors of warfare onto the most innocent and defenseless amongst us, aaaaaaaand that would be a million times worse. If everything we do is not to protect them, then nothing else we do will ever matter.

And it's not a perfect science how you would prevent this either. I think the best way that you could accomplish this... is to foster a culture of disapproval. And then the people would be utterly terrified by even the most vanishingly small possibility that they could ever get caught actually doing something like this. And I guess that would have to be your best option... in a preindustrial world.

So, yeah, you could see how this whole, overpopulation prevention thing could have some really unintended consequences.

Anytime any civilization tries to do some sort of social program, it ALWAYS causes unintended side effects. And then, inevitably, they have to come up with a bunch of *other* programs to solve *those* problems. But then those new programs cause even *more* problems. So, then it's just this endless loop of problems and wasted money. And that is PRECISELY why it is almost always better to just leave a population alone, as a carrot-based society usually does. But we're just discussing these social programs... to see if it's even hypothetically possible to avoid these overpopulation issues.

Alright. So, let's talk about the next complication.

The next occurring problems, in no particular order, are accordingly, travel and homesickness. In a preindustrial world, traveling one sixth of the way across the entire hemisphere is both challenging and heart-wrenching. We instinctively do not want to leave our people. And the people who are occasionally doing this are only doing it out of desperation. So, this prospect for these regional emissaries would be both dangerous and painful.

And the final major issue with this strategy is that this idea would be very very difficult to get everyone to agree to. And if you couldn't get enough people to agree to it, then the plan would simply fail by virtue of disinterest. The plan would have to have a network of allies. ... because every single region that *wasn't* a part of it would need to be unambiguously cognizant that if they every violated the peace, then a force of allied soldiers would descend upon.

Alright. So, we need a bunch of people to agree to it? So, why is this so challenging?

Well, notwithstanding the difficulty of trying to sell everyone on the idea of transcendentalism over materialism, there would also be the problem of ancient rivalries. Yes, indeed. These rivalries run deep. And it doesn't really matter who struck first, because these tit-for-tats are an infinite loop. This has been the hominin condition since the jungle days. ... This is where we get our tribal instincts.

And this wasn't just an issue for the Nauptsèh. ... obviously. This has been a relevant existential issue for the entire world. Our planet has been blanketed in a peppering of regional conflicts since basically forever, and they are just as ferocious now as they have ever been. So, again, the Nauptsèh were by no means an outlier. They were dealing with the same exact problem that even WE have not solved yet!

Alrighty... ...

Well... ...

... uuuuuuuuuuuuuuh...

Then how might they have solved this? How might these preindustrial, medieval, classic era, and Neolithic people have independently solved a problem that even WE cannot figure out apparently?

Well, the first thing that can begin to assist us with this... is to simply improve our communities and raise our standard of living. When people's personal lives are going well for them, they *miraculously* stop obsessing about these rivalries. It's only when the chips are down that we start blaming other people for our misfortunes. ... And, surely, the best way for any group people to increase their standard of living is to allow themselves to organically transition over to a carrot system. So, if you wanna have a lot more peace with your neighbors, then you definitely wanna start by establishing a carrot society.

But, of course, the Nauptseh were already doing this. I mean, they basically invented it. So, they were actually in a pretty good position to start negotiating these peace talks.

But still, the idea of just sitting down and having these peace talks, when you both have been feeling *so much* enmity for *so many* generations, can truly be one of the most difficult things you have ever had to do.

But here's the amazing thing - we don't have to be buddies. We don't have to stop hating each other. We don't have to stop mistrusting one another. We only need to say this simple statement - "yeah, my opinion of you is better left unspoken, but I also have these little ones that I take care of, and so do you, and they are *far* more important than these rivalries we have. So, for THEIR sake... let us affix this one, simple principle above our conflicts – let us establish a better world that protects their innocence."

And if everyone can agree to that, then great.

This is basically that mythological battle between good and evil. It's like, which of these two motivations is more important to you - your hatred for your enemies... or your love for these little innocent things?

Everyone must answer that alone.

And I'll tell you what, if we choose to do the better thing and not obsess over these rivalries, but then we're STIIIILL super bent outta shape, when this whole, rough and rowdy, jacked up, Earth life thing is over, then I *swear* to you on everything, we're gonna go meet up and form a posse in the afterlife, and then go hunt these bastards down and have a shootout... ... ... in the afterlife... ... ... with *guns*... ... ... But for now, there are *far* more important things to focus on... and I'de rather not have to look back at my life and see all of these little, innocent things, that I *could've* been helping, but wasn't, because I was *instead* out wreaking vengeance on my neighbors. ... No way. ...

... That's what my old blind friend Chris used to call it. He'd start cranking up his stereo all loud and be like, "time to make the neighbors pay for their sins." ...

I could put some more Chris quotes in here... ... *but it's all downhill from there.*

Freakin' cripples.

I don't trust any of'm. Whenever I see these supposedly "blind" people, I just start doing air punches really fast, right in front of their face.

FFFHHH'FFFHHH'FFFHHH'FFFHHH'FFFHHH...

... to see if they're like, faking it. ... And they're like, "hey, stop doing that," ... and I'm like, "doing what? ... How would you even know I was doing anything???"

"If you're accusing me of doing something, then maybe I should."

"You wanna *catch these hands*???"

Freakin' cripples. Prob'ly faking it.

One time I saw this like, really young, little blind girl just basically an-nihilate the outside of this dude's Jaguar, trying to figure out where she was going. ... just walking all around it, bashing the holy shit out of it with her cane. *!!WWWACK!! !!WWWACK!! !!WWWACK!! !!WWWACK!! !!WWWACK!!* ... ... And she wasn't even keeping it low. She was just like, slashing right into his windows and stuff. *!!CRACK!! !!CRACK!! !!CRACK!! !!CRACK!! !!CRACK!!*

I was like, "*ooooooohhhhhhh* mmmmmmaaaaaaaan..."

And he was just sitting in his car like, "*what the fU** IS GOIng right now?!?!*" ... ... And he was just like, stuck in traffic, he couldn't go any-where.

Haha...

Oh man...

It was hauntingly beautiful.

But that sucks for him though.

But from that point forward, I just decided I to judge every single blind person from that one individual.

You could do some serious damage like that though, ... if you just like, *pretended* to be a blind person. ... ... like *most* of these fakers.

Alright.

Enough of that.

But anyway, if you wanna get all enlightened about this peace treaty stuff, then there's *ooooooone more* deeper layer that's at least worth mentioning.

We talked earlier in the chapters about the E.L.F. core and the C.A.T. layer, and how if the cat-layer gets unfortunately traumatized then it can eventually go all A.P.E. shnizzle... or even vampire mode.

Well, the vampire isn't just paranoid and aggressive, it is also very ruthless and unthinking. It just blindly wants to dominate everything. And that is the mindless, sociopathic nature of the vampire - Vampiric Aggression and Maladaptive Paranoia with Irrationally Ruthless Entitlement - V.A.M.P.I.R.E. ... - aggressive and paranoid and parasitic.

So, this is what you're dealing with when you're dealing with someone who is getting hostile or sociopathic with you. You're dealing with a vampire. And the vampire has no mind of its own. It's just exclusively speaking the language of its pathology. So, the only way to deal with a vampire is to scare it. You essentially have to appeal to its sense of paranoia. That is the only thing that will actually deter it.

This is where our instincts for revenge come from. This is why movies about retribution are one of the most popular varieties. But if ever we can bring ourselves to not just react for a moment, then there is also a second option which is far more effective.

If you just go and exact your revenge on some aggressor, then that'll undoubtedly make the environment a little bit safer... at least in the short term. ... It'll make them think twice about acting like a maniac. It'll put a little miniature version of you in their head... that'll pop up and try to scare 'm if they ever start to do that again. ... But how much actual benefit is this really providing?

We rarely, if ever, stop to ask ourselves that question. Our societies are all built around threats and retribution and punishment. And, unfortunately, it still has its place in certain situations. But how productive are these knee jerk reactions really? Are we actually fixing the original problem? ... Or are we just diverting it into going after another target? ... Or are we *simply* just making that person a lot more aggressive? ... Because when a person is in vampire mode, attempting to resist that temptation is like trying to resist a blood meal for a vampire. We are never going to fix that craving by simply threatening them.

And so, this isn't really solving any problems... - not on a root cause level, anyway. ... It definitely FEELS good... for sure... ... And sometimes you have no choice, in a deeply sick society... But it isn't actually remediating anything in the long run. We are eventually going to need another solution.

So, this is where a second option presents itself - the elf-core. The elf-core is the absolute antithesis of the vampire. It could not pOSSibly be any more different. Everything you hate about your enemies is not this thing. The elf-core only knows happiness or sadness. This is what every little baby is born as - ... before the world has enshrouded their innocence.

This true self is rational and intelligent and compassionate. And its ONLy potential flaw is a lack of development - it might not have the strength yet to outshine the darkness. And it might not have the wisdom it needs to perceive the good in everyone. ... So, the elf-core might retreat in the face of great difficulty. And then this will expose a vacuum that invites the vampire.

This isn't to say, though, that our normal healthy psyches are just some prancing little ponies without that vampire element activated – they're not – and we still have a rough-and-tumble C.A.T. layer to help assist us with this animal stuff. But rather, this is only to establish, for the purposes of this discussion, that the elf-core, with its longing for inspiration, might not be emboldened enough yet. It might not have the strength to not be smothered.

The vampire, on the other hand, is exclusively born of darkness. It is the reflection of the world around it. And it is protecting the threatened organism in which it is activated. But at what cost? What is the ring of power going to do to your soul in the long run? ... aside from making you feel like "butter scraped across too much toast?"

The vampire is a pretty nasty piece of business. And sometimes that's all you can see in a person. But the vampire is not the executive or the true self - ... the elf-core is, if ever it resolves to summon its courage. ... ... And sometimes... ... just sometimes... ... if you give an enemy civilization a chance... ... ... they may surprise you with how effectively they transcend their own vampires. ... ...

... And you can even encourage this outcome about *a hundred* times faster if you want, by just inspiring those inner elf-cores with your *own* integrity... or even faster *still*, by actually offering them some assistance. ... – assistance that actually helps them to make things *function* better... – assistance that actually helps to improve their *process*. ... ... But either way.

It has been shown... that if you take some regular families, from two opposing cultures, and you just stick'm in a room together, with all of their little babies and their children and stuff, then they all just become really bonded to one another in less than a week or so. ... especially with their little babies in there - those joyful little bundles that know nothing of our sorrows. And so, they see that their enemy culture is just regular people – just regular, struggling people... who are *just as* overwhelmed by life as they are. But of course, it is much much harder to perceive this reality when you cannot actually see them. And it is much much easier to imagine that they are monsters when all you know is what you hear through distant sources.

But either way, even if your blood is *way* too hot right now, and all this jazz about chumming it up with your neighbors is not exactly your cup of jam at the moment, then hopefully we can *still* agree on *one* thing - that the children and the innocents deserve much better. And sometimes we just plainly have to prioritize. Sometimes we just bluntly have to bear it.

It's not always easy though. ... because our world is so distorted now. So, sometimes it can help us to keep a list of things - things that we've encountered that've brought us happiness. ... – real happiness. ... because *that* is what will last into the future. That is what will always spring eternal... when all of this other disfunction has long since faltered.

A lot of people would say that this is silly though. They would say that you should "just stop... hating," ... and that "hate's just for meanies," ... and that "we're having a sale today - half off on all crystals and Portlandia feminist literature."~

Well, I'm sorry to disappoint, folks, but that's just not how we're built. We... are... animals. And we are programmed with an array of survival instincts. Our hate is specifically put there so that we can claw our way out of horrible situations.

Our core may be immortal, but our shell is terrestrial.

Life takes grit... ... - true grit. ... And sometimes you just gotta get to that next level of karate belt color... so your parents don't think you already lost interest... after realizing you can't kick a human being in two pieces.

And so, you just gotta gut up, and cowboy up, and man up... so you can harness that ancient, mystical, oriental, chi force... and then stand there like a boss and put your *hand* through some wood!! HAYAHH-HHH!!!!!!!!!

How you gonna do that if we're all sittin' around singin' campfire songs?

... 't's what I thought.

Can you imagine a world where I have to just be *nice*... to every little anthropologists who comes along? ...

HA!!! ... ...

Steve wishes.

But it's okay though, we don't need to start a *drum circle* with our worst, mortal enemies. We just need some brighter beacon that we can look to. And that beacon can just be goodness - puuuuure and simple goodness. Goodness doesn't judge you. It doesn't condemn you. It doesn't come home and get drunk and hit you! ... It's just goodness.

And that brings us to the crux of this whole discussion. Which is to say, however incredibly difficult it is to establish a peace treaty, the situation is never truly hopeless, and that is because of one enduring principle - time will heal all wounds. ... ... if you let it. ... if you just give it a sporting chance to get its pants on. ... It's like how my buddy Matt and I were friends. We didn't know anything about any old-world insanity. No one ever told us. All we knew is that Percy sucked, and it was funny to mess with him, while he was physically preoccupied. And that's the only thing that really matters in this life - playground trolling Percy and then outrunning 'im.

You know what's funny too... is that when some of these lifelong enemies start getting a little bit older, they often start identifying with one another, even more than their own grandchildren. ... because at least your enemies "actually have some common sense about the world."

You just see these old dudes hangin' out and laughing it up together. And then you find out later that they used to be like, serious mortal enemies or something. But now they're just all old, so they've started identifying with one another more – even more than any younger person that they're related to.

Maybe that's what heaven is. We all just get all old as shit first, then we finally start hanging out again. ... tellin' our old war stories.

"Aaaaah, we got up to some jive back then... uh'heh'heh... Ain't nobody never got up to no jive like that... Let me tell yuh."

"Aaah, you know it, brothuh! We was some bad mamma jammas... uh'heh'heh'heh..."

We're all old black people for some reason.

Heaven is just a bunch of old black people going to church all day.

... just sitting there for eternity.

"... This sUCKS!"

But then occasionally we get to go feed the pigeons in the park for a while.

"... This is better."

Or you could go be a guardian angel for an old, rich, white, bank executive, having pangs of conscience over something.

"Yuh know, suh... sometimes it ain't so much about the golf club we choose... *much as it be duh...*"

I always loved seeing what the ancient cultures thought about heaven. ... or even what modern cultures think about it. People always act like it's just this tiny little spaceship or something. ... like it can only fit just so many people on it.

They're like, "do you think dogs go to heaven? Is there enough room up there? Will the life-support systems be able to cope with all of that?"

It's like, seriously? Are you fu**in' kidding? You think that *that* dimension spawned this one, but then *that* dimension would just be all small and retarded and sh**?! ... ... duFEQ?!?!?!

The distance from Earth to Saturn ALONE... is f****n' mind boggling. ... And that's nothing!!! ... And if we're just this tiny little, itty bitty subset of some much more significant progenitor dimension, then get the french outta here with this, *"is there enough room for dogs to go there?"* ...

The problem wouldn't be *not* having enough room - the problem would be *not* having enough life to fill it up with!

The word is... *"big."*

... very... very... ... ... b'i'i'i'i'i'i'g.

And like man's best friend wouldn't be there. ... ... *Get'the'F...*

So, yeah, like I was sayin', when a sufficient amount of time has gone by, especially in a carrot-based civilization, then these ancient wounds can heal and be rewritten. We can find a better way to understand them. We don't have to be offended by history. We can be like sexy, deep voice Saladin from the Kingdom of Heaven movie, and acknowledge that *"(we... are) not... those... men..."*

All we need to make this happen is just humility. ... - just some good ol' fashion, down home, country humility. That's the secret sauce to make this happen. ... We just need to get down on our knees... and bow our heads... and put our hands together... like a double karate chop... ... and acknowledge that... we... ... ...

are'a'bunch'uh'fuckin'clowns... ... ... That's the secret sauce to getting this party started.

But that's okay though... ... we can still be badass. We don't have to be all epic all the time. We can just be like the drunken master - all stumbling around and homeless looking. ... until someone starts a fight. ... then we just suddenly become all effective. ... - just sloppily diverting all their energy... and making them beat themselves with their own haymakers.

The drunken master doesn't try to act all hard. ... because he knows that he can flow. And he doesn't try to act all heavy... because he knows that he can float.

So, he's capable, and he darn well knows it. And he's not worried about pretending to be all perfect.

Or better yet, he's like Jackie Chan. Jackie doesn't just walk up in a room and start slaughtering everybody. No. Not even close. He always gets kicked in the face like fifty times. But then he still makes it outta there and gets the job done. He still brings home the ambassador's daughter - perfectly safe and sound, and completely unharmed... and also giggling.

Jackie ain't perfect. And he ain't tryin'uh be perfect. But he's still badass. And that's just fine.

So, if we can just perceive the humor in our own ridiculousness, and just admit it to ourselves that we haven't exactly been entirely perfect, then that simple little act of honest humility... can often be the spark of regional healing.

And this future is coming, one way or another... because we live in twilight world."~

After a planet has finally evolved to a certain level, the idea of just attempting to control people becomes untenable. Knowledge and information become ubiquitous. And that awareness then provides them independence from all controlling arbitrary institutions.

Take money for example. The money that exists in the world as we've known it is basically play money. The people who control it can always make more of it. But this waters down the value for all the little people. So, people around the world have had their *life* savings reduced to *nothing*.

This basically happens constantly, on a historical timeline, because the power to make infinite money proves to be far too tempting for all non-hobbits. So, the people who can print it just constantly do it. ... for angels do not rule us, but rather men do.

So, then the average person needs to protects their savings. They can't just leave it in their currency and hope that it survives all of this. So, they need some sort of "store of value" which retains its intrinsic buying power. And thus, this forces the average person to invest in things like land and gold and pop-tarts.

But doing this is a huge pain in the ass. And you can't just transport these assets around as easily as money. And of course, all nations have been known to just straight up confiscate these assets from people whenever they felt like it. And then that causes people to panic, and start burying all their gold in their back yard, and eating all their pop-tarts as fast as possible.

... It helps if you just frantically do both things at once, I've found.

So, then some people said, "screw this. This is bullsh**. We need some sort of money that's quantitatively limited; and that no one can print too much of; and that's indestructible; and that can only be destroyed if a meteor hits the planet; and that's also electronic... so I can just send it and receive it... anywhere in the world... for less than a penny; and so I can travel anywhere in the world and access my money; and no one can ever mess with it; and no one can ever control it; and no one can ever hack it; and not even the quantum computers of the future can ever bother it; and it's totally private, and impossible to spy on... so people can't just go online and find out what my salary is, or what kind of personal items I like to buy for myself... and it isn't pop-tarts. ... and I want it NOW."

And then other people were like, "what? That's impossible. There's no such thing. You can't do that. I suppose you'll be wantin' a magic fairy with that as well..." ... and then the nerds just built it. Yeah, they didn't care. They just went ahead and built it anyway. ... and made our nerd ancestors proud.

What they built is called crypto currency, as I'm sure you're aware. And certain ones can actually do all of that stuff. ... right out of the box. They freed money. And now people can send any amount of money to literally anyone in the world, regardless of what anyone else has to say about it. And they can do it privately and cheaply and sustainably. ... Sure, it isn't exactly convenient yet, but it's getting there. It's *guaranteed* to get there. *That* isn't even a question.

And a lot of different people have had a lot of different opinions about these crypto currencies, but one, hard fact remains... they *freed* money. Money is set free now. And there's no putting that genie back in the bottle.

And information is being set free also. More and more people can do more and more learning on an ever more convenient basis. People can listen to literature while they're jogging. They can learn how to do some cool new task while they're washing dishes or something. Search engines are becoming ever more sophisticated. And the power of the "wisdom of crowds" is just expanding by the minute.

So, once again, after a planet has inevitably reached a certain level, the idea of attempting to control people becomes untenable.

So, what does that mean for us? Well, it means that the stick-based systems are not future-compatible. They are obsoLETE, mister wORDSWORTH!!

Just like the brutal alpha systems of the jungle eventually gave way to our cooperative pair bonding, so too will the older systems of force and domination eventually fade.

Future generations will become more and more collaborative. They will reduce their stress, immensely, and become more prosperous. Then they will start to get away from that ancient sickness. And they will finally have a chance to see more clearly.

So, why wouldn't the aliens just destroy us then? Why would they even care about our species?

Well... ... why wouldn't you destroy a puppy?

Why wouldn't you destroy a toddler?

Why wouldn't you destroy a little old lady at a bus stop?

It's because you see the good in them.

It's *because*... you see their beauty.

Well, maybe a much more advanced race can see this in everyone.

I've actually seen this myself, oddly enough. It's been one of the funny little side effects of having inherited Pomeranians.

My youngest little poma'girl is just this tiny little tater tot - she's about the weight of a ripe green coconut. And she looooooves bein' held. She's just *obsessed* with it. It's one of her favorite things in the world. And she just looooooves being carried around and goin' shoppin'. Seriously. It completely beDAZZles her. She just sits there with this satisfied little look on her face.

So, I'll just be walking around a store or something, carrying her in one arm, and then all of the sudden, some random, total stranger will look over and notice her. And what's funny... is it completely takes 'm off guard. No one's ever expecting to be out shopping... and then suddenly have a little, red panda staring into their soul.

So, it just totally takes 'm off guard.

And it's funny too, because you see their authentic nature in these unguarded moments like this. ... like, when they don't realize anyone else is noticing. And then you see that little sparkle of honest happiness. It's just that fleeting little moment of childlike innocence.

We *all* have a *core* that just wants happiness. None of us were born to be a tyrant. We had to learn that. And that means that darkness is destined to fail, because it is *fighting* our truest nature.

And furthermore, we rose up from the jungles because of teamwork, because the couples started pairing up and working together. And

767 - BARTROLOMEW MC INNCEL

then the families started going into specialties... so that their people could work together... to their mutual advantage.

The trajectory of progress is mutual assistance. It is people being of service to one another. It is a society based on incentives, instead of coercion. And it is people... obtaining fulfillment... to become far more effective.

No other system can compete with this. So, a highly sophisticated civilization would always arc towards these qualities. They would be all of *moral* and *humane* and empathic. And they would also appreciate the value... of this *truest* inner essence of ours. ... this truest inner essence that always smiles when it sees a baby, fox panda.

So, when we finally reach that point where the aliens can come and introduce themselves, they will be likely to come and greet us with a giant shopping mall. ... - not a bunch of *phasors* and dumb sh**.

Why do you think they fly around in those huge, gigantic mother ships with all the flashing lights on the side? Do you really think that they need all that razzle dazzle just for flying through the vacuum of space? ... I don't think so, son. That's a shopping mall.

So, if you ever wanna ask me, "num'num'num'num'num... myeeeeeeh, what's up, doc?" then I'll tell you what's up... doc. It's a spaceship full of women's lingerie and cinnamon rolls. *That's* what's up.

That's about as futuristic as it gets, right there.

There will *always* be something to trade, even if you can assemble your own matter, and even if you can tap into a bandwidth... wherein the energy within its background is essentially unlimited. ... Resources will always be finite. And artwork will forever be inspiring. New discoveries will unceasingly provide a spectacle. And new perspectives will inexhaustibly be considered invaluable. ... ... wink, *wink...*

So, truly, there will always be things of great value... that cultures which can leap spacetime... will be able to trade with one another.

Alright, now brace yourself for this...

The only thing more advanced throughout the cosmos... is just the abstract philosophical concept of unsolicited compassion. If you can reach a certain level of civilization, then a flare for gratuitous altruism is *unavoidable.*

As we discussed earlier, in a previous chapter - advancement is synonymous with learning; and learning is the growth of understanding; understanding is the pathway to unveiling beauty; and then beauty invokes compassion without volition.

In other words, when a civilization becomes sufficiently sophisticated, attempting to avoid compassion is like trying to turn off gravity. Not only is education the pathway to the light side, but a society which is based around incentives is one of service.

Thus, a continuing education leads to altruism, while an incentive-based society does this also. And so, the cosmic binding force is adoration, ... there is *truly* no escape from the pull of levity.

Or, in *OTHer* other words, the aliens would not have destroyed us yet, because they're not that *absurdly* devoid of basic empathy. ... ... generally speaking.

Don't worry though, I'm sure there'll still be a few star wars out there, if you wanna fight some. I mean, you'll probably have to go trekking pretty far across the parsecs to go find'm, but I'm sure they'll still be a few of'm out there, somewhere. ... But it won't be all fun and light swords though. It'll be soldiers coming home in body bags. It'll be jeddis coming home in body bags.

But all *real*-life jeddis will just be nerds though. ... just, like, all different types of heavy set guys with ponytails, and a bunch of comic-con people. And that's because they'll be the only who can't resist the temptation to go messing around with an insanely dangerous plasma sword.

And like, all of'm will just wanna be on the dark side... because it's edgy. So, there'll be no balance in the force. It'll just all be dark side.

And they'll all be itchin' for a fight, so they'll just create a bunch of star wars where there were none.

And I know what you're thinkin'. You're thinking that this is impossible. ... because you can't actually do those things in real-life, that the jeddis do. Well, believe me... they'll find a way. ... They... will *find*... a way. As soon as someone invents a functional plasma saber, they're just gonna completely lose their sh** and start going all bañañas, and it'll be an all out feeding frenzy to go and figure out all the other jeddi powers. ... Believe it. ... They won't stop until they've done it. ... I'm mean... they're nerds... ... that's what they do. ... ... And then we'll all be in trouble.

But there you have it. That's your answer. The final answer to the ultimate question of life, the universe, and everything in between... is... in point of fact... sixty nine... ... ...notwithstanding forty two. It is a six and a nine; It is a yin and a yang; it is tai chi wu de; it is compassion...

spiraling upwards... with courage.

... For courage will escape from its confines, within the dark, fertile soil; and curiosity will emerge from the surface, into the daylight of knowledge; and then this knowledge will impart its warm brilliance, to engender much more growth and expose the beauty; and then this beauty will unfold like a flower, which will inspire... a great compassion... that is irresistible. ... – courage... spiraling upwards... ... with compassion... ... in a helical... yin and yang... across dimensions.

There it is. ... That's your answer.

That... is what is up... ... ... Doc.

That... is what is oriented... upwards... my *dearest*... Doctor... *Watson!*

... Elementary, really.

The future is lit.

Oh, and don't worry about those jeddis. Captain Kurk is gonna go jack those guys up.

Their predictive reflexes won't work on 'im. He's too sloppy and erratic. ... ... Even *he* ... doesn't know what he's about to do.

... goes to shoot one guy and accidentally shoots the one next to him.

Puhtyyywwww...

"Whoo'oops! ... ..."

"I mean... ... ... let that be a lesson... to you!"

... accidentally sets his phasor to full auto...

P''T''T''T''T''T''T''T''T''T''T''T''T''T''T''T''T''T''T...

... trying to hold it with both hands...

"AAAaaAAAAaAAAAAAaaAHHHHHH..."

The jeddis 're all tryin'uh swat 'is phasors back... ... but it just looks like D-Day... ... with jeddis all dazed and confused. ... ...

P''T'F'T'F'T'F'T'F'T'F!!!!!!!!!

Alright. So, what have we learned in this dank a** hitchhiker's guide to humanity? We talked about how nearly everyone alive can boost their IQs up - quite dramatically in most cases; how essentially every single child can be turned into a veritable *super* human, if we just raise'm properly – ... and even adults too, for that matter; how we all can increase our lifespans, and improve our vitality; how we all can understand where we came from, and what our bodies were built for; how we all can eat more paleo, and still be ethical; how we all can enjoy more romance, and do it paleo style; and how our world can protect our children, and our animals, and our environment.

Heck, we even talked about how communities can increase their citizens' buying power by over four hundred percent. And that's just four hundred percent for starters, by the way.

They say that Mother Teresa helped like several thousand people with food and medicine and stuff. ... ... Well... ... we can help like several... ALL... the people... with *all* their *everything*.

... and all the children and the animals too. Because if you help the average person, then you're also helping to save their children and their animals.

And that's the only way to do it really. You can't just extract the children and the animals. You have to save everyone. ... ... And *save them* we can.

We can start our own religion even. We'll just call ourselves The Piglets. ... ... And when you join, we'll call it "coming to the trough." ... ... And you'll have to stand up in front of the chapel, with your arms out, and your head up towards the heavens, and just start squealing as loud as possible... reeeeeeeeeeeeeh!!!

... like Shawshank Redemption meets Deliverance.

Ree!ee!ee!ee!ee!ee!ee!ee!!!!!!!!

And then you'll hafta introduce yourself to everyone... ... in pig latin.
... ... and burping every word of it.

"e'l'l'o'H'a'y'... 'm'y'a'y'... 'a'm'e'n'a'y'...'i's'... J'a'r'v'i's'"'h'a'y'"u'u'u'h..."

Go 'head. Try it out. See if you can do it. ... See if you've got what it
takes to join this religion.

It's okay though, you don't have to do it like, just one single word at
a time. You can try to get as many out in a single burp as possible. ...
We're not gonna be one of those strict churches.

"H'e'l'l'o'm'y'n'a'm'e'i's'J'a'r'i's'U'U'U'H"

Watch, they find this book in like ten thousand years... but just that
part right there out of context. ... ... "Such a strange people, our ancestors..."

Alright, alright... but for real, we can still save everyone. ... - "save"
like... just help'm... ... not like... for religion.

And we don't have to be like, saints or anything – we can just be the
guardians. We can help everyone with everything, and like a million
other things that I can't even talk about yet. We can move from the
age of Pisces to the age of Aquarius. We can move from alpha energy...

into Alfred synergy. We can make the world *actually* fun for people. We can get rid of like, ninety nine point nine nine, nine nine, nine nine, nine nine... nine nine, nine nine, nine nine, *nine* nine percent of things for people to complain about. ... – we can put punk rockers outta business. ... and replace'm with groups like, Lounge Against the Machine."~

Not really. We're always gonna need the punk rockers. We're always gonna need that fresh new energy. And every generation is gonna have to rebel against something. ... And it's like, someone's gonna have to invent sex at some point. ... – four point five billion years and we've just been using storks this whole time. ... It's like, *what the hell*, can't someone just invent something *bETTER already*?! ... ... If only one of these generations could finally be brave enough to try something shameful for once. ... We shouldn't have to keep using a stork to get your mom pregnant.

But anyway, we need to be making love... – not war. ... And we can. We can establish a world of peace... and adopt a strategy for ending world hunger. We can make these dreams come true for every beauty pageant winner. ... And it's *absurdly* simple, even. ... ... All we've ever actually needed was just the right philosophy.

We have all been grasping through the fog of this world with our hands around an elephant, and we have all been arguing about its details and its purpose. So, some have consistently contended that this conundrum is quite characteristic a cargo conveyor... while indeed others have persistently prognosticated that this protuberance is nothing if not a proboscis appendage.

Or is it an elbow or an earflap that makes up this lifeform?

Or is it a liberated or a collectivist society?

Or is it a spiritualism or a materialism that we should turn to?

Or is it a paleo or a highly ethical sort of diet?

Or is it *all* of these things?

Well, the animal which we have here before us is quite assuredly an entire elephant. So, of course, these many features are all connected. We have all been grasping through the fog of this world, trying to suss out what this creature is; and we have all been sensing what we thought was different elements... but is actually...

just...

one...

instinct...

This is an instinct... to consume... our ancient coastal diet; and this is an inkling... to commune... via reputation. This diet... is not just optimal - it is also ethical; and these communities... are not just empowering - they are also emancipated.

This is how we were living in The Garden of Beachin'... - where the firmament rises up from the waters; where our Adams first transcended our alphas; and where our Liliths were finally supplanted by our pairbonders.

Everything we want comes from this paradigm. ... We can have our cake and eat it, if we exploit this nexus union. ... We don't *even* have to agree on how we got there. ... We can all have our own different pathways to this same exact conclusion.

There is a much better world at our fingertips. And we've all been working on this puzzle with one hand tied behind our backs. ... which doesn't really mean much when you're working on a puzzle... ... but we've also been missing a few pieces. ... because those pieces were all *distributed* throughout the world. ... ... So, we have to find them, and then assemble them, and understand them. And it is *then*, and *only* then, that we will see it.

And maybe if the world can ever get this, and our compassion can ever match our great ambitions, then I guess that we could talk about some funner stuff, like what it really was that Nikola Tesla was discovering. ... ... And, no, it's not in any book or any archive. ... ... teehee... ...

There is a lot more in this universe than in our books right now.

But getting back to the whole point of this chapter. Hopefully by now we can at least somewhat begin to appreciate how the Nauptsèh were in a quagmire. The same, epic, coastal geography that enabled their people to become total legends, was also the same exact geography that created these challenges.

And yet... ... ... they persisted. ... ... as still they do. ...

And the last couple of centuries have been anything but kind to them. So, it is *no small* attainment that they yet persist, even now. ... Their culture was almost wiped out completely. The word "holocaust" would not be an overdramatization. ... ... So, it's kings to them for preserving their heritage. ... It hasn't been easy.

The one that really amazes me though is how their people were often criticized for the fact that they really liked to party a lot.

So, it's like... ... their culture actually figured out the secret to happiness... and then some other culture just came along and started criticizing them for that.

... instead of taking notes.

That's too bad.

The one that really gets me *the most* though is what happened to their fluffy little buddies - The Pacific Northwest Wooly Dogs. ... ... The breed was lost forever in all of the chaos. ... And now their memory is only survived by the old histories and photographs.

And there's really nothing that anyone can say that can make these things better. I lost my first dog a long time ago. He was my little buddy. And he went too early - there was a thing he was born with. Of course I missed him. So, I did what any good nerd father would do and did an experiment.

I already knew that there were some investigators who regularly assisted detectives in finding missing persons. And they did so by using some sort of extrasense perception. And this a real phenomenon, by the way. Several universities have already been studying it. It's been validated at far beyond statistical probability.

The studies are real. The people are real. And the phenomenon is real. But I had to see it for myself.

So, I contacted a number of these extrasense investigators, and I solicited their services to help me with my experiment. And they were all completely normal, by the way. You would never even know this fact about them, that they use an actual extrasense to help find missing persons. They don't just go around telling everyone about it, or like, wearing a big hat or something.

But anyway, I got ahold of a few of'm, and I asked 'm if they could get any impressions from "my son." That's all I told 'm, that I was trying to get impressions from my son. I didn't say he was a dog. ... or anything else about 'im. ... And these people had absolutely no idea who I was, by the way, much less any Earthly means of ever knowing anything about me. I've never been one of those people who puts his whole life out there, where everyone can see it and know their story. I was a complete unknown to them, in every possible way.

Well, I'm not gonna get into all of the different experiences, but I will share one of 'm, because I think my little guy will like that.

So, one of these individuals I tracked down was a very sweet little Korean girl. She had been born a twin, and she and her sister had that twin thing, where they were always connected no matter how far apart they were - basically like that superluminal quantum entanglement phenomenon.

Anyway, we started talking, and then she started getting some impressions. Then she started giving me these very specific details, like his exact cause of passing, which in fact was actually kind of confusing to her, because it was stuff that didn't really make sense if you thought you were dealing with a human child, which of course she assumed we were.

But she was a trooper though, and she just kept on going... in a sort of perplexed and cautious manner, still giving me these very personal details - ones that no one could even know, even if they had a bunch of cameras in my house. That's the kind of stuff that gets your attention. And it's very specific too - not like these scam artist that just give you this mentalist, cold reading, deductive extrapolation, vague stuff. No. It'll be like really specific things... like... I want my little brother to have my rock collection in the back yard and my car collection on the side of the house and my pet turtles in the living room, even though no one ever said anything about a little brother, much less any of that other stuff. It's like, seriously, how the fffflip could they even *know* that?

But anyway, so, this very sweet little Korean girl was just busy telling me all this stuff, and then maybe like seven minutes into it, she just stops all the sudden and starts apologizing. She's like "I'm sorry. I'm very sorry to stop the flow. But I just need to ask... and I am sorry... but did your son ever have a little black dog? Because that is the only constant thing he keeps showing me."

Man, I was laughing so hard at that point. She completely broke my poker face. So, I just said screw it and I told her. I was like "yeah, that's actually him," and then she slowly started to realize what I meant. She was like "ooooooohhhhhhhh mmmmmyyyyyyy gooooooooosh! I didn't know I could *talk to* ANIMALS! ... ... He is soooo SMART!"

It was hilarious. Like, just the whole experience.

And this is just, like, one little experience I've had. I could fill a big fat book with all the other ones. But I don't need to. Because people need to see things for themselves - not just to be told about it. And they certainly can. All the resources are available. The science of the multi-verse has been set *free*. The data is now coming in through *millions* of people, especially people who have had these temporary death experiences and come back. And, believe it or not, we know for a fact now... that this is not where we come from. ... and that no one is ever lost.

And to that end, one of the reasons why a lot of people don't really like this sort of loosey-goosey spirituality talk... is because they essentially feel like it's just this license to bullshit. But here's the thing - it's not. Just like crypto currencies are being constantly kept secure through decentralized consensus... so too is this. The truth must be established through the vast majority. That's why I jokingly refer to it as "spiritual crypto."

And this whole field of research was actually started by a Frenchman by the name of Hippolyte Léon Denizard Rivail, nearly *two* hundred years ago - or as he was much more commonly, and *much-less-fun-to-sayilly* known, from his research, Allan Kardec. And what that means for us is this... no, *one*, single, solitary individual, or even group of individuals for that matter, *ever* gets to just go around acting like they've got the exclusive scoop on everything. The consensus about this progenitor dimension is now coming in through *millions*.

We've all had our important pieces over the centuries. And the future is about all of us. And no one, single group of people will get to say, "I told you so." ... because everyone will get to say, "I told you so." ... to some degree. And everyone will have a few different places where they just simply didn't know about something - like, modern germ theory, or electricity for example. And I don't think that this was an accident. ... because a future that gives the power to the people... is a future that values *everyone* and their contributions.

But either way, as far as this current research is concerned, the truth must be presumed through mass consensus. Much like any other science.

And I always have to be careful in how I talk about this, though, because I've already known about it for so long... that I just talk about it all nonchalantly. And that's how it'll be in the future. It'll just be one more ordinary aspect of reality, and people won't even be all that amazed by it anymore. Same thing with aliens. It'll be all flabbergasting for like six months, but then it'll just be ordinary again, and people'll think it's *boring* if you still wanna talk about it. They'll be like chewing their bubble gum and twirling their hair with their finger and rolling their eyes at you like, "yeah, yeah, yeah, aliens exist or whatever. Quit obsessing over it, stalker."

Seriously... people'll just be bored with it after a few months. You're probably already bored of it *right now* and it hasn't even happened yet. That's just the way human psychology works. We always trivialize everything once it finally becomes established. ... "Yeah, yeah, yeah, there's like a spiritual dimension or whatever. Why don't you marry it if you love it so much?"

If you took any ancient human from pretty much any former era, and you just plopped 'm right down into our current period, they would be absolutely sTUPEFied with our technology. They would be utterly and inconsolably dumbstruck. ... But for us it's just normal. For us it's just boring. We're already surrounded by miracles that no one even pays attention to. ... unless they get high a little. ... but that's another story. And that's why I don't shy away from this stuff - I know in the not-too-distant future it'll all be ordinary.

And that's because the truth is now coming in like a tidal wave, and *no one*... will be able to stop it. It's too late. So, it's just *a million* times better... for us to just go mount up... and let's surf this thing.

But, either way, the truth about many ancient mysteries is coming to light now. Everything is becoming decentralized. Crypto coins and metal coin certificates have decentralized money. Carrots have decentralized power. And clairvoyance has decentralized, what you might call, the extra-dimensional realities.

So, now we've got Carrots, Coins, and Clairvoyance - A Recipe for Sovereign Independence.

The power is going back to the people. ... if you want it. It's up to you. Some communities will, and it'll be astonishing to witness. They won't just *merit* a better world, they will actually *build* it. And other communities won't, and that's just fine too. It's all about self-determination. Or they might even just want a hybrid model. But either way.

But like I said, anyone can easily verify this transdimensional stuff. The resources are out there. The challenge is not in the proof, but rather in our psychology. Our animal brain rejects it. It's like, "what?! Whach'you talkin' 'bout immortality?! ... u'duh co' personality?!" ... Our subconscious brain is an old black guy too. "What's all this flamma jamma bibbity bop about immortality?! We goen die, son! ... We BOTH goen die! ... I goen die; you goen die; ey'bauy goen die. ... We all dien'! ... ... You ain't goen *fly* away like no angel!"

You know, that's actually funny too, because that really *did* happen to my mom. She just flew away out of her body... while she was giving birth to me of all people. It was one of those water births too... with the gravity assist. But the stress was just too unbearable, because apparently my head was like, at a weird angle or something. But, yeah, the stress was just too unbearable, and then she just, *swoosh*, popped right out. ... And she could see all around the area, in like ultra high definition... and see like all these little details in the surrounding structure... that you wouldn't've ever seen otherwise.

It's a really boring story, though, that only takes like sixty seconds to tell. She's only told it like, twice, because it's so boring, and we just shame her into silence, because of its boringness.

"Boo'oo'oo'oo... you should'a flown around or something... boo'oo'oo'oo..."

But, yeah, she just *flew* away for a second. And what's interesting too... is that these temporary death experiences sort of crack our inter-dimensional doorways open a little bit. And then the experiencer becomes a little "sensitive." ... and they can pick up on "impressions" a little bit... that would be otherwise imperceptible. It sort of depends on how deep they went. So, people could actually recreate this, and study it, if they were brave enough to try it. I'm not saying it would be easy, but it *would* be scientific.

Maybe people could do *that* instead of like, donating plasma, or like, testing drugs or whatever. ... If you need some extra cash, you could just go and volunteer to get killed for a little bit. ... ... and then studied for your new abilities.

That's prob'ly how the *nerds* are gonna get their jeddi powers. They'll just keep killing themselves until they're strong enough. ... ... SEE?!?! ... I told you they would find a way!! ... ... Well... I guess I just gave it to'm. ... ... *Ehh*... ... they would'a found it anyway.

But don't worry though... the jeddis'll be no match for cold, hard Vulcan logic. ... and Spock's paralyzing shoulder pinch. He'll just *drop'm* to their knees until they release their glow sword.

Then he'll just reach down with his free hand and pick it up and *wack* their head off.

Then HE'll be the jeddi!!

Oh shit... Woe... Wait a minute... I just had a freakin' lightbulb go off. Dude. I think I'm freakin' one of these people. For real though. For really real in real life. Yeah, because there might be a pattern... that whenever this happens to a pregnant mother, and she has a temporary death experience, there's also a really high probability that it'll have the same exact effect on her unborn baby. ... Holy shit. That might be me. My freakin' jaw is hanging open. ... like, figuratively. Not literally. ... Literally my eyes are *bugged* out... and I'm *amazed!* ... and *frozen*. ... just looking at my little dog sleeping... and she looks like a little baby seal... with feet instead of flippers.

But, holy shnazzle. I might be one of those people. Maybe that's why I've had these little experiences... that none of my siblings ever had. ... and so, they think I'm crazy. They were all born in the open air of a hospital with no deaths.

Haha.

Chumps.

But, yeah... maybe *that's* why I saw my grandma crossing over, and *they* didn't. ... - I told that story earlier in the book, in case you skipped ahead... like a cheat.

Wow. That's insane. I'm glad I wrote this book. I was just lookin' for some random backdrop to do a bunch of riffing, but now we've *actu-*

ally come up with some good stuff in here. And some of it might even be accurate. And now I might even be a little bit jeddi it turns out. ... like, maybe one or two parts per million.

I hope Spock doesn't find out.

But... wow! ... I'm really amazed by this. And it's also sweet that we can initiate this scientifically. I mean, I'm not sayin' it's easy. I'm just sayin'... you could do it. But it would probably be a whole lot easier, though, to just go and round up some of these tens of millions of people who've already had these experiences. ... and then study the ones with the strongest clairvoyant abilities.

And, yes, it's millions. ... It's a lot. We have a loooooooooooot of data to pull from.

And *I* might even be one of these people apparently. ... But not really. I don't want that ability. I've got enough thoughts of my own without some extrasense perception pouring in. The only one *I* want is that one jeddi one, where they can see the future, like a couple seconds in advance, and then it seems like they have these really fast reflexes. ... That's the one I want. I mean, it wouldn't help with Kurk at all... but... yuh know, still... it'd be a good one.

And get this, I know this sounds completely insane, but *that*... is actually possible. ... - seeing the future, I mean. It's *possible*. It happens all the time. For real. ... Future probabilities can *actually* be known somehow. ... I wouldn't be saying that if I hadn't actually seen it. ... myself. ... *numerous* times. ... in aLARMing degrees of detail.

Like, I don't know if I should tell this... buuuuut... yeah, I won't tell it. But, yeah, there're other examples of how this happens. Like, for example, if a mothman"- just called you up on the telephone, and prophesized the statement "Denver 99," and you were like, "what? What the hell's that supposed to mean?" and then the next day you were watching the news, and it said that a plane went down in Denver with 99 people on board, you would know without a doubt... that whoever had told you that... somehow *knew* that that was coming.

And that's how they do it. They give you just enough information to see that they predicted it, but then *not* enough information so that you could actually *stop* it. It's like they don't want us messing around with the timeline and stuff. ... They just want *you* to know... that *they* know. ... and that these sorts of things *can* be known. ... and that you can at least take *comfort* in that. ... – that every single timeline is something they know about. ... and that everything that happens is for some reason... ... and that apparently there's a method to the madness... ... and that eventually everything ends well, and it'll all be worth. ... ... - according to them. ... ... for what it's worth.

We're not really given to know what that method is though, but what I *do* know is this - learning is a pathway to loving, and there's no better way to learn than by *immersion*. No one appreciates the water like a person from the desert. And no one else will ever have that character, ... until they *themselves* have also lived that hardship. ... ... So, apparently, getting your ass kicked helps you love more. ... ... Apparently, living through *insanity* makes you more beautiful. ... ... *eventually*.

But the mothmen, and the dearly departed, are always telling us that the cheat code to this safari ride... is "love." ... Just love. ... And they're really big on second chances. ... and not always conflating the actions of an individual with their first and final essence.

And people sometimes ask them what the one true path is, and they just say, "whatever brings you closer to love." And I guess, in light of everything we've been talking about, *that actually makes a lot uh sense.*

... because, as we've been discussing in this chapter, the glue that can bind the universe is literally learning how to love more. And according to the mothmen and the departed, the secret to the multiverse is *also*.

Wow.

I don't know how we could top that.

Everything *really is* connected... isn't it.

It's the unified philosophy theory.

Woe.

How 'bout *them* apples?

Some people aren't quite sure about these mothmen though. They say, "how do you know you can trust 'm?"

And to that I simply say, I believe there's an old quote about this - "you shall know a (moth)man by his deeds." ... Or something like that. ... ... But seriously though, we don't - we don't know that we can trust them. But that's okay. We don't need to. We've *always* had the tools to find our *own* way. If our philosophy... is consistently... informed... by both compassion... and courage, then we will instinctively... return... to the path... of existential fulfillment.

But that's not the point, really – the point is not to trust what they've been saying, necessarily. The point is that there IS a "they" to say something. ... The point is that there IS a continuation of consciousness. ... *That's* the point. ... We'll iron out the details whenever we get over there - right now we've got enough to worry about.

But some people don't believe in these mothmen though. They say it defies the laws of physics within our universe. And to that I say, yes, that is precisely correct – wwwwwwwithin'our'universe. But what is the definition of magic? ... according to me? ... It is the penetration of influences into *this* universe, which originate from *another*, and thus conform to its tenets. No laws are being broken. It's just a blend across the octaves.

The unified field theory of physics... will never be complete... until it can learn to sing like this.

But we're not gonna talk about that little guy. Maybe better days will come and we could talk about that one.

Until then, mainstream science *still* cannot explain things like quantum entanglement, and dark matter, and dark energy, and apparent red shift, and about a million other, heretofore, inexplicable phenomena, that just simply don't make the headlines.

So, we are not exactly in a position to be telling *ANYONE*... what IS... and what is NOT...

Seriously.

There is far more going on than meets the eye, out there. And none of this is any late-breaking news to a traditional Nauptsèh community. Culturally, they are already acutely aware of what my wonderful little Korean friend was so beside herself to discover - that the immortal essence of all life is far more intelligent than the roles we play in this world.

Their little wooly dogs are still with them. They're all around them, throughout the forests, and amongst the ancestors. They have never left their sides... through all these ages. They still bark at red canoes... and play with children.

That's it.

Book's over.

I'm done.

),x

I'm freakin' DONE!

SHOW'S OVER!!

GET OUTTA HERE!!

What?

Bill said what?!

PFFFFFFFF!!!!!! ... ... Well, you can tell Bill he can go shove it up his...
... What? ... ... aNOTHER book?!

HAAAAAAAA!!! ... I just barely finished THIS one!! ... Is Bill IN-

SANE?! ... ...Wai'wai'wait, don't answer that... don't answer that. I think we both know the answer to that.

What?

PFFFFFFFFFFFFFFFFFFFFFFFFFFFF!!!!!!!!

HAAAA!!!

Get a load of this guy with his "*second book*" over here.

Fuck outa here with this "second book" shit. No, fuck that. I wanna talk to my family.

No. Fuck that. Bill said when I finished *this* one I get to talk to my family.

No! ... I wanna talk to my family... ... *right*... ... now!!!

And I want some better FOOD DOWN HERE!!!

... or I ain't puttin' your *damn* lotion anywhere.

www.ingramcontent.com/pod-product-compliance
Lightning Source LLC
Chambersburg PA
CBHW021839020426
42334CB00013B/128